INTERFEROGRAM ANALYSIS FOR OPTICAL TESTING

OPTICAL ENGINEERING

Series Editor

Brian J. Thompson

Distinguished University Professor
Professor of Optics
Provost Emeritus
University of Rochester
Rochester, New York

Additional Volumes in Preparation

INTERFEROGRAM ANALYSIS FOR OPTICAL TESTING

DANIEL MALACARA
MANUEL SERVÍN
ZACARIAS MALACARA
Centro de Investigaciones en Optica, A.C.
León, Mexico

MARCEL DEKKER, INC.   NEW YORK · BASEL · HONG KONG

Phys

Library of Congress Cataloging-in-Publication Data

Malacara, Daniel.
Interferogram analysis for optical testing / Daniel Malacara, Manuel Servín, Zacarias Malacara.
p. cm. — (Optical engineering; 61)
Includes index.
ISBN 0-8247-9940-2 (alk. paper)
1. Optical measurements. 2. Interferometry. 3. Interferometers. 4. Diffraction patterns—Data processing. I. Servín, Manuel. II. Malacara, Zacarias. III. Title. IV. Series. V. Series: Optical engineering (Marcel Dekker, Inc.); v. 61.
QC367.M25 1998
681'.25—dc21
98-24471
CIP

This book is printed on acid-free paper.

Headquarters
Marcel Dekker, Inc.
270 Madison Avenue, New York, NY 10016
tel: 212-696-9000; fax: 212-685-4540

Eastern Hemisphere Distribution
Marcel Dekker AG
Hutgasse 4, Postfach 812, CH-4001 Basel, Switzerland
tel: 44-61-261-8482; fax: 44-61-261-8896

World Wide Web
http://www.dekker.com

The publisher offers discounts on this book when ordered in bulk quantities. For more information, write to Special Sales/Professional Marketing at the headquarters address above.

Current printing (last digit)
10 9 8 7 6 5 4 3 2 1

PRINTED IN THE UNITED STATES OF AMERICA

From the Series Editor

Interferometry continues to be a major tool in the fields of inspection, testing, and measurement not only for the optical industry but for an extremely wide variety of industrial and scientific purposes. Optical science and engineering provide a very important set of enabling technologies for many advancing fields of endeavor and for the commercialization of new products. Interferometry is certainly one of the major components of a lengthening list of enabling optical technologies.

The power of interferometry changed dramatically with the introduction of automated methods of detection and analysis of fringe patterns. We have come a long way since the early days of interferometry, when we relied on visual detection and interpretation of fringe patterns. In turn, the automated analysis has allowed for the introduction of new and expanded techniques.

Interferogram Analysis for Optical Testing describes in detail the fringe analysis and interpretation techniques that are currently available and in use. The descriptions will allow for the ready application of these methods.

Brian J. Thompson

Preface

This book is intended as a continuation of the effort initiated years ago with the publication of the first edition of *Optical Shop Testing*. In that book, and later in its second edition, the main interest was to give a detailed description of the wide variety of interferometers being used in the optical shop. Little attention was given to computational methods for analyzing the fringe patterns obtained, except for the phase shifting interferometry technique. This volume presents the other side of the coin; that is, the most important optical interferometers are briefly reviewed and then classical as well as very recent methods for fringe pattern analysis are discussed.

This book focuses mainly on the analysis of static fringe patterns but phase shifting algorithms have also been analyzed. This work is intended for the optical shop practitioner interested in the computational methods of fringe analysis as well as for newcomers to the exciting research field of digital fringe pattern analysis.

The 12 chapters cover many classical techniques in fringe analysis as well as some very recent ones. Chapter 1 gives a general review and comparison of the major interferometric systems. Twyman-Green, Fizeau, common path, and lateral shear interferometers are studied as well as the principles of phase stepping interferometry. The objective of Chapter 1 is to make this book as self-contained as possible.

Chapter 2 covers various general principles of Fourier theory and digital image processing as applied to the analysis of digitized interferograms. Chapter 3 describes standard methods of digital image enhancement and filtering. Also, there is a section that discusses regularized iterative filtering of digitized images that are bounded by a pupil. Chapter 4 on fringe contouring and polynomial fitting covers the most ancient methods of fringe analysis that nevertheless are still very useful in the optical shop. Chapter 5 describes the theoretical foundations of the methods of signal phase detection. Chapter 6 studies the different algorithms that have been devised for phase detection.

Chapter 7 describes the popular phase shifting methods. Chapter 8 covers the subject of linear and circular carrier fringe analysis. Here we describe the Fourier, direct, phase-locked loop, spatial phase shifting techniques, and quadrature methods of wavefront estimation. Chapter 9 studies the analysis of interferograms with moiré methods. Chapter 10 describes techniques to determine wavefront deformations by measuring slopes and local average curvatures. These techniques range from the ancient but still useful Hartmann test to newer methods based on two defocused images. Chapter 11 describes the important subject of phase unwrapping, giving the most important and easiest techniques for both path-dependent and the more recent path-independent phase unwrapping methods. Finally, Chapter 12 describes the techniques used to test aspheric wavefronts and how the various interferometers reviewed in Chapter 1 can be modified to introduce optical compensators to reduce the number of interfering fringes over the CCD camera used to analyze and measure these wavefronts.

We gratefully acknowledge the careful reading of the manuscript and suggestions to improve the book provided by many friends and colleagues, especially Prof. Malgorzata Kujawinska, Dr. Joanna Schmit, Prof. Johannes Schwider, Jorge García-Márquez, and Efrain Hernández.

Daniel Malacara
Manuel Servín
Zacarias Malacara

Contents

1

Review and Comparison of Major Interferometric Systems

1.1 TWO-WAVE INTERFEROMETERS

Two-wave interferometers produce an interferogram by superimposing two wavefronts, one of which is typically a flat reference wavefront and the other a distorted wavefront whose shape is to be measured. There are many descriptions of interferometers in the literature (Malacara 1992; Creath 1987). Here we describe just some of the more important aspects.

An interferometer can measure small wavefront deformations with a high accuracy, of the order of a fraction of the wavelength. The accuracy in a given interferometer depends on many factors such as the optical quality of its components, the measuring methods, and the light source properties, and also on disturbing external factors such as atmospheric turbulence and mechanical vibrations. Kafri (1989) showed that there is, however, a limit to the accuracy of any interferometer. He proved that if everything else is perfect, a short coherence length and a long sampling time improve the accuracy. Unfortunately, both a short coherence length and a large measuring time make the instrument more sensitive to mechanical vibrations. In conclusion, the uncertainty principle imposes a fundamental limit to the accuracy. This limit depends on several parameters, but it is of the order of one-thousandth of a wavelength.

To study the main principles of interferometers, let us consider a two-wave interferogram with a flat wavefront which has a positive tilt about the y axis and a wavefront under test whose deformations with respect to a flat surface without

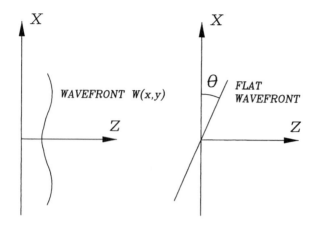

Figure 1.1   Two interfering wavefronts shown separately.

tilt are given by $W(x, y)$. This tilt is said to be positive when the wavefront is as in Fig. 1.1b. The complex amplitude in the observing plane, where the two wavefronts interfere, is the sum of the complex amplitudes of the two waves as follows.

$$E_1(x, y) = A_1(x, y) \exp[ikW(x, y)] + A_2(x, y) \exp[i(kx \sin \theta)] \qquad (1.1)$$

where A_1 is the amplitude of the light beam at the wavefront under test, A_2 is the amplitude of the light beam with the reference wavefront, and $k = 2\pi/\lambda$. Hence, the irradiance is

$$E_1(x, y)E_1^*(x, y) = A_1^2(x, y) + A_2^2(x, y)$$
$$+ 2A_1(x, y)A_2(x, y) \cos k[x \sin \theta - W(x, y)] \qquad (1.2)$$

where the asterisk (*) denotes the complex conjugate of the electric field. Here, an optional tilt θ about the y axis between the two wavefronts has been introduced. The irradiance function $I(x, y)$ may then be written as

$$I(x, y) = I_1(x, y) + I_2(x, y)$$
$$+ 2\sqrt{I_1(x, y)I_2(x, y)} \cos k[x \sin \theta - W(x, y)] \qquad (1.3)$$

where $A_1(x, y)$ and $A_2(x, y)$ are the irradiances of the two beams and the phase difference between them is given by $\phi = k[x \sin \theta - W(x, y)]$. This function is graphically shown in Fig. 1.2.

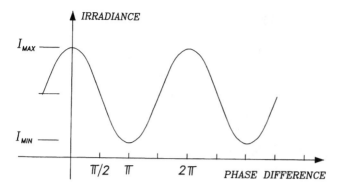

Figure 1.2 Irradiance as a function of the phase difference along the light path, between the two waves.

For convenience Eq. 1.3 is frequently written as

$$I(x, y) = a(x, y) + b(x, y) \cos k[x \sin \theta - W(x, y)] \tag{1.4}$$

Assuming that the variation in the values of $a(x, y)$ and $b(x, y)$ inside the interferogram aperture are smoother than the variations of the cosine term, the maximum irradiance in the vicinity of the point (x, y) in this interferogram is given by

$$
\begin{aligned}
I_{max}(x, y) &= (A_1(x, y) + A_2(x, y))^2 \\
&= I_1(x, y) + I_2(x, y) + 2\sqrt{I_1(x, y)I_2(x, y)}
\end{aligned}
\tag{1.5}
$$

and the minimum irradiance in the same vicinity is given by

$$
\begin{aligned}
I_{min}(x, y) &= (A_1(x, y) - A_2(x, y))^2 \\
&= I_1(x, y) + I_2(x, y) - 2\sqrt{I_1(x, y)I_2(x, y)}
\end{aligned}
\tag{1.6}
$$

The fringe visibility $v(x, y)$ is defined by

$$v(x, y) = \frac{I_{max}(x, y) - I_{min}(x, y)}{I_{max}(x, y) + I_{min}(x, y)} \tag{1.7}$$

hence, we may find

$$v(x, y) = \frac{2\sqrt{I_1(x, y)I_2(x, y)}}{I_1(x, y) + I_2(x, y)} = \frac{b(x, y)}{a(x, y)} \tag{1.8}$$

Hence, using the fringe visibility, Eq. 1.3 is sometimes also written as

$$I(x, y) = I_0(x, y)\{1 + v(x, y) \cos k[x \sin \theta - W(x, y)]\} \qquad (1.9)$$

where $I_0(x, y) = a(x, y)$ is the irradiance for a fringe free field when the two beams are incoherent to each other. This irradiance as a function of the phase difference between the two interfering waves is shown in Fig. 1.2.

1.2 INTERFEROMETER CONFIGURATIONS USED IN OPTICAL TESTING

There are several basic interferometric configurations used in optical testing procedures, but almost all of them are two-wavefront systems. Both wavefronts come from a single light source and their paths are separated by amplitude division. Furthermore, most modern interferometers use a helium-neon laser as the light source. The main advantage of using a laser as the source of light is that fringe patterns may be easily obtained without any problem because of its great coherence. In fact, this advantage is also a serious disadvantage, since spurious diffraction patterns and secondary fringe patterns are also easily obtained. Special precautions must be taken to have a clean interference pattern. In this chapter we review some of these interferometers. More details about these systems may be found in many books (Malacara 1992).

1.3 TWYMAN-GREEN INTERFEROMETER

The Twyman-Green interferometer was invented by F. Twyman and A. Green in 1916. The basic configuration of the Twyman-Green interferometer is illustrated in Fig. 1.3.

The fringes in a Twyman-Green interferometer are of equal thickness. The light from the laser is expanded and collimated by means of a telescopic system. This telescope is usually formed with a microscope objective and a collimator. To have a clean wavefront without diffraction rings on the field, the optical components must be as clean as possible. To clean the beam even more, a spatial filter (pinhole) may be used at the focal plane of the microscope objective. The quality of the wavefront produced by this telescope does not need to be extremely high, because its deformations will appear on both interfering wavefronts without producing any fringe deviations. If the optical path difference between the two interfering beams is large, the tolerance on the wavefront deformations in the illuminating telescope may be drastically reduced and then the illuminating wavefront must be quite flat, within a fraction of a wavelength.

If the beam splitter is nonabsorbing, the main interference pattern is complementary to the one returning to the source, because of the conservation of

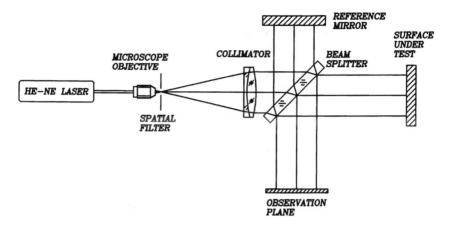

Figure 1.3 Basic configuration of a Twyman-Green interferometer.

energy principle, even though the optical path difference is the same for both patterns. Phase shifts upon reflection on dielectric interfaces may explain this complementarity.

The beam splitter must be of high quality—not only its surfaces but also its material, which must be extremely homogeneous. The best surface must be the reflecting one, which must be flat with an accuracy of about twice the required interferometer accuracy. The quality of the nonreflecting surface may be relaxed by a factor of 4 with respect to the other face. The nonreflecting surface must not reflect any light, to avoid spurious interference fringes. This may be accomplished in many ways, for example, by coating the surface with an antireflection multilayer coating. Another possible method is to have an incidence angle on the beam splitter with a magnitude equal to the Brewster angle and properly polarize the incident light beam. This solution, however, substantially increases the size of the beam splitter, making it more difficult to construct and hence more expensive.

Many different optical elements may be tested using a Twyman-Green interferometer, as described by Malacara (1992). For example, a plane parallel plate of glass may be tested as in Fig. 1.4a.

The *optical path difference OPD* introduced by this glass plate is

$$OPD = 2(n - 1)t \qquad (1.10)$$

where t is the plate thickness and n is the refractive index. The interferometer is first adjusted so that no fringes are observed before introducing the plate into the light beam, so that all the fringes that appear are due to the plate. If the field

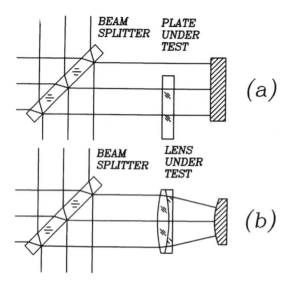

Figure 1.4 Testing (a) a glass plate and (b) a lens in a Twyman-Green interferometer.

remains free of fringes after the plate is introduced, we can say that the quantity $(n-1)t$ is constant over all the plate aperture. If the fringes are straight, parallel, and equidistant and we may assume that the glass is perfectly homogeneous so that n is constant, then the fringes are produced by a small angle between the two flat faces of the plate.

If the fringes are not straight but distorted, we may conclude that either the refractive index is not constant or the surfaces are not flat, or both. We can only be sure that $(n - 1)t$ is not constant. To measure the n and t separately, we must augment this test with measurement made in a Fizeau interferometer, which measures the values of nt.

The optical arrangements in Fig. 1.4b can be used to test a convergent lens. A convex spherical mirror with its center of curvature at the focus of the lens is used for lenses with long focal lengths, and a concave spherical mirror for lenses with short focal lengths. A small, flat mirror at the focus of the lens can also be employed. The portion of the flat mirror being used is so small that its surface does not need to be very accurate. However, the wavefront is rotated 180° and thus the spatial coherence requirements are stronger and odd aberrations cancel.

Concave or convex optical surfaces may also be tested using a Twyman-Green interferometer, using the configurations in Fig. 1.5. Even large astronomical

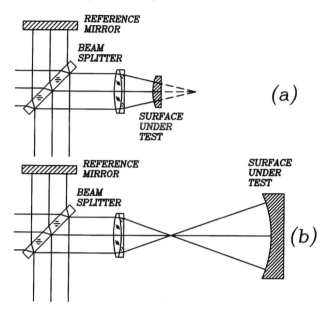

Figure 1.5 Twyman-Green interferometer configurations to test (a) a convex or (b) a concave optical surface.

mirrors can be tested. For this purpose, an unequal-path interferometer for optical shop testing was designed by Houston *et al.* (1967).

When the beam-splitter plate is at the Brewster angle, there is a wedge angle of 2 to 3 minutes of arc between the surfaces. The reflecting surface of this plate is located to receive the rays returning from the test specimen in order to preclude astigmatism and other undesirable effects. A two-lens beam diverger can be placed in one arm of the interferometer. It is made of high index glass, all the surfaces being spherical, and has the capability for testing a surface as fast as $f/1.7$.

1.4 FIZEAU INTERFEROMETERS

Like the Twyman-Green, the Fizeau interferometer is a two-beam instrument with equal-thickness fringes. Its basic configuration is illustrated in Fig. 1.6.

The optical path difference *OPD* introduced when testing a plane parallel glass plate placed in the light beam is

$$OPD = 2nt \tag{1.11}$$

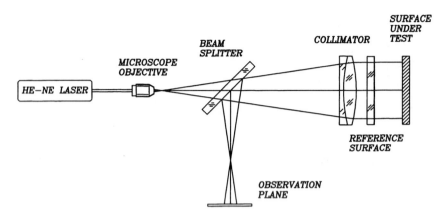

Figure 1.6 Basic Fizeau interferometer configuration.

which, as we may notice, is different from the corresponding expression for the Twyman-Green interferometer. In this sense the two interferometers are complementary, so that the constancy of the thickness t and of the refractive index n may be tested only when both interferometers are used.

A large concave optical surface may also be tested with a Fizeau interferometer, as shown in Fig. 1.7. If the concave surface is aspherical, the spherical aberration may be compensated if the converging lens has the opposite aberration. The reference surface is placed between the collimator and the converging lens.

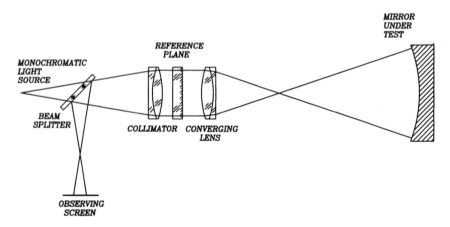

Figure 1.7 Fizeau interferometer being used to test a concave surface.

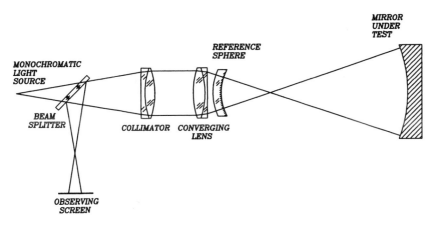

Figure 1.8 Fizeau interferometer being used to test a concave spherical surface using a concave reference surface.

When the reference surface is flat as in Fig. 1.7, no off-axis configuration appears when the concave mirror under test is tilted to introduce many tilt fringes (linear carrier). However, a perfect focusing lens is required, because it is located inside the cavity and thus the wavefront under test passes through this lens but the reference wavefront does not. Then, any error in the focusing lens appears in the interferogram. A second possible source of errors exists when a flat reference is used, because the reference wavefront returns to the collimator lens at an angle with respect to the optical axis. Then the collimator has to be corrected for some field angle.

A spherical reference surface is sometimes used, as shown in Fig. 1.8. In this case the linear carrier may be introduced by tilting the concave spherical surface under test or the reference sphere. This arrangement avoids the need for any optical elements inside the interferometer cavity, between the reference surface and the surface under test, relaxing the requirements for good focusing and collimating optics. These lenses still have to be corrected for some small field angle, but their degree of correction does not need to be very high. Even better, if the whole optical system formed by the focusing lens and the collimator is made symmetrical, the correction of coma is automatic. In this configuration some wavefront aberrations may appear when the linear carrier is introduced, due to the large tilt in the spherical mirror, besides the well-known primary astigmatism.

With this arrangement an off-axis configuration results when a large tilt is applied to the interferometer in order to introduce a linear carrier with more than 200 fringes in the interferogram (Kuchel 1990). The linear carrier is obtained by tilting the reference. The surface being tilted may be the concave mirror

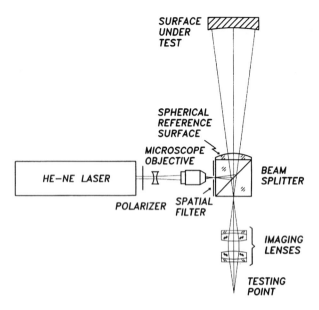

Figure 1.9   Shack-Fizeau interferometer.

under test or the spherical reference. We have seen that in addition to the well-known astigmatic aberration introduced (due to off-axis testing), spherical and high order (ashtray) astigmatism are also generated. However, we may see that even for a large number of fringes the wavefront aberration remains small for all practical purposes, and thus we may introduce as many fringes as desired.

Another source of wavefront errors in the spherical cavity configuration, when testing a high aperture optical element, may be introduced by large axial displacements of the concave surface under test with respect to the spherical reference sphere. Besides the expected defocusing, spherical aberration is introduced in the wavefront.

A common variation of the Fizeau interferometer to test a large concave surface using a spherical reference surface is the Shack-Fizeau interferometer shown in Fig. 1.9.

1.5   TYPICAL INTERFEROGRAMS IN TWYMAN-GREEN AND FIZEAU INTERFEROMETERS

The interferograms produced by primary aberrations have been described by Kingslake (1926). An aberrated wavefront with primary aberrations, as measured with respect to a sphere with its center of curvature at the Gaussian image point, is given by

$$W(x, y) = A(x^2 + y^2)^2 + By(x^2 + y^2) + C(x^2 - y^2)$$
$$+ D(x^2 + y^2) + Ex + Fy + G \qquad (1.12)$$

where

$A = $ spherical aberration coefficient

$B = $ coma coefficient

$C = $ astigmatism coefficient

$D = $ defocusing coefficient

$E = $ tilt about the y axis coefficient (image displacement along the x axis)

$F = $ tilt about the x axis coefficient (image displacement along the y axis)

$G = $ piston or constant term

This expression may also be written in polar coordinates (θ, ρ). For simplicity, in computing typical interferograms of primary aberrations, a normalized entrance pupil with unit semidiameter can be taken.

Some typical interference patterns are shown in Fig. 1.10. A more complete set of pictures may be found in Malacara (1992).

Pictures of typical interferograms can be simulated in a computer by means of fringes of equal inclination on a Michelson interferometer (Murty 1964) using the *OPD*s introduced by a plane parallel plate and cube corner prisms instead of mirrors, or by electronic circuits on a CRT (Geary 1979).

Twyman-Green interferograms were analyzed by Kingslake (1926–1927) by measuring the optical path difference at several points by fringe sampling. Then, solving a system of linear equations, he computed the *OPD* coefficients A, B, C, D, E, F. A similar method for analyzing a Twyman-Green interferogram was proposed by Saunders (1965). He found that the measurement of nine appropriately chosen points is sufficient to determine any of the three primary aberrations. The points were selected as in Fig. 1.11, and then the aberration coefficients were calculated as

$$A = \frac{128}{81r^2}[W_1 - W_9 + 2(W_8 - W_7)] \qquad (1.13)$$

$$B = \frac{128}{3r^2}[W_2 - W_4 + 2(W_6 - W_5)] \qquad (1.14)$$

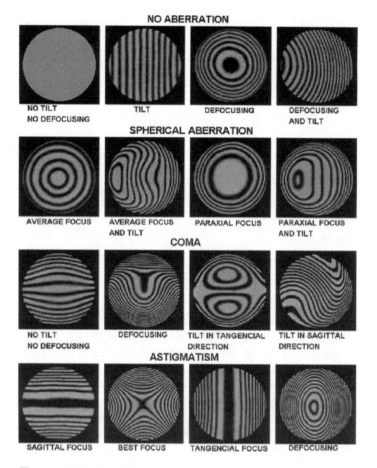

Figure 1.10 Some Twyman-Green interferograms.

and

$$C = \frac{1}{4r^2}[W_2 + W_4 - W_1 - W_3] \tag{1.15}$$

where W_i is the estimated wavefront deviation at the point i.

The aberration coefficients can be determined by direct reading on the interferogram setting, looking for interference patterns with different defocusing settings and tilts.

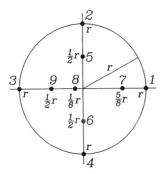

Figure 1.11 Selected points for evaluation of primary aberrations.

1.6 LATERAL SHEAR INTERFEROMETERS

A lateral shear interferogram does not need any reference wavefront. Instead, the interference takes place between two identical aberrated wavefronts laterally sheared with respect to each other as shown in Fig. 1.12.

The optical path difference *OPD* is

$$OPD = W(x, y) - W(x - S, y) \tag{1.16}$$

where S is the lateral shear in the sagittal (x) direction.

Let us now assume that the lateral shear S is sufficiently small that the wavefront slopes in the x direction may be considered almost constant in an

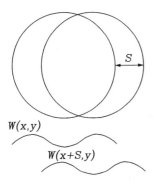

Figure 1.12 Two laterally sheared wavefronts.

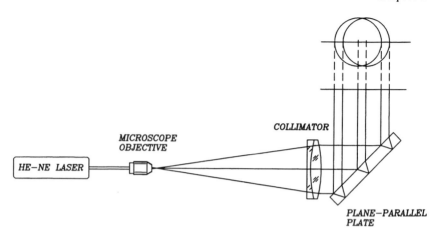

Figure 1.13 Murty's lateral shear interferometer.

interval S. This is equivalent to the condition that the fringe spatial frequency in the x direction is almost constant in an interval S. Then we may expand in a Taylor series, obtaining

$$OPD = W(x + S, y) - W(x, y) = \frac{\partial W(x, y)}{\partial x} S \qquad (1.17)$$

A bright fringe occurs when

$$OPD = \frac{\partial W(x, y)}{\partial x} S = \frac{TA_x(x, y)}{r} = m\lambda \qquad (1.18)$$

where m is an integer number and $TA_x(x, y)$ is the transverse aberration of the ray perpendicular to the wavefront, measured at a plane containing the center of curvature of the wavefront.

This result may be interpreted by saying that a lateral shear interferometer does not measure the wavefront deformation $W(x, y)$ in a direct manner, but its slope or transverse aberration in the direction of the lateral shear. To measure the two components of the transverse aberrations it is necessary to take two laterally-sheared interferograms in perpendicular directions.

There are many practical configurations for lateral shear interferometers. The most popular due to its simplicity is the Murty interferometer (Murty 1964), illustrated in Fig. 1.13.

1.6.1 Primary Aberrations

The lateral shear interferograms for the primary aberrations may be obtained with the expression for the primary aberrations in Eq. 1.12, as will now be described.

Defocus

The interferogram with a defocused wavefront is given by

$$2DxS = m\lambda \tag{1.19}$$

This is a system of straight, parallel, and equidistant fringes. These fringes are perpendicular to the lateral shear direction.

When the defocusing is large, the spacing between the fringes is small. On the other hand, when there is no defocus, there are no fringes in the field.

Spherical Aberration

In this case the interferogram is given by

$$4A(x^2 + y^2)xS = m\lambda \tag{1.20}$$

If this aberration is combined with defocus we may write instead

$$[4A(x^2 + y^2)x + 2Dx]S = m\lambda \tag{1.21}$$

Then the interference fringes are cubic curves.

Coma

In the case of the coma aberration the interferogram is given by

$$2BxyS = m\lambda \tag{1.22}$$

when the lateral shear is S in the *sagittal* direction. If the lateral shear is T in the *tangential* (y) direction, the fringes are given by

$$B(x^2 + 3y^2)T = m\lambda \tag{1.23}$$

Primary Astigmatism

In the case of astigmatism, when the lateral shear is S in the sagittal (x) direction, the fringes are given by

$$(2Dx + 2Cx)S = m\lambda \tag{1.24}$$

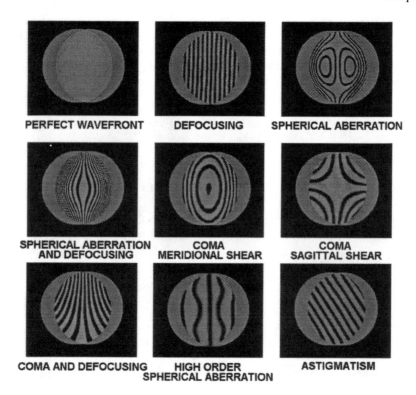

Figure 1.14 Some lateral shear interferograms.

and for the lateral shear T in the tangential (y) direction we have

$$(2Dy - 2Cy)T = m\lambda \tag{1.25}$$

The fringes are straight and parallel as in the case of defocus, but with a different separation for the two interferograms.

Some lateral shear interferograms for primary aberrations are shown in Fig. 1.14.

1.6.2   Wyant-Rimmer Method to Evaluate Wavefronts

The Wyant-Rimmer method performs a polynomial interpolation while determining the wavefront shape from a set of lateral shear interferogram sampled points. The wavefront represented by $W(x, y)$ may be expressed by the x, y polynomial of degree k

$$W(x, y) = \sum_{n=0}^{k} \sum_{m=0}^{n} B_{nm} x^m y^{n-m} \tag{1.26}$$

with $N = (k + 2)(k + 1)/2$ coefficients B_{nm}. The expression for the wavefront laterally sheared a distance S in the x direction is

$$W(x + S, y) = \sum_{n=0}^{k} \sum_{m=0}^{n} B_{nm} (x + S)^m y^{n-m} \tag{1.27}$$

and similarly, the wavefront sheared a distance T in the y direction is

$$W(x, y + T) = \sum_{n=0}^{k} \sum_{m=0}^{n} B_{nm} x^m (y + T)^{n-m} \tag{1.28}$$

On the other hand, the Newtonian binomial theorem is

$$(x + S)^m = \sum_{j=0}^{m} \binom{m}{j} x^{m-j} S^j \tag{1.29}$$

where

$$\binom{m}{j} = \frac{m!}{(m - j)! j!} \tag{1.30}$$

Thus, Eqs. 1.27 and 1.28 may be written

$$W(x + S, y) = \sum_{n=0}^{k} \sum_{m=0}^{n} \sum_{j=0}^{m} B_{nm} \binom{m}{j} x^{m-j} y^{n-m} S^j \tag{1.31}$$

and

$$W(x, y + T) = \sum_{n=0}^{k} \sum_{m=0}^{n} \sum_{j=0}^{n-m} B_{nm} \binom{n - m}{j} x^m y^{n-m-j} T^j \tag{1.32}$$

Hence, subtracting Eq. 1.26 from Eq. 1.31 we have

$$\Delta W_S = W(x + S, y) - W(x, y) = \sum_{n=0}^{k-1} \sum_{m=0}^{n} C_{nm} x^m y^{n-m} \tag{1.33}$$

and subtracting Eq. 1.26 from Eq. 1.32,

$$\Delta W_T = W(x, y + T) - W(x, y) = \sum_{n=0}^{k-1} \sum_{m=0}^{n} D_{nm} x^m y^{n-m} \tag{1.34}$$

with $k(k+1)/2$ coefficients C_{nm} and the same number of coefficients D_{nm} given by

$$C_{nm} = \sum_{j=1}^{k-n} \binom{j+m}{j} S^j B_{j+n,j+m} \tag{1.35}$$

and

$$D_{nm} = \sum_{j=1}^{k-n} \binom{j+n-m}{j} T^j B_{j+n,m} \tag{1.36}$$

The values of C_{nm} and D_{nm} are obtained from the two laterally sheared interferograms in orthogonal directions by means of a two-dimensional least squares fit to the measured values of ΔW_S and ΔW_T. Then the values of all coefficients B_{nm} are calculated by solving the system of linear equations defined by Eqs. 1.35 and 1.36, each with a matrix of dimensions $N \times M$.

1.6.3 Saunders Method to Evaluate Interferograms

In the evaluation of an unknown wavefront it is possible to determine its shape from a lateral shear interferogram.

To illustrate the method proposed by Saunders (1961), let us consider Fig. 1.15. To begin, let us assume that $W_1 = 0$. Then we may write

$$W_1 = 0$$
$$W_2 = \Delta W_1 + W_1$$
$$W_3 = \Delta W_2 + W_2 \tag{1.37}$$
$$\cdots\cdots\cdots\cdots$$
$$W_N = \Delta W_{N-1} - W_{N-1}$$

The main problem with this method is that the wavefront is evaluated only at points separated by a distance S. Intermediate values are not measured and have to be interpolated.

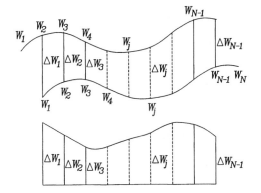

Figure 1.15 Saunders method to obtain the wavefront in a lateral shear interferogram.

Orthogonal polynomials, as described in chapter 4 in this book, may also be used with advantage to represent the wavefront in a lateral shear interferometer. The accuracy of this mathematical representation was studied by Wang and Ling (1989).

1.6.4 Spatial Frequency Response of Lateral Shear Interferometers

Unlike Twyman-Green interferometers, lateral shear interferometers have a non-uniform response to spatial frequencies (Fourier components) in the wavefront deformations function. This response may be analyzed as illustrated in Fig. 1.16.

The spatial frequency content of the lateral shear optical path difference function, which is the interferometer output *OPD*, is given by

$$F\{OPD\} = F\{W(x, y) - W(x - S, y)\} \tag{1.38}$$

or

$$F\{OPD\} = F\{W(x, y)\} - F\{W(x - S, y)\} \tag{1.39}$$

Figure 1.16 The lateral shear interferometer considered as an electronic system.

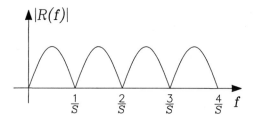

Figure 1.17 Lateral shear interferometer sensitivity as a function of the spatial frequency.

where F{g} is the Fourier transform of g. Using the lateral displacement theorem of Fourier theory, this expression is transformed into

$$F\{OPD\} = F\{W(x, y)\} - F\{W(x, y)\} \exp(-i2\pi f S) \tag{1.40}$$

where f is the spatial frequency of a Fourier component, or

$$F\{OPD\} = F\{W(x, y)\}[1 - \exp(-i2\pi f S)] \tag{1.41}$$

from which we may obtain

$$F\{OPD\} = 2i \sin(\pi f S) F\{W(x, y)\} \exp(-i\pi f S) \tag{1.42}$$

The spatial frequency sensitivity of the interferometer, $R(f)$, may now be defined as

$$R(f) = \frac{F\{OPD\}}{F\{W(x, y)\}} = 2i \sin(\pi f S) \exp(-i\pi f S) \tag{1.43}$$

which may also be written as

$$R(f) = 2 \sin(\pi f S) \exp\left[-i\pi \left(f S - \frac{1}{2}\right)\right] \tag{1.44}$$

This function has zeros at $\pi f S = m\pi$. Thus, the lateral shear interferometer is not sensitive to spatial frequencies given by

$$f = \frac{m}{S} \tag{1.45}$$

where m is an integer, as shown in Fig. 1.17.

This result implies that the wavefront deformations $W(x, y)$ are not obtained with the same precision for all spatial frequencies. The recovery of the spatial frequency components close to the zeros in Eq. 1.44 will have a larger uncertainty in the calculation.

1.7 MOIRÉ METHODS

Moiré methods are not really interferometric. Nevertheless, their fringe analysis is so similar that a description of these methods is convenient. Whenever two slightly different periodic structures are superimposed, a beating between the two structures is observed in the form of another periodic structure with a lower spatial frequency. These fringes are called moiré fringes.

Moiré techniques have been used in metrology for a long time, as reviewed by Sciammarella (1982), Reid (1984), and Patorski (1988), with many different configurations and purposes. The most common basic configurations are

1. *Fringe projection.* A periodic ruling is projected on a solid body. Then the image of this body with the fringes over its surface is imaged over another periodic ruling, to form the moiré fringes.
2. *Shadow moiré.* A periodic ruling is placed over a nearly flat surface and illuminated from the top. The moiré fringes are formed between the ruling structure and its own shadow over the surface.
3. *Talbot interferometry.* A ruling is illuminated with a collimated, convergent, or divergent beam of light. The shadow of the ruling is projected upon a screen placed at some distance from the ruling, where another ruling is placed to form the moiré.

Moiré patterns are studied in more detail in Chap. 9, mainly as a tool for the analysis of interferograms. Here, we briefly describe the basic moiré configurations.

1.7.1 Fringe Projection

The shape of a solid body can be measured by projecting a periodic structure or ruling over the body (Idesawa *et al.* 1977; Takeda 1982; Doty 1983; Gåsvik 1983; Creath and Wyant 1988). The fringes may be projected on the body by a lens or slide projector (Takasaki 1970, 1973; Parker 1978; Pirodda 1982; Halioua *et al.* 1993; Yatagai and Idesawa 1982; Reid *et al.* 1984; Cline *et al.* 1984; Varman 1984; Gåsvik 1983; Suganuma and Yoshisawa 1991). In another method the interference fringes produced by two tilted flat wavefronts are projected over the body (Brooks and Heflinger 1969).

A slightly different method called shadow moiré produces the moiré fringes between a Ronchi ruling and the shadow of the ruling projected over a solid

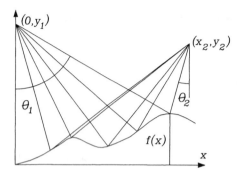

Figure 1.18 Projecting fringes of a periodic structure over a solid body to measure its shape.

body located just behind the ruling. This method makes it possible to find the shape of nearly flat surfaces (Jaerisch and Makosch 1973; Pirodda 1982).

Let us consider that a straight fringe is projected from point **A** at height z_a to point **C** on the plane $z = 0$, as shown in Fig. 1.18. This fringe is observed from point **B** at height z_b over the plane $z = 0$. If a surface to be measured is located over the plane $z = 0$, this surface will intersect the fringe at point **D.** As observed from point **B,** the fringe appears to be at point **E** on the plane $z = 0$. The separation between the points **E** and **C** allows us to calculate the object height over the plane $z = 0$. Obviously the lines **AC** and **BE** are on a common plane since they intersect at **D.** Nevertheless, this plane is not necessarily perpendicular to the plane $z = 0$.

This geometry is completely general. The shape of the body is determined if the tridimensional coordinates of point **D** are calculated from measurements of the coordinates of point **E** on the plane $x = 0$ for many positions on the projected fringes.

This is the general configuration for fringe projection, but a simpler analysis can be made if both the lens projector and the observer are optically placed at infinite distances from the body to be measured, as shown in Fig. 1.19. The observer is located in a direction parallel to the z axis. In this case the object heights are given by

$$f(x, y) + \frac{x}{\tan \theta} - \frac{s}{\sin \theta} = \frac{md}{\sin \theta} \tag{1.46}$$

where m is the fringe number, with the fringe $m = 0$ being located at the origin $(x = 0)$. The angle θ is the inclination of the illuminator. The distance d is the fringe period in a plane perpendicular to the illuminating light beam.

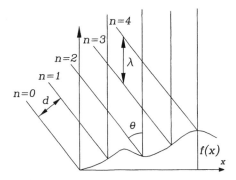

Figure 1.19 Projecting a periodic structure over a solid body to measure its shape, with both the projector and the observer at infinity.

The equivalent two-beam interferometric expression for the wavefront deformation $W(x)$ is

$$W(x, y) + \frac{\lambda}{p}x = m\lambda \tag{1.47}$$

Hence, the surface deformation $f(x, y) = 2W(x, y)$ when tested in a Fizeau interferometer is

$$f(x, y) + \frac{2\lambda}{p}x + a = 2m\lambda \tag{1.48}$$

where n is the order of interference, p is the fringe period introduced by tilting the reference wavefront, and a is a constant. By comparing these two expressions we see that we may consider fringe projection with this geometry as Fizeau interferometry with a wavelength λ given by

$$\lambda = \frac{d}{2 \sin \theta} \tag{1.49}$$

These projected fringes may then be considered Fizeau fringes with a large linear carrier (tilt) introduced. This body with the fringes or interferogram is imaged on the observing plane by means of an optical system, photographic camera, or television camera. This interferogram with tilt may be analyzed with any of the traditional methods, but one common method is with moiré techniques, as will be described in Chap. 9. Then the image is superimposed on a linear ruling with approximately the same frequency as the fringes on the inter-

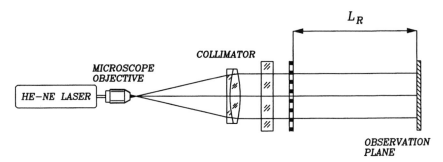

Figure 1.20 Talbot autoimage formation of a ruling illuminated with a collimated beam of light.

ferogram. This linear ruling may be real or software-generated on the computer analyzing the image.

1.7.2 Talbot Interferometry

Another method commonly used to measure wavefront deformations uses the Talbot autoimaging procedure illustrated in Fig. 1.20.

Talbot (1836) discovered that when a linear ruling is illuminated with a collimated beam of light, perfect images of this ruling are formed without any lenses, at distances which are an integer multiple of a distance called the Rayleigh (1881) distance L_R, as shown in Fig. 1.20.

If the illuminating wavefront is not flat but spherical or distorted, the fringes in the autoimage are not straight but distorted. The interferometric explanation assumes that the diffracted wavefronts produce a lateral shear interferogram as in Fig. 1.21a. On the other hand, the geometric interpretation considers the fringes to be shadows of the ruling lines, projected in a direction perpendicular to the wavefront, as in Fig. 1.21b. The two models are equivalent.

When the moiré pattern between the fringe image represented by the autoimage and a superposed linear ruling is formed, we speak of a Talbot interferometer. Talbot interferometry has been described by many researchers, including Yokoseki and Susuki (1971a, 1971b), Takeda and Kobayashi (1984), and Rodríguez-Vera *et al.* (1991). These authors interpreted the fringe with an interferometric model like multiple beam lateral shear interferometry. Ten years later, Kafri (1981, 1989) rediscovered this method from the geometrical point of view, calling it moiré deflectometry. Glatt and Kafri (1988), Stricker (1985), and Vlad *et al.* (1991) described this method and some applications. The interferometric and geometric interpretations may be proved to be equivalent, as

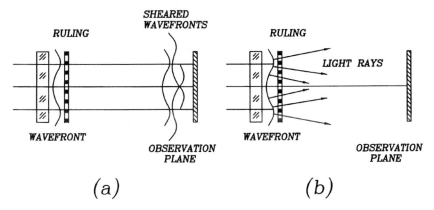

Figure 1.21 Formation of autoimages with a distorted or spherical wavefront. (a) Shadows interpretation and (b) laterally sheared wavefronts interpretation.

pointed out by Patorski (1988). This procedure has a close analogy with the Ronchi test (Cornejo 1992).

1.8 COMMON LIGHT SOURCES USED IN INTERFEROMETRY

By far the most common light source in interferometry is the helium-neon laser. The great advantage of this light source is its large coherence length and its monochromaticity. However, these characteristics are sometimes a great problem, because many spurious fringes are also formed unless great precautions are taken to avoid them.

When a laser light source is used, extremely large *OPD*s can be introduced. As shown in Fig. 1.22, the light emitted by a gas laser usually consists of several

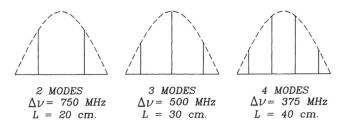

Figure 1.22 Line spectra of the light (longitudinal modes) from a gas laser for three values of *L*, the laser cavity length.

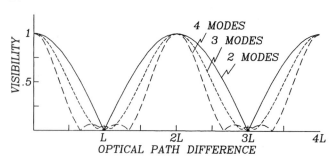

Figure 1.23 Visibility in a Twyman-Green interferometer using a He-Ne laser, as a function of the optical path difference, for three different lengths of the laser cavity.

equally spaced spectral lines (longitudinal modes) with a frequency separation equal to

$$\Delta v = \frac{c}{2L} \tag{1.50}$$

where L is the laser cavity length.

If the cavity length L of the laser changes because of thermal expansion or contraction or mechanical vibrations, the lines move in preserving their relative separations along the frequency scale, but with their intensities inside the dashed envelope (power gain curve) as in Fig. 1.22.

Single-mode or single-frequency lasers produce a perfectly monochromatic wavetrain, but because of instabilities in the cavity length, the frequency may be unstable. By the use of servomechanisms, single-frequency lasers with extremely stable frequencies are commercially produced. They are the ideal source for interferometry because an *OPD* as long as desired can be introduced without any loss in contrast.

The fringe visibility in an interferometer using a laser source with several longitudinal modes is a function of the optical path difference. To have good contrast, the *OPD* has to be an integral multiple of $2L$ (as shown in Fig. 1.23). A laser with two longitudinal modes is sometimes stabilized to avoid contrast changes by a method recommended by Bennett *et al.* (1973), Gordon and Jacobs (1974), and Balhorn *et al.* (1972).

Another type of laser frequently used in interferometers is the laser diode. Creath (1985), Ning *et al.* (1989), and Onodera and Ishii (1996) studied the most important characteristics of these lasers for use in interferometers. Their low coherence length, of the order of 1 mm, is a great advantage in many applications. Other advantages are their low price and small size.

REFERENCES

Brooks R. E. and L. O. Heflinger, "Moiré Gauging Using Optical Interference Fringes," *Appl. Opt.*, **8**, 935–939 (1969).

Cline H. E., W. E. Lorensen, and A. S. Holik, "Automatic Moiré Contouring," *Appl. Opt.*, **23**, 1454–1459 (1984).

Cornejo, A., "Ronchi Test," in *Optical Shop Testing*, D. Malacara, Ed., John Wiley and Sons, New York, 321, 1992.

Creath K., "Interferometric Investigation of a Laser Diode," *Appl. Opt.*, **24**, 1291–1293 (1985).

Creath K., "Wyko Systems for Optical Metrology," *Proc. SPIE*, **816**, 111–126 (1987).

Creath K. and J. C. Wyant, "Comparison of Interferometric Contouring Techniques," *Proc. SPIE*, **954**, 174–182 (1988).

Doty J. L., "Projection Moiré for Remote Contour Analysis," *J. Opt. Soc. Am.*, **73**, 366–372 (1983).

Gåsvik K. J., "Moiré Technique by Means of Digital Image Processing," *Appl. Opt.*, **22**, 3543–3548 (1983).

Geary J. M., "Real Time Interferogram Simulation," *Opt. Eng.*, **18**, 39–45 (1979).

Glatt I. and O. Kafri, "Moiré Deflectometry—Ray Tracing Interferometry," *Opt. and Lasers in Eng.*, **8**, 227–320 (1988).

Halioua M., R. S. Krishnamurthy, H. Liu and F. P. Chiang, "Projection Moiré with Moving Gratings for Automated 3-D Topography," *Appl. Opt.*, **22**, 805–855 (1983).

Idesawa M., T. Yatagai, and T. Soma, "Scanning Moiré Method and Automatic Measurement of 3-D Shapes," *Appl. Opt.*, **16**, 2152–2162 (1977).

Jaerisch W. and G. Makosch, "Optical Contour Mapping of Surfaces," *Appl. Opt.*, **12**, 1552–1557 (1973).

Kafri O., "High Sensitivity Moiré Deflectometry Using a Telescope," *Appl. Opt.*, **20**, 30980–3100 (1981).

Kafri O., "Fundamental Limit on the Accuracy in Interferometers," *Opt. Lett.*, **14**, 657–658 (1989).

Kingslake, R., "The Interferometer Patterns Due to the Primary Aberrations," *Trans. Opt. Soc.*, **27**, 94- (1926).

Kuchel M., "The New Zeiss Interferometer," *Proc. SPIE.* **1332**, 655–663 (1990).

Malacara D., Ed., *Optical Shop Testing*, 2nd Edition, John Wiley and Sons, New York, 1992.

Murty M. V. R. K., "The Use of a Single Plane Parallel Plate as a Lateral Shearing Interferometer with a Visible Gas Laser Source," *Appl. Opt.*, **3**, 531–351 (1964).

Ning Y., K. T. V. Grattan, B. T. Meggitt, and A. W. Palmer, "Characteristics of Laser Diodes for Interferometric Use," *Appl. Opt.*, **28**, 3657–3661 (1989).

Onodera R. and Y. Ishii, "Phase-Extraction Analysis of Laser-Diode Phase-Shifting Interferometry That Is Insensitive to Changes in Laser Power," *J. Opt. Soc. Am. A*, **13**, 139–146 (1996).

Parker R. J., "Surface Topography of Non Optical Surfaces by Oblique Projection of Fringes from Diffraction Gratings," *Opt. Acta*, **25**, 793–799 (1978).

Patorski K., "Moiré Methods in Interferometry," *Opt. and Lasers in Eng.*, **8**, 147–170 (1988).

Pirodda L., "Shadow and Projection Moiré Techniques for Absolute and Relative Mapping of Surface Shapes," *Opt. Eng.*, **21**, 640–649 (1982).

Rayleigh Lord, *Phil. Mag.*, **11**, 196 (1881).

Reid G. T., "Moiré Fringes in Metrology," *Opt. and Lasers in Eng.*, **5**, 63–93 (1984).

Reid G. T., R. C. Rixon and H. I. Messer, "Measurements of the Three-Dimensional Shape by Phase-Measuring Moiré Topography," *Opt. and Laser Tech.*, **16**, 315–319 (1984).

Rodriguez-Vera R., D. Kerr, and F. Mendoza-Santoyo, "3-Dimensional Contouring of Diffuse Objects by Talbot Projected Fringes," *J. Mod. Opt.*, **38**, 1935–1945 (1991).

Saunders J. B., "Measurement of Wavefronts Without a Reference Standard: The Wavefront Shearing Interferometer," *J. Res. Natl. Bur. Stand.*, **65B**, 239 (1961).

Saunders J. B., "Precision Method for Evaluating Primary Aberrations of Lenses with a Twyman Interferometer," *J. Res. Nat. Bur. Stand.*, **69C**, 251 (1961).

Sciammarella C. A., "The Moiré Method. A Review," *Exp. Mech.*, **22**, 418–433 (1982).

Stricker J., "Electronic Heterodyne Readout of Fringes in Moiré Deflectometry," *Opt. Lett.*, **10**, 247–249 (1985).

Suganuma M. and T. Yoshisawa, "Three Dimensional Shape Analysis by Use of a Projected Grating Image," *Opt. Eng.*, **30**, 1529–1533 (1991).

Takasaki H., "Moiré Topography," *Appl. Opt.*, **9**, 1467–1472 (1970).

Takasaki H., "Moiré Topography," *Appl. Opt.*, **12**, 845–850 (1973).

Takeda M., "Fringe Formula for Projection Type Moiré Topography," *Opt. and Lasers in Eng.*, **3**, 45–52 (1982).

Takeda M. and S. Kobayashi, "Lateral Aberration Measurements with a Digital Talbot Interferometer," *Appl. Opt.*, **23**, 1760–1764 (1984).

Talbot W. H. F., "Facts Relating to Optical Science," *Phil. Mag.*, **9**, 401 (1836).

Twyman F. and A. Green, British Patent No. 103832 (1916).

Varman P. O., "A Moiré System for Producing Numerical Data of the Profile of a Turbine Blade Using a Computer and Video Store," *Opt. and Lasers in Eng.*, **5**, 41–58 (1984).

Vlad V., D. Popa, and I. Apostol, "Computer Moiré Deflectometry Using the Talbot Effect," *Opt. Eng.*, **30**, 300–306 (1991).

Wang G.-Y. and X.-P. Ling, "Accuracy of Fringe Pattern Analysis," *Proc. SPIE*, **1163**, 251–257 (1989).

Yatagai T. and M. Idesawa, "Automatic Fringe Analysis for Moiré Topography," *Opt. and Lasers in Eng.*, **3**, 73–83 (1982).

Yokoseki S. and T. Susuki, "Shearing Interferometer Using the Grating as the Beam Splitter," *Appl. Opt.*, **10**, 1575–1580 (1971a).

Yokoseki S. and T. Susuki, "Shearing Interferometer Using the Grating as the Beam Splitter. Part 2," *Appl. Opt.*, **10**, 1690–1693 (1971b).

2

Fourier Theory Review

2.1 INTRODUCTION

Fourier theory is an important mathematical tool for the digital processing of interferograms. Hence, it is logical to begin this chapter with a review of this theory. Extensive treatments of this theory may be found in many textbooks, for example, those of Bracewell (1986) and Gaskill (1978). The topic of digital processing of images has also been treated in several textbooks, for example, Gonzales and Wintz (1987), Jain (1989), and Pratt (1978).

2.1.1 Complex Functions

Complex functions are very important tools in Fourier theory. Before beginning the study of Fourier theory let us make a brief summary of complex functions. A complex function may be plotted in a so-called *phasor diagram*, where the real part of the function is plotted on the horizontal axis and the imaginary part on the vertical axis. A complex function may be written as

$$g(x) = \text{Re}\{g(x)\} + i\,\text{Im}\{g(x)\} \tag{2.1}$$

where $\text{Re}\{g\}$ stands for the real part of g and $\text{Im}\{g\}$ stands for the imaginary part of g.

The phase of this complex number is the angle, with respect to the horizontal axis, of the line from the origin to the complex function value being plotted. Thus, the phase of any complex function $g(x)$ may be obtained as

$$\phi = \tan^{-1}\left[\frac{\text{Im}\{g(x)\}}{\text{Re}\{g(x)\}}\right] \tag{2.2}$$

However, this phase has a wrapping effect, because if both the real and the imaginary parts are negative, the ratio is the same as if both quantities were positive. Thus, this phase is within the limits $0 \leq \phi \leq \pi$. The magnitude of this complex number is defined by

$$g(x) = \left[(\text{Re}\{g(x)\})^2 + (\text{Im}\{g(x)\})^2\right]^{1/2} \tag{2.3}$$

which is always positive. This complex number may also be written as

$$g(x) = |g(x)|\exp(i\phi) \tag{2.4}$$

where the phase ϕ may have any value between 0 and 2π.

Sometimes this representation has some disadvantages. Let us consider, for example, the case of the real function $g(x) = x$. Using this expression, it has to be written as $g(x) = |x|$ for $x \geq 0$ and as $g(x) = |x|\exp(\pi)$ for $x \leq 0$. To avoid this discontinuity, both in the derivative of the function and in the phase, we may define the *amplitude* Am of a complex function by the expression

$$g(x) = \text{Am}\{g(x)\}\exp(i\phi) \tag{2.5}$$

where the derivative of the function $g(x)$ and the phase are now continuous over the complex plane. This amplitude is the equivalent of the radial coordinate in polar coordinates. A change in the sign of the amplitude is equivalent to a change of π in the phase.

The phase, as plotted in the phasor diagram, of a periodic real function like the functions $\sin\phi$ and $\cos\phi$ is zero because the function is real. However, there is a concept of phase ϕ associated with a real sinusoidal function. Frequently, we refer to these real functions as stationary waves, and their phase in the phasor diagram is zero. On the other hand, a function $\exp(i\phi) = \cos\phi + I\sin\phi$ on the phase diagram plots a unit circle, and its phase may be represented there. For this reason, this function is sometimes called a traveling wave. These two phases, the phase of a complex function and the phase of a real periodic function, are slightly different concepts but quite related to each other. In general, it is not necessary to specify which phase we are considering, because normally that is clear from the context.

2.2 FOURIER SERIES

A real, infinitely extended periodic function with fundamental frequency f_1 may be decomposed into a sum of real (stationary) sinusoidal functions with frequencies that are multiples of the fundamental, called *harmonics*. Thus, we may write

$$g(x) = \frac{a_0}{2} + \sum_{n=1}^{\infty} [a_n \cos(2\pi n f_1 x) + b_n \sin(2\pi n f_1 x)] \tag{2.6}$$

The coefficients a_n and b_n are the amplitudes of each of the sinusoidal components. If the function $g(x)$ is real, these coefficients are also real. Multiplying this expression first by $\cos(2\pi m f_1 x)$ and then by $\sin(2\pi m f_1 x)$ and making use of the well-known orthogonality properties for the trigonometric functions, we may easily obtain, after integrating for a full period, an analytical expression for the coefficients, which may be calculated from $g(x)$ by

$$a_n = \frac{1}{x_0} \int_{-x_0}^{x_0} g(x) \cos(2\pi n f_1 x) \, dx \tag{2.7}$$

and

$$b_n = \frac{1}{x_0} \int_{-x_0}^{x_0} g(x) \sin(2\pi n f_1 x) \, dx \tag{2.8}$$

where the fundamental frequency is equal to twice the inverse of the period length $2x_0 (f_1 = 1/2x_0)$. We may see that the frequency components have a constant separation equal to the fundamental frequency f_1. If the function is symmetrical, i.e., $g(x) = g(-x)$, only the coefficients a_n may be different from zero, but if the function is antisymmetrical, i.e., $g(x) = -g(-x)$, only the coefficients b_n may be different from zero. If the function is asymmetrical, both coefficients a_n and b_n may be different from zero. The coefficients a_n and b_n always correspond to positive frequencies. Figure 2.1 shows some common periodic functions and their Fourier transforms.

Fourier series may also be written in terms of complex functions. The periodic functions just described are represented by a sum of real (stationary) sinusoidal functions. In order to describe complex functions, the coefficients a_n and b_n have to be complex. An equivalent expression in terms of complex (trav-

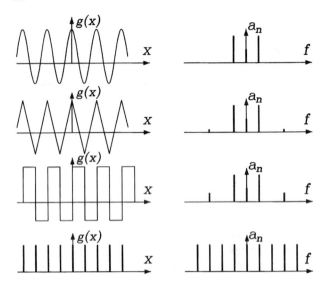

Figure 2.1 Some periodic functions and their spectra.

eling) sinusoidal functions $\exp(i2\pi nf_1x)$ and $\exp(-i2\pi nf_1x)$ using complex exponential functions instead of real trigonometric functions is

$$g(x) = \sum_{n=-\infty}^{\infty} c_n e^{i2\pi nf_1x} \tag{2.9}$$

where the coefficients c_n may be real, imaginary, or complex. These exponential functions are also orthogonal, like the trigonometric functions. The coefficients may be calculated with

$$c_n = \int_{-x_0}^{x_0} g(x)e^{-i2\pi nf_1x}\,dx \tag{2.10}$$

In this case the coefficients c_n correspond to positive frequencies (phase increasing in the negative direction of x) as well as to negative frequencies (phase increasing in the positive direction of x). Thus, the number n may be positive as well as negative. In general, the coefficients c_n are complex. If the function $g(x)$ is symmetrical, the coefficients c_n are real, with $c_n = c_{-n} = 2a_n$. On the other hand, if the function $g(x)$ is antisymmetrical, the coefficients c_n are imaginary, with $c_n = -c_{-n} = -2ib_n$. Table 2.1 shows some periodic functions and their coefficients a_n and b_n.

Table 2.1 Some Periodic Functions and Their Coefficients a_n and b_n

Function	Coefficients
Cosinusoidal $g(x) = A + B\cos(2\pi f_1 x)$	$a_0 = 2A$ $a_1 = B;\ b_n = 0$ $a_n = 0;\ n \geq 2$
Triangular $g(x) = A + B(1 - 4f_1 x);\ -x_0 \leq x \leq 0$ $g(x) = A + B(1 - 4f_1 x);\ 0 \leq x \leq x_0$	$a_0 = 2A;\ b_n = 0$ $a_n = \dfrac{2B}{n^2\pi};\ n$ odd $a_n = 0;\ n$ even
Square $g(x) = A - B;\ -x_0 \leq x \leq 0$ $g(x) = A + B;\ 0 \leq x \leq x_0$	$a_0 = 2A$ $b_n = \dfrac{2B}{n\pi};\ n$ odd $b_n = 0;\ n$ even
Comb $g(x) = \displaystyle\sum_{n=-\infty}^{\infty} \delta(x - nx_0)$	$a_0 = \dfrac{\delta(f)}{2};\ b_n = 0$ $a_n = \delta(f - nf_1);\ n \neq 0$

2.3 FOURIER TRANSFORMS

If the period of the function $g(x)$ is increased, the separation between sinusoidal components decreases. In the limit when the period becomes infinity, the frequency interval among harmonics tends to zero. Any nonperiodic function may be regarded as a periodic function with an infinite period. Thus, a nonperiodic continuous function may be represented by an infinite number of sinusoidal functions, transforming the series in Eq. 2.5 into an integral, where the frequency separation f_1 becomes df. This leads us to the concept of the Fourier transform.

Let $g(x)$ be a continuous function of a real variable x. Then the *Fourier transform* of $g(x)$ is $G(f)$ defined by

$$G(f) = \int_{-\infty}^{\infty} g(x)e^{-i2\pi f x}\ dx \tag{2.11}$$

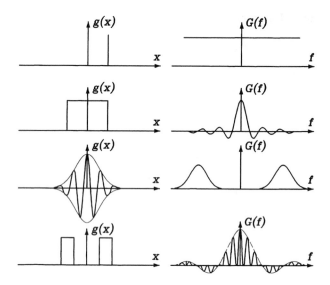

Figure 2.2 Some Fourier transform pairs. Each $G(f)$ at the right is the amplitude spectrum of the function $g(f)$ shown to its left.

This Fourier transform function $G(f)$ is also called the *amplitude spectrum* of $g(x)$, and its magnitude is the *Fourier spectrum* of the function $g(x)$. The Fourier transform of $g(x)$ may also be represented by $F\{g(x)\}$. For example, a perfectly sinusoidal function $g(x)$ without any constant term added has a single frequency component. The spectrum is a pair of Dirac delta functions located symmetrically with respect to the origin. Given $G(f)$, the function $g(x)$ may be obtained by its *inverse Fourier transform* defined by

$$g(x) = \int_{-\infty}^{\infty} G(f)e^{i2\pi fx}\, df \qquad (2.12)$$

We may notice that Eq. 2.10 is similar to Eq. 2.11 and that Eq. 2.12 is similar to Eq. 2.9 when the fundamental frequency tends to zero. Here, x is the space variable and its domain is called the *space domain*. On the other hand, f is the frequency variable and its domain is called the *frequency* or *Fourier domain*. A *Fourier transform pair* is defined by Eqs. 2.11 and 2.12. Both functions $g(x)$ and $G(f)$ may be real or complex. Figure 2.2 and Table 2.2 show some examples of Fourier transform pairs.

Table 2.2 Some Fourier Transform Pairs

Space domain function	Frequency domain function
Dirac delta (impulse) function $g(x) = \delta(x - x_0)$	Constant: $G(f) = Ae^{-i2\pi f x_0}$
Square function $g(x) = A; \; \lvert x \rvert \leq a$ $g(x) = 0; \; \lvert x \rvert > a$	Sinc function: $G(f) = 2Aa \dfrac{\sin(2\pi f x_0)}{2\pi f x_0}$
Gaussian modulated wave $g(x) = A \cos(2\pi f_0 x) e^{-x^2/a^2}$	Gaussian function: $G(f) = \dfrac{Aa\sqrt{\pi}}{2} e^{-2\pi a^2 (f - f_0)^2/4}$ $\quad + \dfrac{Aa\sqrt{\pi}}{2} e^{-2\pi a^2 (f + f_0)^2/4}$
Pair of square functions $g(x) = A; \; b - a \leq \lvert x \rvert \leq b + a$ $g(x) = 0; \; \lvert x \rvert < b - a \wedge \lvert x \rvert b + a$	Sinc modulated wave: $G(f) = 4Aa \cos(2\pi f b) \dfrac{\sin(2\pi f a)}{2\pi f a}$

The magnitude $\lvert G(f) \rvert$, as we mentioned before, is called the *Fourier spectrum* of $g(x)$, and the square of this magnitude is the *power spectrum*, sometimes also called the *spectral density*.

The phase ϕ, at the origin ($x = 0$), of a real cosinusoidal function $\cos(\omega_S x + \phi)$ is equal to the complex phase, at the origin, of its spectral component $\exp[I(\omega_S x + \phi)]$, which in turn is equal to the complex phase of the Fourier transform $[\delta(\omega - \omega_S) \exp(i\phi)]$ of the cosine function at the frequency $\omega = \omega_S$. An important and useful conclusion is that the phase of the real cosinusoidal Fourier components of a real function is equal to the complex phase of its Fourier transform at the frequency of that component.

2.3.1 Parseval Theorem

Another important theorem is the Parseval theorem, which may be written as

$$\int_{-\infty}^{\infty} \lvert g(x) \rvert^2 \, dx = \int_{-\infty}^{\infty} \lvert G(f) \rvert^2 \, df \tag{2.13}$$

This theorem may be described by saying that the total power in the space domain is equal to the total power in the frequency domain.

2.3.2 Central Ordinate Theorem

From Eq. 2.11 we may see that

$$\left[\int_{-\infty}^{\infty} g(x) e^{-i2\pi fx}\, dx \right]_{f=0} = G(0) = \int_{-\infty}^{\infty} g(x)\, dx \tag{2.14}$$

Thus, the integral of a function is equal to the central ordinate of the Fourier transform. An immediate consequence is that since any lateral translation of the function $g(x)$ does not change the area, the central ordinate value does not change either.

2.3.3 Translation Property

Another useful property of the Fourier transform is the translation property, which states that a translation of the input function $g(x)$ changes the phase of the transformed function as follows:

$$F\{g(x + x_0)\} = G(f) \exp(i2\pi f x_0) \tag{2.15}$$

or, in the frequency domain,

$$G(f + f_0) = F\{g(x) \exp(i2\pi f_0 x)\} \tag{2.16}$$

A consequence of this theorem is that the Fourier transform of any function, with any kind of symmetry, can be made to be real, imaginary, or complex, by means of a proper translation of the function $f(x)$.

2.3.4 Derivative Theorem

If $g'(x)$ is the derivative of $g(x)$, the Fourier transform of this derivative is given by

$$\int_{-\infty}^{\infty} g'(x) \exp(-2\pi i f x)\, dx = \int_{-\infty}^{\infty} \lim_{\Delta x \mapsto 0} \frac{g(x + \Delta x) - g(x)}{\Delta x}$$
$$\times \exp(-i2\pi f x)\, dx$$
$$= \lim_{\Delta x \mapsto 0} \frac{\exp(i2\pi f \Delta x) G(f) - G(f)}{\Delta x}$$
$$= i2\pi f G(f) \tag{2.17}$$

or

$$g'(x) = F^{-1}\{i2\pi f G(f)\} \tag{2.18}$$

Thus, the Fourier transform of the derivative of a function $g(x)$ is equal to the Fourier transform of the function, multiplied by $i2\pi f$. Now, using the convolution expression in Eq. 2.25 (see Sec. 2.4), we may write

$$g'(x) = F^{-1}\{G(f)H(f)\} = g(x) * h(x) \tag{2.19}$$

with

$$h(x) = F^{-1}\{i2\pi f\} = F^{-1}\{H(f)\} \tag{2.20}$$

This means that the derivative of $g(x)$ may be calculated with the convolution of this function with the function $h(x)$. By taking the inverse Fourier transform, this function $h(x)$ is equal to

$$h(x) = \frac{2f}{x} \cos(2\pi f_0 x) - \frac{1}{\pi x^2} \sin(2\pi f_0 x)$$

$$= \lim_{f_0 \mapsto \infty} 2f_0 \frac{d}{dx} [\text{sinc}(2\pi f_0 x)] \tag{2.21}$$

where $\text{sinc}(\alpha) = \sin(\alpha)/\alpha$.

2.3.5 Symmetry Properties of Fourier Transforms

A function $g(x)$ is symmetrical or even if $g(x) = g(-x)$, antisymmetrical or odd if $g(x) = -g(-x)$, or asymmetrical if it is neither symmetrical nor antisymmetrical. An asymmetrical function may always be expressed by the sum of a symmetrical function plus an antisymmetrical function. A complex function is Hermitian if the real part is symmetrical and the imaginary part antisymmetrical. For example, the function $\exp(ix)$ is Hermitian. The complex function is anti-Hermitian if the real part is antisymmetrical and the imaginary part is symmetrical. These definitions are illustrated in Fig. 2.3.

The Fourier transform has many interesting properties as shown in Table 2.3.

The fact that the Fourier transform of a real asymmetrical function is Hermitian is called the *Hermitian property* of the spectrum of real functions.

A few more properties of Fourier transforms, derived from their symmetry properties, are

1. If the function $g(x)$ is complex, of the form $\exp[i\phi(x)]$ where $\phi(x)$ is positive for all values of x (imaginary part has the same sign as the real part for all values of x), then the spectral function $G(f)$ is different from zero only for positive values of f.
2. If the function $g(x)$ is complex, of the form $\exp i\phi(x)$ where $\phi(x)$ is negative for all values of x (the imaginary part is of opposite sign to the

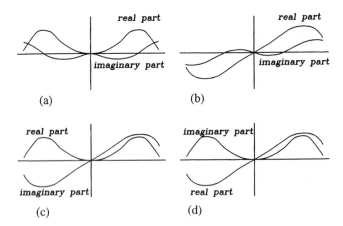

Figure 2.3   Possible symmetries of a function. (a) Symmetrical; (b) antisymmetrical; (c) Hermitian; (d) anti-Hermitian.

real part for all values of x), then the spectral function $G(f)$ is different from zero only for negative values of f.

3.   It is easy to show that for any complex function $g(x)$,

$$F\{g^*(x)\} = G^*(-f)$$ (2.22)

where the asterisk signifies the complex conjugate.

Table 2.3   Symmetry Properties of Fourier Transforms

	$g(x)$		$G(f)$
Real	Symmetrical	Real	Symmetrical
	Antisymmetrical	Imaginary	Antisymmetrical
	Asymmetrical	Complex	Hermitian
Imaginary	Symmetrical	Imaginary	Symmetrical
	Antisymmetrical	Real	Antisymmetrical
	Asymmetrical	Complex	Anti-Hermitian
Complex	Symmetrical	Complex	Symmetrical
	Antisymmetrical	Complex	Antisymmetrical
	Hermitian	Real	Asymmetrical
	Anti-Hermitian	Imaginary	Asymmetrical
	Asymmetrical	Complex	Asymmetrical

A particular and important case is when the function $g(x)$ is real. Then we may write

$$G(f) = G^*(-f); G(-f) = G^*(f) \tag{2.23}$$

which implies that

$$|G(f)| = |G^*(-f)| \tag{2.24}$$

From this expression we may conclude that if the function $g(x)$ is real, as in any image to be digitized, then the Fourier transform is Hermitian and the Fourier spectrum (or magnitude) $|G(f)|$ is symmetrical.

2.4 THE CONVOLUTION OF TWO FUNCTIONS

The convolution operation of the two functions $g(x)$ and $h(x)$ is defined by

$$g(x) * h(x) = \int_{-\infty}^{\infty} g(\alpha)h(x - \alpha) \, d\alpha \tag{2.25}$$

where the symbol $*$ denotes the convolution operator. It may be seen that the convolution is commutative, that is,

$$g(x) * h(x) = h(x) * g(x) \tag{2.26}$$

A property of the convolution operation is that the Fourier transform of the product of two functions is equal to the convolution of the Fourier transforms of the two functions:

$$F\{g(x)h(x)\} = G(f) * H(f) \tag{2.27}$$

or

$$F^{-1}\{G(f)H(f)\} = g(x) * h(x) \tag{2.28}$$

Conversely, the Fourier transform of the convolution of two functions is equal to the product of the Fourier transforms of the two functions, as follows:

$$F\{g(x) * h(x)\} = G(f)H(f) \tag{2.29}$$

or

$$F^{-1}\{G(f) * H(f)\} = g(x)h(x) \tag{2.30}$$

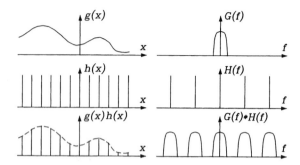

Figure 2.4 A function $g(x)$, a comb function $h(x)$, and their product, and the convolution of their Fourier transforms.

Figure 2.4 shows the product of the function $g(x)$ and the comb function $h(x)$ and, on the right side, the convolution of the Fourier transforms of these functions. A comb function is formed by an infinite array of equally spaced Dirac deltas.

The convolution may be interpreted in several equivalent ways. Two different models that can be used to interpret the convolution are described here.

The first interpretation may be explained as a sequence of the following three steps, as illustrated in Fig. 2.5a:

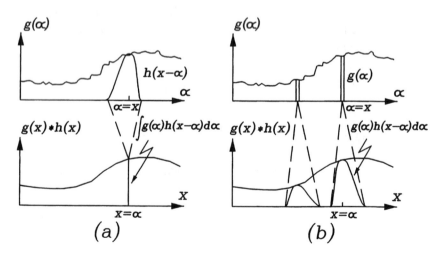

Figure 2.5 Two models used to interpret the convolution of two functions. See text for descriptions.

1. A value of x is selected in the domain of the convolution.
2. At the point $\alpha = x$ in the functions space, the function $h(\alpha)$ is placed with a reversed orientation, obtaining $h(x - \alpha)$.
3. An average of the function $g(\alpha)$, weighted by the function $h(x - \alpha)$ is obtained by first multiplying the function $g(\alpha)$ by the function $h(x - \alpha)$ and then integrating. The result of the integration is the value of the convolution at the point x.

The second interpretation of the convolution operation is explained by means of the following four steps, as shown in Fig. 2.5b:

1. The α axis is divided into many extremely narrow intervals of equal width $d\alpha$. The narrow interval at any position α is selected.
2. At the corresponding point $x = \alpha$ in the convolution space, the function $h(x)$ is placed without reversing it, to obtain the function $h(x - \alpha)$. Then its height is made directly proportional to the value of $g(x)$ by multiplying the two functions.
3. The preceding steps are repeated for all narrow intervals in the functions space.
4. All functions $g(x)h(x - \alpha)\, d\alpha$ in the convolution space are added by integration.

A property of the convolution is that the extent of the convolution is equal to the sum of the two function bases being convolved.

2.4.1 Filtering by Convolution

An important application of the convolution operation is the low pass, band pass, or high pass filtering of a function $g(x)$ by means of a filter function $h(x)$. This filtering property of the convolution operation may be easily understood if we write, from Eqs. 2.27 and 2.25,

$$\overline{g}(x) = \mathrm{F}^{-1}\{G(x)H(x)\} = \int_{-\infty}^{\infty} g(\alpha)h(x - \alpha)\, d\alpha \tag{2.31}$$

We see that the filtering or convolution operation is equivalent to multiplying the Fourier transform of the function to be filtered by the Fourier transform of the filtering function and then obtaining the inverse Fourier transform of the product. If the filtering function $h(x)$ has a great content of low frequencies and no high frequencies, we have a low pass filter. On the other hand, if the filtering function $h(x)$ has a great content of high frequencies and no low frequencies, we have a high pass filter. This convolution process with its associated low pass filtering is illustrated in Fig. 2.5.

Let us consider the special case of the convolution of a sinusoidal real function $g(x)$ formed by the sum of a sine function and a cosine function, with a filter function $h(x)$. Then, we obtain the filtered function $\overline{g}(x)$:

$$\overline{g}(x) = \int_{-\infty}^{\infty} (a\sin(2\pi n f\alpha) + b\cos(2\pi n f\alpha))\, h(x - \alpha)\, d\alpha \tag{2.32}$$

This expression, which is a function of x, must have a zero value for all values of x. The value of this function at the origin $(x = 0)$ is

$$\overline{g}(0) = \int_{-\infty}^{\infty} [a_n \sin(2\pi n f\alpha) + b_n \cos(2\pi n f\alpha)]\, h(-\alpha)\, d\alpha \tag{2.33}$$

The real sinusoidal function $g(x)$ with frequency f has two Fourier components, one with frequency f and the other with frequency $-f$. If only the first term (sine) is present in $g(x)$, the signal is antisymmetrical and the two Fourier components have the same magnitudes but opposite signs. In this case, if the signal is filtered with a filter function with symmetrical values at the frequency to be filtered, we see that the desired zero value is obtained at the origin but not at all values of x.

If only the second term (cosine) is present in $g(x)$, the signal is symmetrical and the two Fourier components have the same magnitudes and the same signs. In this case, if the signal is filtered with a filter function with antisymmetrical values at the frequency to be filtered, the correct filtered value of zero is again obtained only at the origin.

In the most general case, when both the sine and cosine functions are present in $g(x)$, the magnitudes and signs of the two Fourier components may be different. In the general case the filtering function must have zero values at both Fourier components.

2.5 THE CROSS-CORRELATION OF TWO FUNCTIONS

The cross-correlation operation of the two functions $g(x)$ and $h(x)$ is similar to the convolution, and it is defined by

$$g(x) \otimes h(x) = \int_{-\infty}^{\infty} g(\alpha)h(x + \alpha)\, d\alpha \tag{2.34}$$

where the symbol $\otimes$ denotes cross-correlation. This operation is not commutative but satisfies the relation

$$g(x) \otimes h(x) = h(-x) \otimes g(-x) \tag{2.35}$$

A property of the cross-correlation operation is that the Fourier transform of the product of the two functions is equal to the cross-correlation of the Fourier transforms:

$$F\{g(-x)h(x)\} = G(f) \otimes H(f) \tag{2.36}$$

Conversely, the Fourier transform of the cross-correlation is equal to the product of the Fourier transforms:

$$F\{g(x) \otimes h(x)\} = G(-f)H(f) \tag{2.37}$$

The cross-correlation is related to the convolution by

$$g(x) \otimes h(x) = g(-x) * h(x) \tag{2.38}$$

As the convolution operation, the cross-correlation may be used to remove high frequency Fourier components from a function $g(x)$ by means of a filter function $h(x)$.

2.6 SAMPLING THEOREM

Let us consider a band-limited real function $g(x)$ whose spectrum is $G(f)$. The width Δf of this spectrum is equal to the maximum frequency contained in the function.

To sample the function $g(x)$ we need to multiply it by a comb function $h(x)$, whose spectrum $H(f)$ is also a comb function, as shown in Fig. 2.4. The fundamental frequency of the comb function $h(x)$ is defined as the sampling frequency. A direct consequence of the convolution theorem is that the spectrum of this sampled function (product of the two functions) is the convolution of the two Fourier transforms $G(f)$ and $H(f)$.

In Fig. 2.6 we can see that if the sampling frequency of the function $h(x)$ decreases, the spectral elements in the convolution of the functions $G(f)$ and $H(f)$ get closer to each other. If these spectral elements are completely separated without any overlapping (Fig. 2.6a), the inverse Fourier transform recovers the original function with full details and frequency content. If the spectral elements overlap each other as in Fig. 2.6c, the process is not reversible. The original function may not be fully recovered after sampling if the spectral elements do overlap or even touch each other. Thus, the sampling theorem requirements have just been violated when the spectral elements are just touching each other as in Fig. 2.6b.

The total width $2\Delta f$ of the base of the spectral elements is smaller than twice the maximum frequency f_{max} in the signal or function being sampled, as

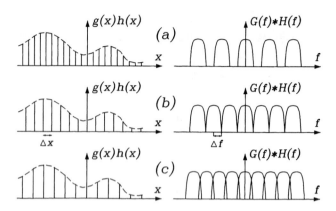

Figure 2.6 Sampling of a function with sampling frequencies (a) above the Nyquist frequency; (b) just below the Nyquist frequency; (c) below the Nyquist frequency.

defined by its Fourier transform. On the other hand, the frequency separation between the peaks in the Fourier transform of the comb function is equal to the sampling frequency. Hence, the sampling frequency $f_S = 1/\Delta x$ must be greater than half the maximum frequency f_{max} contained in the signal or function to be sampled,

$$f_S > 2f_{max} \tag{2.39}$$

This condition is known as the *Whittaker-Shannon sampling theorem*. The minimum sampling frequency is known as the *Nyquist frequency* (Nyquist 1928). Alternatively, we can say that once a signal has been sampled, the maximum frequency in this sampled signal is equal to half the sampling frequency.

If the spectral elements overlap, the recovery of the sampled function is not perfect and a phenomenon called *aliasing* occurs.

In this discussion we have assumed that the sampling function $h(x)$ extends from $-\infty$ to $+\infty$ and that the sampled function is band-limited. In most practical cases neither of these is true. Later we examine this in more detail.

If the sampling extends only from $-x_0$ to x_0, we may consider instead, for simplicity, that the sampling points, that is the function $h(x)$, extend from $-\infty$ to $+\infty$ but that the function to be sampled $g(x)$ is multiplied by a window function $w(x)$ as shown in Fig. 2.7. Then, by the convolution theorem, the spectrum of the product of these two functions is the convolution of its Fourier transforms. The Fourier transform of the window function is the *sinc function*, whose spectrum extends from $-\infty$ to $+\infty$. Thus, the spectrum elements of the windowed sampled function necessarily have some overlap. The important

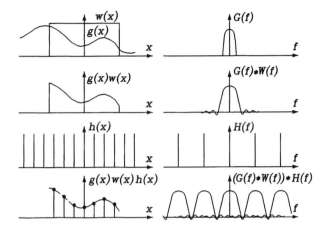

Figure 2.7 Illustration of the sampling theorem with a limiting aperture (window). The left side shows the product of the function $g(x)$ being sampled and the right side shows the Fourier transform.

conclusion is that a sampling-bounded function (or an interval-limited sampling) is always imperfect, since a perfect recovery of the function is not possible.

2.7 SAMPLING OF A PERIODIC FUNCTION

There is only one important case in which a limited sampling may lead to a perfect recovery of the function. That is when the function is periodic (not necessarily sinusoidal) and band-limited (a highest order harmonic frequency must exist) with a fundamental spatial period equal to the length of the total sampling interval. The reason will be seen in the next paragraph.

Let us assume that the function is periodic and band-limited. Then it may be represented by a Fourier series with a finite number of terms. Due to the periodicity of the function we may assume that the sampling pattern repeats itself outside the sampling interval, as in Fig. 2.8. Even if the sampling points are equally spaced if they are not uniformly distributed in the interval and the sampling pattern is repeated, the whole distribution of virtual sampling points (empty points) is not uniform (see Fig. 2.8a). However, if we take the N sampling points uniformly and equally spaced, as shown in Fig. 2.8b, with phases ϕ_n given by

$$\phi_n = \frac{2\pi(n-1)}{N} + \phi_0 \tag{2.40}$$

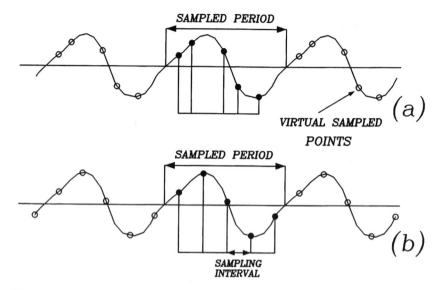

Figure 2.8 Sampling of a periodic function with a finite sampling interval. See text for details.

where ϕ_0 is the phase at the first sampling point ($n = 1$), then the virtual sampling points in the whole infinite interval will be equally distributed. Then a sampling in an interval of length equal to the period of the fundamental is enough to obtain full recovery of the function. Of course, this is assuming that the sampling frequency is greater than twice the maximum frequency contained in the function.

The advantage of extrapolating the function in this manner outside the sampling interval is that the sampling may be mathematically considered as extending throughout the whole interval from $-\infty$ to $+\infty$. Then we may be sure that the sampling theorem is strictly satisfied.

An interesting particular case of a periodic and bandwidth-limited function is a pure sinusoidal function. If we sample a sinusoidal function, the sampling theorem requires a greater sampling frequency than twice the frequency of the sinusoidal function (equal is not acceptable). Two sampling points in the period length makes the sampling frequency equal to twice the frequency of the sampled function. If the sampling interval is much larger than one period, we could sample with a frequency just slightly greater than this required minimum of two points per period. However, if the sampling interval is just one period (as in most phase shifting algorithms), we need a minimum of three sampling points per period.

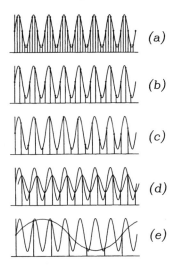

Figure 2.9 Sampling of a periodic function with a finite sampling interval. (a) Frequency higher than twice the frequency of the function. (b) Three points per period. (c) Smaller sampling frequency, satisfying the sampling theorem. (d) Sampling frequency equal to 2. (e) Sampling frequency less than twice the frequency of the sinusoidal function.

Figure 2.9a shows a sinusoidal signal sampled with a frequency f_S much higher than twice the frequency f of this signal. Figure 2.9b shows the sampling with three points per period.

Figure 2.9c shows a smaller sampling frequency, but one that still satisfies the sampling theorem requirements.

Figure 2.9d has a sampling frequency equal to 2, just outside the sampling theorem requirements. We see that there are several possibilities in the function reconstruction (two of them are illustrated).

Finally, Fig. 2.9e has a sampling frequency less than twice the frequency of the sinusoidal function, with the aliasing effect clearly shown. With aliasing, instead of reproducing the signal with frequency f, a false signal with a frequency $f_S - f$ and the same phase at the origin as the signal appears. Since the requirements of the sampling theorem were violated, the frequency of this aliased wave is smaller than the signal frequency.

Another way to visualize these concepts is by analyzing the same cases in the Fourier space, as shown in Fig. 2.10. These spectra correspond to the same cases as in Fig. 2.9.

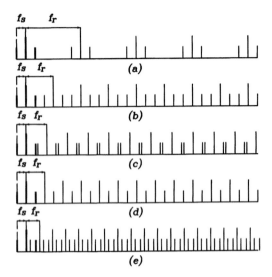

Figure 2.10 Spectra when sampling a periodic function with a finite sampling interval, as in Fig. 2.9.

2.7.1 Sampling of a Periodic Function with Interval Averaging

We have studied the sampling of a periodic function using a detector that measures the signal at one value of the phase. However, most real detectors cannot measure the phase at one value of the phase but take the average value in one small phase interval. This may be the case in space signals as well as in time signals.

In the case of a time-varying signal, as in phase shifting interferometry, the phase may be continually changing while the measurements are being taken. Thus, the number being read is the average of the irradiance during the time spent in the measurement. This method is frequently called *bucket integration.*

In the case of a space-varying signal, as when digitizing the image of sinusoidal interference fringes with a detector array, the detector may have a significant size compared with the separation between the detector elements. In this case the measurements are also the average of the signal over the detector extension.

Let us consider this signal averaging as in Fig. 2.11, where the signal $s(x)$ is measured in an interval centered at x and extending from $x - x_0/2$ to $x + x_0/2$,

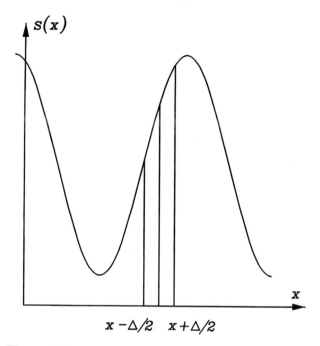

Figure 2.11 Signal averaging when measuring a sinusoidal signal in a phase interval from $x - x_0/2$ to $x + x_0/2$.

where x_0 is the size of the interval. Then the average signal on this interval is given by

$$\bar{s}(x) = \frac{\int_{x-x_0/2}^{x+x_0/2} s(x)\, dx}{x_0} = \frac{\int_{x-x_0/2}^{x+x_0/2} (a + b\cos x)\, dx}{x_0} \qquad (2.41)$$

thus obtaining

$$\bar{s}(x) = a + b\, \text{sinc}(x_0/2)\cos x \qquad (2.42)$$

This result tells us that the effect of this signal averaging just reduces the contrast of the fringes with the filtering function sinc $(x_0/2)$. As is to be expected, for an infinitely small averaging interval ($x_0 = 0$) there is no reduction in contrast. However, for finite size intervals the contrast is reduced.

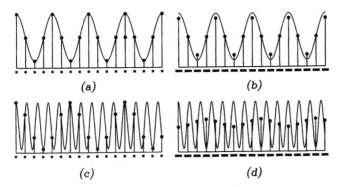

Figure 2.12 Contrast of a detected signal for a finite size of integration. (a) Below the Nyquist limit and small integration interval; (b) below the Nyquist limit and large integration interval; (c) above the Nyquist limit and small integration interval, showing aliasing; (d) below the Nyquist limit and large integration interval, showing reduction and inversion in the contrast.

The sinc function has zeros at $x_0 = 2m\pi$ with m being an integer. Thus, the first zero occurs at $x_0 = 2\pi$. If the sampling detectors have a size equal to their separation, so that there is no space between them, as in most practical CCD detectors, this corresponds to half the sampling frequency allowed by the sampling theorem. In other words, when increasing the signal frequency, the Nyquist frequency is reached before the first zero of the contrast. Hence, at these values of x_0, when the averaging interval is a multiple of the wavelength of the signal (spatial or temporal), the contrast is reduced to zero and no signal, but the DC component is detected. For averaging intervals between π and 2π the contrast is reversed. These contrast changes are illustrated in Fig. 2.12.

When the signal is sampled at equally spaced intervals, there is an upper limit for the size of the averaging interval, when the averaging intervals just touch each other. Then, the averaging interval size is equal to the inverse of the sampling frequency, that is, $x_0 = 1/f_s$. With this detector, at the Nyquist limit (sampling frequency equal to twice the signal frequency) the integration interval is equal to half the period of the signal ($x_0 = \pi$) and the contrast reduction is $2/\pi = 0.6366$. The contrast is zero when the sampling frequency is equal to the signal frequency f.

In digitization of images, this frequency-selective contrast reduction (filtering) is sometimes an advantage because it reduces the aliasing effect. However, in some interferometric applications, as will be described later in this book, the aliasing effect may be useful.

2.8 FAST FOURIER TRANSFORM

The numerical computation of a Fourier transform takes an extremely long time even in modern powerful computers. Several algorithms have been designed by various authors since the beginning of the century, but the work of Cooley and Tukey (1965) was the first to become very well known. It was called the *fast Fourier transform* (FFT). Tukey devised an algorithm to compute the Fourier transform in a relatively short time by eliminating unnecessary calculations, and Cooley did the programming. Their work was not published, but it aroused enough interest to be used by several researchers. Richard L. Garwin was in need of this algorithm and went to see Cooley to ask about his work. Cooley's answer was that he did not publish it because he considered the algorithm quite elementary. However, he agreed to work again on the problem. Eventually, the Tukey-Cooley algorithm was published, to be later also known as the fast Fourier transform. Now this method may be found in several books and publications (e.g., Brigham 1974; Hayes 1992). Code for programs in C language (Press *et al.* 1988) or in Basic (Hayes 1992) for the numerical computation may also be found in the literature.

Since the Fourier transform is to be calculated with a computer, the function to be transformed needs to be sampled by means of a comb sampling function. Then the integral becomes a discrete sum. The *discrete Fourier transform* (DFT) pair is defined by

$$G_k = \sum_{l=0}^{N-1} g_l e^{-i2\pi kl/N} \tag{2.43}$$

and

$$g_l = \frac{1}{N} \sum_{k=0}^{N-1} G_k e^{i2\pi kl/N} \tag{2.44}$$

The first expression may be written as

$$G_k = \sum_{l=0}^{N-1} g_l W^{kl} \tag{2.45}$$

where

$$W = e^{-i2\pi/N} \tag{2.46}$$

We see that the sampled function g_l to be Fourier transformed has a bounded domain contained in an array of N points. The Fourier transform G_k is calculated

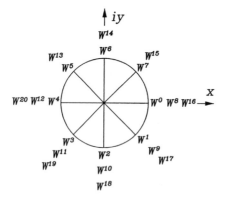

Figure 2.13   Phasor diagram representing the values of W^{kl} for $N = 8$.

at another array of N points in the frequency space. Thus, N multiplications have to be carried out for each G_k. To calculate the whole Fourier transform set of numbers G_k, N_2 multiplications are necessary. This operation, written in matrix notation (Iisuka 1987), is equivalent to

$$
\begin{pmatrix} G_0 \\ G_1 \\ G_2 \\ \cdot \\ \cdot \end{pmatrix} = \begin{pmatrix} W^0 & W^0 & W^0 & W^0 & \cdot & W^0 \\ W^0 & W^1 & W^2 & W^3 & \cdot & W^{N-1} \\ W^0 & W^2 & W^4 & W^6 & \cdot & W^{2(N-1)} \\ \cdot & & \cdot & & \cdot & \cdot \\ W^0 & W^{(N-1)} & \cdot & \cdot & \cdot & W^{(N-1)(N-1)} \end{pmatrix} \begin{pmatrix} g_0 \\ g_1 \\ g_2 \cdot \\ \cdot \end{pmatrix} \qquad (2.47)
$$

Hence the discrete Fourier transform may be regarded as a linear transform. If there are N points to be sampled, there are N points in the transform. The elements of the matrix are given by Eq. 2.47.

This matrix has some interesting characteristics that may be used to reduce the time used in the matrix multiplication. The fast Fourier transform is just an algorithm to reduce the number of operations. This matrix multiplication involves $N \times N$ multiplications and $N \times (N - 1)$ additions.

The values of W^{kl} may be represented in a phasor diagram in the complex plane, as shown in Fig. 2.13. All values fall in a unit circle, and we may see that there are only N different values. We may also notice using Eq. 2.46 that values at opposite sides of the circle differ only in their sign. Points symmetrically placed with respect to the x axis have the same real part, and their imaginary parts differ only in sign. Points symmetrically placed with respect to the y axis have the same imaginary part, and their real parts differ only in sign.

The key property that allows us to reduce the number of numerical operations when calculating this Fourier transform is that a discrete Fourier transform of length N can be expressed as the sum of two discrete Fourier transforms of length $N/2$. One of the two transforms is formed by the odd points and the other by the even points, as follows:

$$
\begin{aligned}
G_k &= \sum_{l=0}^{N-1} g_l e^{-i2\pi kl/N} \\
&= \sum_{l=0}^{N/2-1} g_{2l} e^{-i2\pi k(2l)/N} + \sum_{l=0}^{N/2-1} g_{2l+1} e^{i2\pi k(2l+1)/N} \\
&= \sum_{l=0}^{N/2-1} g_{2l} e^{-i2\pi kl/(N/2)} + W^k \sum_{l=0}^{N/2-1} g_{2l+1} e^{-i2\pi kl/(N/2)}
\end{aligned}
\tag{2.48}
$$

where we have assumed that N is even. This property is called the *Danielson-Lanczos lemma*. Thus, we may also write

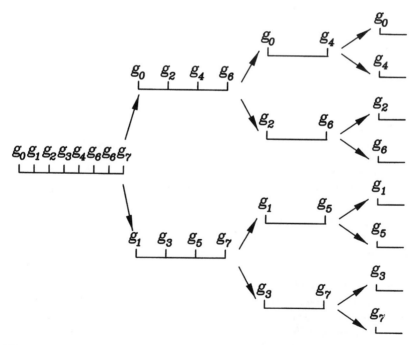

Figure 2.14 Fragmentation of a digitized signal with eight values into two parts in a successive manner until we have eight single values.

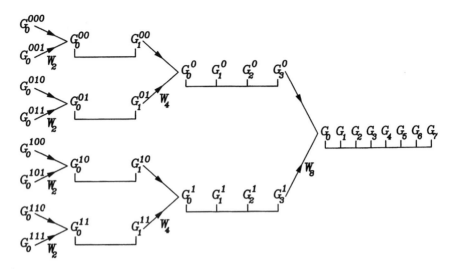

Figure 2.15 Calculation of the fast Fourier transform by grouping.

$$G_k = G_k^{\text{even}} + W^k G_k^{\text{odd}} \tag{2.49}$$

where each of these two Fourier transforms is of length $N/2$. So, now we have two linear transforms half the size of the original, and the total number of multiplications has been reduced to one-fourth.

The wonderful thing is that this principle can be used recursively. It is only necessary that in each step the number of points be even. Then it is ideal if the total number of points is $N = 2^M$, where M is an integer. The result is that the number of multiplications has been reduced from N^2 to $N \log_2 N$.

As an example of the procedure to find the fast Fourier transform, let us consider Fig. 2.14, where we have a signal with eight digitized values g_i. These values are first divided into two groups, one with the odd sampled values and another with the even sampled values. Again, each group is divided in two, and so on, until we have eight groups with a single value.

The next step is to find the Fourier transform of each of the single values, which is trivial. Then, with the procedure described earlier, the Fourier transforms of larger groups of signal values are calculated until we obtain the desired Fourier transform at eight frequency values, as shown in Fig. 2.15.

As an example, Fig. 2.16 illustrates the positions of the sampling points in the space domain as well as the calculated points in the frequency domain for a rectangular function. It is interesting to notice that if the sampling points are located only over the top of the rectangular function, the calculated points do

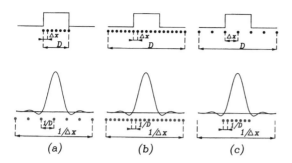

Figure 2.16 Location of sampling points in the transformed function and the location of the calculated points in frequency space.

not have enough resolution to give the shape of the expected sinc function. A solution is to sample a larger space in the function domain with additional points with zero values on both sides of the aperture.

The details of the fast Fourier transform algorithms have been described by several authors (e.g., Hayes 1992; Iizuka 1987; Press *et al.* 1988).

REFERENCES

Bracewell R. N., *The Fourier Transform and Its Applications*, 2nd Edition, McGraw-Hill, New York, 1986.

Brigham E. O., *The Fast Fourier Transform*, Prentice-Hall, Englewood Cliffs, NJ, 1974.

Cooley J. W. and J. W. Tukey, "An Algorithm for the Machine Calculation of Complex Fourier Series," *Math. Comput.*, **19** (No. 90), 297–301 (1965).

Gaskill J. D., *Linear Systems Fourier Transforms and Optics*, John Wiley and Sons, New York, 1978.

Gonzales R. C. and P. Wintz, *Digital Image Processing*, 2nd Edition, Addison-Wesley, Reading, MA, 1987.

Hayes J., "Fast Fourier Transforms and Their Applications," in *Applied Optics and Optical Engineering*, Vol XI, p. 56, Academic Press, New York, 1992.

Iisuka K., *Optical Engineering*, 2nd Edition, Springer Verlag, Berlin, 1987.

Jain A. K., *Fundamentals of Digital Image Processing*, Prentice-Hall, Englewood Cliffs, NJ, 1989.

Nyquist H., "Certain Topics in Telegraph Transmission Theory," *AIEE. Trans.*, **47**, 817–844 (1928).

Pratt W. K., *Digital Image Processing*, John Wiley and Sons, New York, 1978.

Press W. H., B. P. Flannery, S. A. Teukolsky, and W. T. Vetterling, *Numerical Recipes in C*, Cambridge University Press, Cambridge, 1988.

3
Digital Image Processing

3.1 INTRODUCTION

Digital image processing is a very important field by itself that has been treated in many textbooks (Gonzales and Wintz 1987; Jain 1989; Pratt 1978) and chapter reviews (Morimoto 1993). To digitize an image, it is separated into an array of small image elements called pixels. Each of these pixels has a different color and irradiance (gray level). The larger the number of pixels in the image, the greater the definition and sharpness of this image. Interferograms, as described in Chap. 1, may be analyzed using digital processing techniques. In this case, however, color information is not necessary. This is clearly illustrated in the images in Fig. 3.1.

The great advantage of digital image processing is that the image may be improved or analyzed using many different techniques. These techniques may be applied to the analysis of interferograms, as described by many authors over the last twenty years or so (e.g., Kreis and Kreitlow 1979).

When digitizing an image, different gray levels (irradiance) are digitized and transformed into numbers in the computer. These numbers are internally represented by binary numbers with only ones and zeros, called bits. A quantity written as a series of 8 bits is called a byte. A quantity may be represented by one, two, or even three bytes. Thus, the total number of bits being used to digitize an image determines the number of possible gray levels that may be used to represent this luminance level, as shown in Table 3.1.

Figure 3.1 Digitized images of the same picture, with different pixel separations, (a) 256 × 256 pixels; (b) 128 × 128 pixels; (c) 64 × 64 pixels; (d) 32 × 32 pixels.

Table 3.1 Gray Levels According to the Number of Bits

Number of unsigned bytes	Number of bits	Number gray levels
1	8	256
2	16	65,536

a

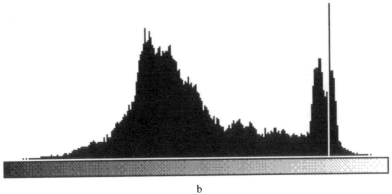

b

Figure 3.2 (a) A digitized image and (b) its gray level histogram.

a

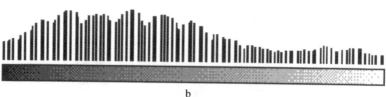

b

Figure 3.3 (a) Increased contrast in a digitized image and (b) its modified histogram.

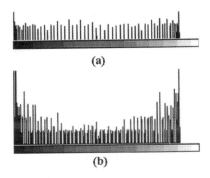

Figure 3.4 Histograms for two digitized interferograms, the first (a) with 20 pixels per fringe period and the second (b) with 200 pixels per fringe period.

3.2 HISTOGRAM AND GRAY SCALE TRANSFORMATIONS

One of the most important properties of a digitized image is the relative population of different gray levels. We may plot this information in a diagram where the x axis represents the luminance of the pixel and the y axis represents the number of pixels in an image with that value of the gray level. This diagram is called a *histogram*. Gray levels have only discrete quantized values, according to the number of bits used to represent them. Thus, the histogram is not a continuous curve, but a set of vertical line segments. Figure 3.2 shows a digitized image and its histogram.

The contrast of an image is also reflected in the histogram. This may be observed in Fig. 3.3, where the same image as in Fig. 3.2 but with a much greater contrast is shown with its histogram.

It is interesting to note that the image of a digitized interferogram with perfectly sinusoidal fringes, without noise, has more dark and clear pixels than pixels with intermediate gray levels. There are two histogram maxima. The first maximum corresponds to the gray level at the top of the clear fringes, and the second maximum corresponds to the gray level at the top of the dark fringes. If noise is present, the height of the first peak in the histogram is reduced. The aspect of the histogram depends on the number of pixels per fringe period, as shown in Fig. 3.4.

3.3 SPACE AND FREQUENCY DOMAIN OF INTERFEROGRAMS

When digitizing or sampling an interferogram, the selection of the sampling points is extremely important. This may be seen from a study of the effect of sampling points on the frequency domain. Womack (1983, 1984) described the properties of this frequency domain of interferograms.

Let us consider the interferogram of an aberrated wavefront with a large tilt (linear carrier) as in Fig. 3.4a. Let us assume that the irradiance signal in this interferogram may be written as

$$s(x, y) = a(x, y) + b(x, y) \cos[k(x \sin \theta - W(x, y))] \tag{3.1}$$

This irradiance has been represented here by $s(x, y)$ instead of $I(x, y)$, so that the Fourier transform becomes $S(f_x, f_y)$. The variable θ represents the tilt angle introducing the linear carrier, k is equal to $2\pi/\lambda$, and $W(x, y)$ is the wavefront deformation. We may also write this irradiance as

$$s(x, y) = a(x, y) + b(x, y) \cos[2\pi f_0 x - kW(x, y)] \tag{3.2}$$

where f_0 is the spatial frequency introduced in the interferogram by the tilt. This expression may also be written as

$$s(x, y) = a(x, y) \left[1 + \frac{b(x, y)}{a(x, y)} \cos(2\pi f_0 x - kW(x, y)) \right]$$
$$= a(x, y)[1 + v(x, y) \cos(2\pi f_0 x - kW(x, y))] \tag{3.3}$$

where $v(x, y)$ is the fringe visibility. If we define a function $u(x, y)$, sometimes called the complex fringe visibility, as

$$u(x, y) = v(x, y)e^{-ikW(x,y)} \tag{3.4}$$

we obtain

$$s(x, y) = a(x, y) + 0.5a(x, y)[u(x, y) \exp(i2\pi f_0 x)$$
$$+ u^*(x, y) \exp(-i2\pi f_0 x)] \tag{3.5}$$

Then, using the convolution theorem and Eq. 2.15, the Fourier transform of this function $s(x, y)$ is

$$S(f_x, f_y) = A(f_x, f_y) + 0.5A(f_x, f_y)$$
$$* [U(f_x - f_0, f_y) + U^*(-f_x - f_0, -f_y)] \tag{3.6}$$

where the symbol $*$ represents the convolution operation. Thus, this spectrum would be concentrated in three circles (lobes), a small one at the origin and two

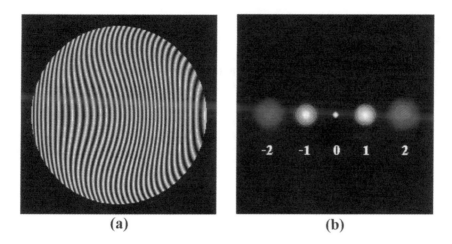

(a)   (b)

Figure 3.5 Interferogram and its frequency domain space image. (a) Interferogram with tilt; (b) spectrum. The second-order lobes are due to nonlinearities.

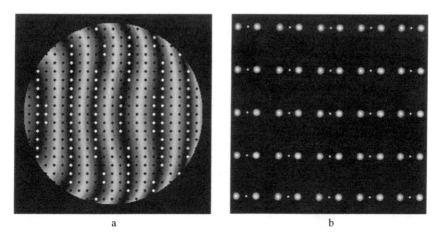

a   b

Figure 3.6 (a) Interferogram sampled with a rectangular array of points; (b) spectrum.

larger ones centered at f_0 and $-f_0$, with a radius equal to the frequency cut-off of $U(f)$.

The image in the frequency domain space (spectrum) of an interferogram without any tilt is a bright spot at the center in the frequency space. If tilt is added to the interferogram, as shown in Fig. 3.5a, the spectrum splits into several orders, as illustrated in Fig. 3.5b, but the three brightest components are the zero, -1, and $+1$ orders. The central bright peak is at the center, with two smaller lobes corresponding to the two first (± 1) orders on each side. If the tilt is increased, the separation between these lobes also increases.

If the interferogram is sampled with a rectangular array of points, as in Fig. 3.6a, the spectrum looks like Fig. 3.6b. To separate the different orders of diffraction and to be able to reconstruct the image of the interferogram, according to the sampling theorem, the sampling points must have a spatial frequency higher than twice the maximum spatial frequency present in the interferogram.

3.4 DIGITAL PROCESSING OF IMAGES

In a digital image or interferogram, some types of spatial characteristics must sometimes be detected, reinforced, or eliminated. Also, some kinds of noise may need to be removed by means of some type of averaging or spatial filtering. The general procedures used in the digital processing of images will now be described.

Digital processing of the image is performed by means of a *window* or *mask*, sometimes also called a *kernel*, represented by a matrix of $N \times N$ pixels. This mask is placed over the image to be processed. Then each value h_{nm} in the mask is multiplied by the corresponding number of pixels with signal (gray level) s_{nm} on the image as shown in Fig. 3.7, and all these products are added to obtain the result s' as follows:

$$s'_{00} = \sum_{n=-M}^{M} \sum_{m=-M}^{M} h_{nm} s_{nm} \tag{3.7}$$

where $M = (N-1)/2$.

The result s' of this operation is used to define a new number to be inserted in the processed new image, at the pixel corresponding to the center of the window. After this, the mask is displaced to the next pixel in the image being processed, and the preceding operations are repeated for the new position. In this manner the whole image is scanned. In the next sections some of the main image operations that can be performed will be described.

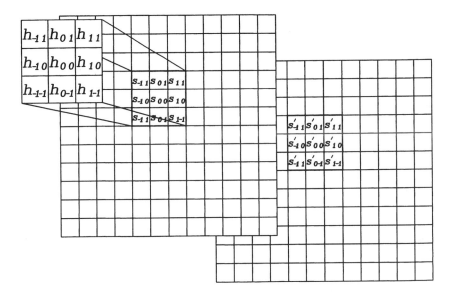

Figure 3.7 Image processing with window or mask.

3.4.1 Point and Line Detection

The simplest operation is the detection of a pixel with a gray level too different from those of its surrounding pixels. To do this we take the average signal of the eight pixels surrounding the one being considered. If this average is very different from the signal at the pixel being considered, a point has been identified. This operation may be carried out with the mask in Fig. 3.8a. A point is said to be detected if

$$s' > T \tag{3.8}$$

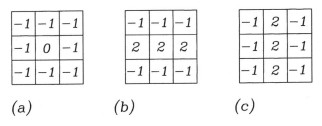

Figure 3.8 Masks for point and line detection. (a) Point detection; (b) horizontal line detection; (c) vertical line detection.

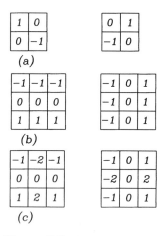

Figure 3.9 Masks for evaluating derivatives. (a) Roberts operators; (b) Prewitt operators; (c) Sobel operators.

where T is a predefined threshold value. If s' is close to zero, the point is not different from its surroundings.

The next degree of complexity is the detection of a line. To detect a horizontal line, the average cf the pixels above and below the line being considered is compared with the average of the pixels on this line. This is accomplished with the masks in Figs. 3.8b and 3.8c. The criterion in Eq. 3.8 is also used to decide if a line has been detected.

3.4.2 Derivative and Laplacian Operators

The partial derivatives of the signal values with respect to x and y may be estimated if we calculate the difference in the signal values of two adjacent pixels, as

$$\frac{\partial s}{\partial x} \propto s_{10} - s_{00} \tag{3.9}$$

The 2×2 Roberts masks shown in Fig. 3.9a are used to evaluate the partial derivatives in the diagonal directions. An important problem encountered with these operators is their large susceptibility to noise. For this reason they are seldom used.

The 3×3 Prewitt operators illustrated in Fig. 3.9b evaluate the partial derivatives in the x and y directions. They are less sensitive to noise than the Roberts operators because they take the average of three pixels in a line to evaluate these derivatives.

0	−1	0
−1	4	−1
0	−1	0

Figure 3.10 Laplacian operator.

The 3×3 Sobel operators in Fig. 3.9c also evaluate the partial derivatives in the x and y directions. They differ from the Prewitt operators in that in the averaging they give more weight to the central points.

The Laplacian of a function s is given by

$$\nabla^2 s = \frac{\partial^2 s}{\partial x^2} + \frac{\partial^2 s}{\partial y^2} \tag{3.10}$$

The value of the Laplacian is directly proportional to the average of the curvatures of this function s in the directions x and y. The 3×3 Laplacian operator is shown in Fig. 3.10. A problem with this operator is that it is quite sensitive to noise.

Figure 3.11 shows an interferogram processed with some of these operators.

3.4.3 Spatial Filtering by Convolution Masks

A *filtering mask* represents the filtering function $h(x, y)$ with a matrix of $N \times N$ pixels. As we have seen in Chap. 2, a function may be filtered by convolving it with a filter function. The Fourier transform of the filter function is called the *frequency response function* of the filter.

The filtering function with a mask with $N \times N$ pixels may be written as

$$h(x, y) = \sum_{n=-M}^{M} \sum_{m=-M}^{M} h_{nm} \delta(x - n\alpha, y - m\alpha) \tag{3.11}$$

where $M = (N - 1)/2$. The Fourier transform (or frequency response) of this filter is

$$H(f_x, f_y) = \sum_{n=-M}^{M} \sum_{m=-M}^{M} h_{nm} \exp[-i2\pi\alpha(nf_x + mf_y)] \tag{3.12}$$

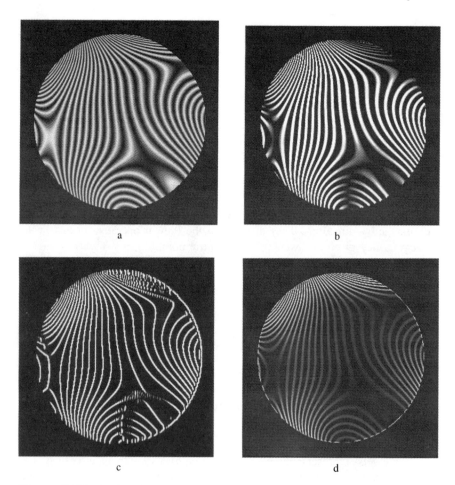

Figure 3.11 An interferogram processed by some operators. (a) Original interferogram; (b) after processing with horizontal Sobel operator; (c) result after four passes with horizontal Sobel operator; (d) after processing with the Laplacian.

where α is the separation between two consecutive pixels. Hence, we may write the sampling frequency as $f_S = 1/\alpha$.

The kernel or mask may be of any size $N \times N$. The larger the size, the greater the control over the functional form of the filter. This size has to be decided based on the spatial frequencies in the image that need to be filtered. A small 3×3 size is the most common. The mask may be asymmetrical or symmetrical. A symmetrical mask has a real Fourier transform and is thus called a zero phase mask. In this case we have $h_{-11} = h_{-1-1} = h_{1-1} = h_{11}, h_{-10} = h_{10}$ and $h_{0-1} = h_{01}$. Thus, in this particular case we may write

$$H(f_x, f_y) = h_{00} + 2h_{10} \cos\left(2\pi \frac{f_x}{fs}\right) + 2h_{01} \cos\left(2\pi \frac{f_y}{fs}\right)$$
$$+ 4h_{11} \cos\left(2\pi \frac{f_x}{fs}\right) \cos\left(2\pi \frac{f_y}{fs}\right) \tag{3.13}$$

As pointed out before, when sampling a digital image it is assumed that it is band-limited and that the conditions of the sampling theorem are not violated. Hence, the maximum values that f_x and f_y may have are equal to half the sampling frequency. This filter function along the x axis is

$$H(f_x, 0) = h_{00} + 2h_{01} + 2(h_{10} + 2h_{11}) \cos\left(2\pi \frac{f_x}{fs}\right) \tag{3.14}$$

The coefficients h_{nm} are frequently normalized so that the filter frequency response at zero frequencies $H(0, 0)$ is equal to 1, in order to preserve the DC level of the image. In this case we have

$$H(0, 0) = h_{00} + 2h_{10} + 2h_{01} + 4h_{11} = 1 \tag{3.15}$$

that is, the sum of all elements in the kernel should be equal to 1. In some other kernels, for example in the Laplacian, this sum of coefficients is made equal to zero in order to eliminate the DC level of the image. Examples of some common filtering masks are illustrated in Fig. 3.12. The frequency responses for some of these filters are shown in Fig. 3.13. The frequency responses are plotted only up to the highest frequency in the image, which is half the sampling frequency. For some of these filters the response at some frequencies may become negative, so that the contrast is reversed for these frequency components.

The main application of the low pass filters is to reduce the noise level in an image. The low pass kernel in Fig. 3.12a is quite effective in reducing Gaussian noise, which covers the whole image at random and seriously degrades its quality. The frequency response of this filter is shown in Fig. 3.13a. We see that the first zero of this filter is at approximately 0.31 of the sampling frequency. In other words, the period of the first zero is at 3.2 times the pixel separation, which is approximately the full mask size (3 pixels). A low pass filter with its first zero at a lower spatial frequency requires a larger mask. Thus, a rule of thumb is that the period of the first zero is about the mask size.

Applying the low pass filter reduces the noise but also reduces the high frequency content of the image. Another common consequence is that the image

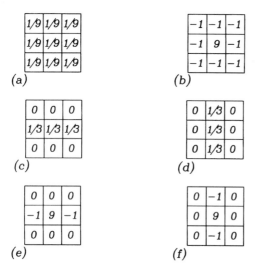

Figure 3.12   Some typical 3 × 3 kernels used to filter images. (a) Low pass bidirectional; (b) high pass bidirectional; (c) low pass horizontal; (d) low pass vertical; (e) high pass horizontal; (f) high pass vertical.

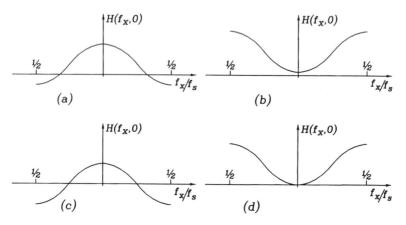

Figure 3.13   Frequency responses of some 3 × 3 kernels used to filter images. (a) Low pass; (b) high pass; (c) vertical edge detection; (d) Laplacian mask.

contrast is reduced. The filter may be applied to the image several times to reduce the noise even more, but always at the expense of reducing the image sharpness.

However, this is not the only type of noise that may affect an image. There is also shot or binary noise. This type of noise affects some isolated pixels, which may have the maximum brightness, independently of the surrounding pixels. It does not in general degrade the image definition, but it produces the appearance of speckles. In this case the low pass filter reduces the image definition without suppressing the binary noise. A much better filter for binary noise is the so-called *median filter*, which reduces the binary noise without reducing the image definition. In the median filter the value to be inserted at the center of the kernel is not the average value of the surrounding pixels. Instead, it is the median value of these pixels. The median value is obtained by sorting the surrounding pixels in order of decreasing or increasing value. Then the value of the pixel at the center is taken. If the kernel side is odd, as in the 3 × 3 just considered, the number of pixels around the central one is even. In this case the median is the average of the two pixels in the middle after sorting. It is interesting to know that the median filter performs very poorly with Gaussian noise. Figures 3.14 and 3.15 show images with binary and Gaussian noise, respectively, and their filtered versions using the two noise filters.

A high pass filter is illustrated in Fig. 3.12b and its frequency response in Fig. 3.13b. An example of filtering with this filter is given in Fig. 3.16.

3.4.4 Edge Detection

It is possible to detect fringe edges by means of a derivative, as shown in Fig. 3.17, where the location of the edge is defined as the points with maximum slopes. At the maximum slope locations the second derivative is zero, as shown in the same figure.

We have seen in Chap. 2 that the derivative of a function may be found by convolving it with a filtering function whose Fourier transform is linear with the frequency. This is possible only for a large mask. However, as we have already seen, a good approximation may be obtained with some 3 × 3 masks.

Then the edges may be detected by calculating the partial derivatives in order to obtain the gradient, defined by a vector with two components,

$$\nabla s = \left(\frac{\partial s}{\partial x}, \frac{\partial s}{\partial y} \right) \tag{3.16}$$

Then edges are located where the gradient has a maximum value with an orientation perpendicular to the gradient. The Laplacian is not often used for edge detection due to its large sensitivity to noise. However, it is useful to help to determine which side of the edge the dark or clear zone is on. Figure 3.18 shows an example of edge detection.

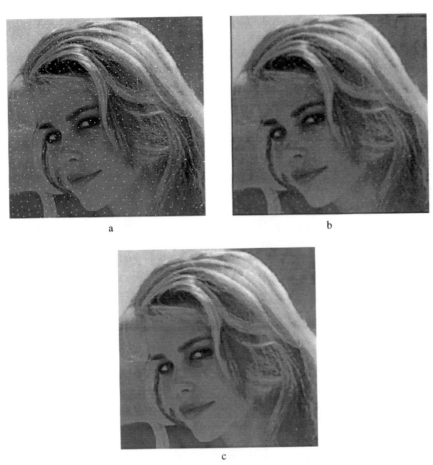

Figure 3.14 An image (a) with binary noise, (b) filtered with a low pass filter, and (c) filtered with a median filter.

3.4.5 Smoothing by Regularizing Filters

We have seen how we can use small convolution matrices to filter images. In fringe analysis one often needs to low pass filter a fringe pattern which has a finite extent. This finite extent may be due to the pupil of the optical instrument under analysis. The main drawback of using low pass convolution filters is that at the edges of the fringe pattern the fringe pattern is mixed with the illumination background. In other words, the information is cross talking at the fringe boundary between the background illumination and the fringe pattern.

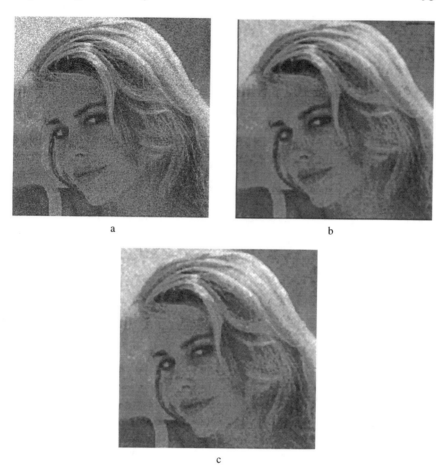

Figure 3.15 An image (a) with Gaussian noise, (b) filtered with a low pass filter, and (c) filtered with a median filter.

This causes problems for phase detection near the boundary. The phase distortion at the edge introduced by a convolution filter may be very important when testing, for example, a large telescope mirror.

One way of filtering that alleviates this cross talk problem is with the use of so-called regularized filters (Marroquin 1993). These filters are obtained as minimizers of quadratic cost functionals. The basic principle behind them is to assume that neighborhood pixels of the filtered image must have similar values, while their processed values also resemble the raw image data, i.e., large changes among neighboring pixels are penalized. A merit function U may be defined as

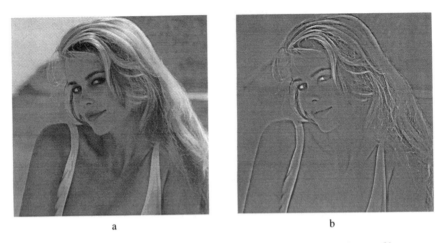

a   b

Figure 3.16 (a) An image and (b) its filtered version with a high pass filter.

$$U = \sum_{i,j} [(s'_{i,j} - s_{i,j})^2 m_{i,j} + \eta_x (s'_{i,j} - s'_{i-1,j})^2 m_{i,j} m_{i-1,j}$$
$$+ \eta_y (s'_{i,j} - s'_{i,j-1})^2 m_{i,j} m_{i,j-1}] \tag{3.17}$$

where the field signal $s_{i,j}$ is the image being filtered and $s'_{i,j}$ is the filtered field signal. A mask field $m_{i,j}$ is equal to 1 in the region of valid image data and zero otherwise. The first term in the quadratic merit function defined by

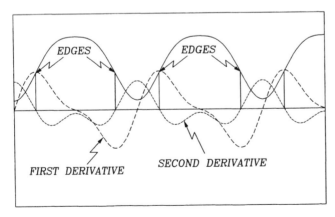

Figure 3.17 Edge detection with first and second derivatives.

a   b

Figure 3.18 An image and its filtered version with an edge detection filter.

this expression is the fidelity to the observations term. The constants η_x and η_y penalize large gray level changes of the filtered field signals $s'_{i,j}$ along the i and j directions, respectively. We need to specify a mask field $m_{i,j}$ over the image being filtered by setting on the valid region a value $m_{i,j} = 1$ and on the background a value $m_{i,j} = 0$. This field mask therefore denotes the region where we want to filter the field $s_{i,j}$ to obtain a filtered field $s'_{i,j}$. The filtered field then will be the one that minimizes the above cost functional for each pixel. This field may be found by deriving the cost functional U with reference to the filtered field $s'_{i,j}$ and making this derivative equal to zero, that is,

$$\frac{\partial U}{\partial s'_{i,j}} = (s'_{i,j} - s_{i,j})m_{i,j}$$

$$+ \eta_x[(s'_{i,j} - s'_{i-1,j})m_{i,j}m_{i-1,j} - (s'_{i+1,j} - s'_{i,j})m_{i+1,j}m_{i,j}]$$

$$+ \eta_y[(s'_{i,j} - s'_{i,j-1})m_{i,j}m_{i,j-1} - (s'_{i,j} - s'_{i,j1})m_{i,j}m_{i,j-1}] = 0 \tag{3.18}$$

This expression represents a linear set of simultaneous equations that must be solved for the $s'_{i,j}$ field. One simple iterative method that may be used to solve Eq. 3.18, thus minimizing the merit function, is by gradient descent:

$$s'^{k+1}_{i,j} = s'^{k}_{i,j} - \tau \frac{\partial U}{\partial s'_{i,j}} \tag{3.19}$$

where τ is a damping parameter. Coding this equation into a digital computer is very simple, but it is not an efficient method. One may instead use the conjugate gradient for a much more efficient method.

One may also use the Fourier method to analyze this kind of filter. The Fourier method of analyzing these filters assumes that the region of valid image data is very large, that is, the indicating mask field $m_{i,j}$ is equal to 1 over the whole (i, j) plane. With this in mind, Eq. 3.18 may be rewritten as

$$\frac{\partial U}{\partial s'_{i,j}} = s'_{i,j} - s_{i,j} + \eta_x[-s'_{i-1,j} + 2s'_{i,j} - s'_{i+1,j}]$$

$$+ \eta_y[-s'_{i,j-1} + 2s'_{i,j} - s'_{i,j+1}] \tag{3.20}$$

Taking the Fourier transform on both sides of Eq. 3.20 one may obtain the frequency response of the system as

$$H(\omega) = \frac{F\{s'_{i,j}\}}{F\{s_{i,j}\}} = \frac{1}{1 + 2\eta_x[1 - \cos(\omega_x)] + 2\eta_y[1 - \cos(\omega_y)]} \tag{3.21}$$

This transfer function represents a low pass filter with bandwidth controlled by the parameter constants η_x and η_y.

3.5 SOME USEFUL SPATIAL FILTERS

We will now describe some of the filters most commonly used in interferogram analysis, with their associated properties.

3.5.1 Square Window Filter

One common filter function is a square function, with a width x_0, defined by

$$h(x) = 1.0 \qquad \text{for } |x| < \frac{x_0}{2}$$

$$= 0 \qquad \text{elsewhere} \tag{3.22}$$

The spectrum of this filter, shown in Fig. 3.19a, is the sinc function, illustrated in Fig. 3.19b and given by

$$H(f) = \frac{\sin(\pi f x_0)}{\pi f x_0} = \sin(\pi f x_0) \tag{3.23}$$

The first zero of the spatial frequency is for the frequency f_0 given by

$$f_0 = 1x_0 \tag{3.24}$$

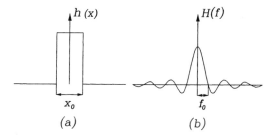

Figure 3.19 A one-dimensional square filter (a) and its spectrum (b).

This filter is equivalent to averaging the irradiance over all pixels in a window of 1 pixel height by N pixels width. This width is selected so that the row of N pixels just covers the window width x_0, defined by the desired low pass cutting point f_0 for the spatial frequency. In other words, the length of the filtering window should be equal to the period of the signal to be filtered out.

The height of the first secondary (negative) lobe is equal to 0.2172 times the height of the main lobe (central peak). Hence, the amplitude of this secondary maximum is -7.63 Db (decibels) below the central peak.

We may also use a window with a sinc profile. Thus, the spectrum would be a square function.

3.5.2 Hamming and Hanning Window Filters

The square filter just described is not the ideal because it leaves unfiltered some high frequencies due to the secondary maxima in the spectrum of the sinc function. A better filtering function is the Hamming function defined by

$$h(x) = 0.54 + 0.46\cos\frac{2\pi x}{x_0} \qquad \text{for } |x| < \frac{x_0}{2}$$
$$= 0 \qquad \text{elsewhere} \tag{3.25}$$

This function and its spectrum are illustrated in Fig. 3.20. The Fourier transform of this filter is given by

$$H(f) = 1.08 \ \text{sinc}(\pi f x_0) + 0.23 \ \text{sinc}(\pi f x_0 + \pi)$$
$$+ 0.23 \ \text{sinc}(\pi f x_0 - \pi) \tag{3.26}$$

The first zero for the spatial frequency of this filter is

$$f_0 = \frac{1}{2x_0} \cdot \tag{3.27}$$

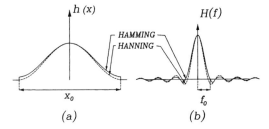

Figure 3.20   (a) Hamming and Hanning filters and (b) their Fourier transforms.

The height of the first secondary lobe (negative) has a height equal to 0.0063 times that of the main lobe or 22 Db below, which is a much lower value than for the square filter.

The Hanning filter is very similar to the Hamming filter, and it is defined by

$$h(x) = 0.5 \left(1 + \cos \frac{2}{x_0} \right) \qquad \text{for } |x| < \frac{x_0}{2}$$

$$= 0 \qquad \text{elsewhere} \tag{3.28}$$

This function and its spectrum are also illustrated in Fig. 3.20. The Fourier transform of this filter is given by

$$H(f) = 1.00 \ \text{sinc}(\pi f x_0) + 0.25 \ \text{sinc}(\pi f x_0 + \pi) + 0.25 \ \text{sinc}(\pi f x_0 - \pi) \tag{3.29}$$

The differences between the Hamming and Hanning filters are the relative heights of the secondary lobes with respect to the main lobe and the main lobe widths.

3.5.3   Cosinusoidal and Sinusoidal Window Filters

Combined cosinusoidal and sinusoidal window filters are band pass, not low pass, filters. The cosinusoidal filter may be expressed as the product of a Hamming filter and a cosinusoidal function as follows:

$$h(x) = \left(0.54 + 0.46 \cos \frac{2\pi x}{x_0} \right) \cos(2\pi f_R) \qquad \text{for } |x| < \frac{x_0}{2}$$

$$= 0 \qquad \text{elsewhere} \tag{3.30}$$

and it is illustrated in Fig. 3.21. The half-width of each band is the same as in the Hamming filter, and their separation from the origin is equal to f_R.

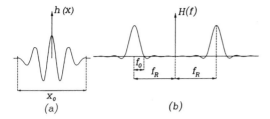

Figure 3.21 (a) A cosinusoidal window filter and (b) its spectrum.

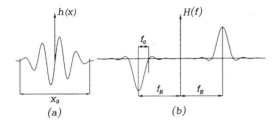

Figure 3.22 (a) A sinusoidal window filter and (b) its spectrum.

The disadvantage of this filter is that it has two symmetrical pass bands, and hence a single sideband cannot be isolated. The solution is to complement its use with a sinusoidal filter, defined by

$$h(x) = \left(0.54 + 0.46 \cos \frac{2\pi x}{x_0}\right) \sin(2\pi f_R) \qquad \text{for } |x| < \frac{x_0}{2}$$
$$= 0 \qquad \qquad \text{elsewhere} \qquad (3.31)$$

This filter and its spectrum are shown in Fig. 3.22. We may see that the two pass bands are now of opposite sign. Then, with a combination of the two filters any of the sidebands may be isolated.

A combination of the two filters is known as a quadrature filter.

3.6 EXTRAPOLATION OF FRINGES OUTSIDE OF THE PUPIL

In order to avoid some errors in phase detection, as suggested by Roddier and Roddier (1987), the Gerchberg (1974) method may be used to extrapolate the fringes in an interferogram with a large tilt (spatial carrier), outside the pupil boundary. Let us assume that the irradiance signal in the interferogram with a large spatial carrier may be written as

$$s(x, y) = p(x, y)a(x, y)[1 + v(x, y)\cos(2\pi f_0 x - kW(x, y))] \qquad (3.32)$$

where $p(x, y)$ is the domain over which the interferogram extends, as follows:

$$
\begin{aligned}
p(x, y) &= 1 \qquad \text{inside the pupil} \\
p(x, y) &= 0 \qquad \text{outside the pupil}
\end{aligned}
\qquad (3.33)
$$

Now we define the continuum as the interferogram irradiance when there are no fringes, which is equal to $a(x, y)$. This continuum may be measured by several different procedures, as described by Roddier and Roddier (1987). If we divide the irradiance by the continuum and subtract the pupil domain function, we obtain a function $g(x, y)$:

$$
\begin{aligned}
g(x, y) &= \frac{s(x, y)}{a(x, y)} - p(x, y) \\
&= p(x, y)v(x, y)\cos(2\pi f_0 x - kW(x, y))
\end{aligned}
\qquad (3.34)
$$

If we use the complex fringe visibility $u(x, y)$ defined in Eq. 3.4, we obtain

$$g(x, y) = \frac{p(x, y)}{2}[u(x, y)\exp(i2\pi f_0 x) + u^*(x, y)\exp(-i2\pi f_0 x)] \qquad (3.35)$$

The Fourier transform of this function $g(x, y)$, obtained by using the convolution theorem in Eq. 2.30, is

$$G(f_x, f_y) = 0.5P(f_x, f_y) * [U(f_x - f_0, f_y) + U^*(f_x - f_0, -f_y)] \qquad (3.36)$$

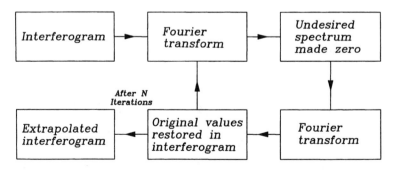

Figure 3.23   Algorithm used to extrapolate the fringes in an interferogram.

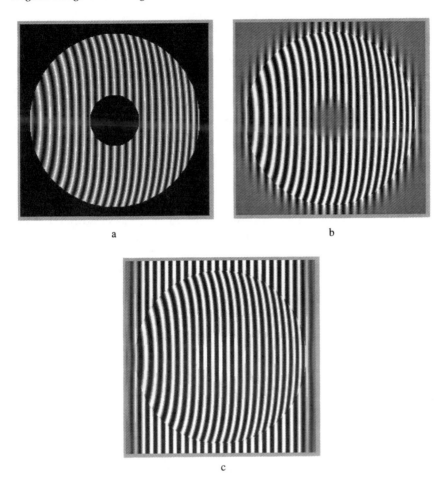

Figure 3.24 Interferogram and its extrapolated interferogram using the Gerchberg method and filtering with a Gaussian filter. (a) Original interferogram, (b) after 10 passes, (c) after 60 passes.

Thus, if the interferogram has no pupil boundaries, this spectrum would be concentrated in two circles with radius equal to the frequency cutoff of $U(f)$ centered at f_0 and $-f_0$. Due to the pupil's circular boundary, these circles increase in size as the pupil size decreases. The extrapolation of the fringes is easily achieved if the size of these two spots is reduced by cutting around them and then taking the inverse Fourier transform. This cut, however, distorts the fringes a little. Then the original fringe pattern inside the pupil area is recovered by inserting it back into the extrapolated fringe pattern. This process is repeated several times.

This algorithm to extrapolate the fringes outside of the pupil's boundary is illustrated in Fig. 3.23.

Figure 3.24 shows an example of fringe extrapolation using this method.

If there is no noise in the interferogram and the interferogram boundary is well defined, this algorithm works quite well, producing clean and continuous fringes. An improved version of this algorithm when there is some noise present was proposed by Kani and Dainty (1988).

3.7 LIGHT DETECTORS USED TO DIGITIZE IMAGES

Modern instrumentation to digitize images is of many different types, and it is rapidly evolving and changing. A description of these instruments is bound to be obsolete in a relatively short time. Nevertheless, a brief and superficial description may be useful for people beginning to work in the field of interferogram analysis.

Microcomputer systems for the acquisition and processing of interferogram video images may have many different configurations. One of these systems has been described by Oreb *et al.* (1982).

3.7.1 Image Detectors and Television Cameras

Image detectors are of many different types, depending on several factors such as wavelength, resolution, and price. For example, Stahl and Koliopoulos (1987) reported the use of pyroelectric vidicons as detectors of interferograms produced with infrared light. Prettyjohns (1984) described the use of CCD arrays.

A television camera is one of the most common image detectors used to digitize interferograms (Hariharan 1985). The most important characteristic from the point of view of this application is its resolving power. Here we will study in some detail the most important television systems.

The typical image detector is a coupled charge device (CCD), illustrated in Fig. 3.25 and described in many places in the scientific literature (e.g., Tredwell 1995).

There are many different television systems. The NTSC (National Television Systems Committee) system for color television and the EIA for black and white television is used in the United States, Canada, Mexico, and Japan. The PAL (phase alternating line) system, which is used in Germany, the United Kingdom, parts of Europe, South America, and part of Asia and Africa; and the SECAM (sequential couleur à mémoire) system is used in France, eastern Europe, and Russia. Table 3.1 shows the typical image resolutions for these three systems.

The image is formed by a series of horizontal lines. A complete scan of an image is called a *frame*. Frequently, to avoid flickering, the odd-numbered lines are scanned first and then the even-numbered lines, in a alternate manner, as

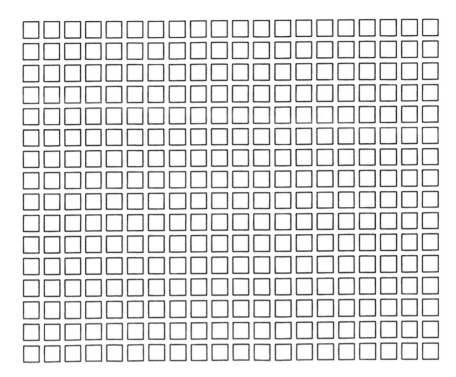

Figure 3.25 Television CCDs.

in Fig. 3.26. The set of all odd-numbered lines is the *odd field*, and the set of all even-numbered lines is the *even field*. This manner of scanning is called *interlaced scanning*. The total number of lines per frame is 525 in the NTSC system. In interlaced scanning the two alternate fields each have 263.5 lines.

Not all the lines in the frame contribute to the image. Approximately 41 lines are blanked out because they are either retrace lines or are at the extreme top or

Table 3.1 Image Resolution in Vertical Lines for Main Television Systems

	NTSC	EIA	PAL	SECAM
Resolution				
Vertical	340	340	400	400
Horizontal	330	360	390	470

Figure 3.26 Interlaced lines in a television frame.

bottom of the frame. Subtracting these lines from the total number in the whole frame, we are left with about 484 visible lines.

The aspect ratio of a standard television image is 4:3 (1.33:1). However, broadcast television images have an aspect ratio of 1.56:1, based on an unofficial standard for professional digital television equipment, as illustrated in Fig. 3.27.

The main characteristics of the two main television systems, i.e., NTSC and PAL in Table 3.2.

The vertical resolution depends on the number of scanning lines. A line covers a row of pixels on the CCD, as illustrated in Fig. 3.28. Hence, a CCD array must have 485 pixels or more in the vertical direction. Then the maximum vertical resolution is 486 TV lines.

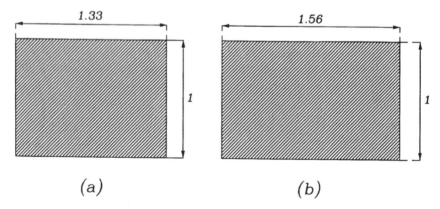

Figure 3.27 Aspect ratios in television images. (a) Standard and (b) professional digital television.

Table 3.2 Characteristics of Some Commercial TV Cameras

Specifications	Monochrome	Color	Color (high resolution)
Signal format	EIA	NTSC	NTSC
Horizontal resolution	570 TV lines	330 TV lines	470 TV lines
Picture elements	768 H × 494 V	510 H × 492 V	768 H × 494 V
Sensing area H (mm) × V (mm)	6.2 × 4.6	6.2 × 4.6	6.3 × 4.7
Interlaced	Optional	Yes	Yes

The signals from each row (image line) in the CCD detector are transformed into an analog signal. The horizontal detail—that is, the number of image elements in the horizontal line—is defined by the bandwidth of the television signal, which is approximately 4.0 MHz, but it may vary, as shown in Table 3.2. If the horizontal resolution is equal to the vertical resolution, we say that the horizontal resolution is equal to 484 TV lines. However, since the aspect ratio is equal to 4:3, the horizontal resolution is equivalent to having $484 \times 4/3 = 645$ lines. The horizontal resolution specified in TV lines is variable, depending on the number of pixels on the CCD. The frequency bandwidth in the electronics of the camera is constructed to fit the horizontal resolution of the CCD detector. Thus, the horizontal resolution may be higher than the vertical res-

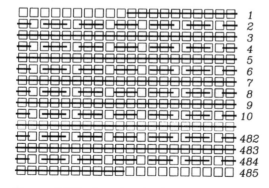

Figure 3.28 Scanning the image from a CCD detector in a TV camera. Continuous odd-numbered lines show the first field, while dashed even-numbered lines show the second field.

olution. Table 3.2 shows the resolution characteristics for some commercial TV cameras.

In color television cameras, dichroic RGB (red-green-blue) color filters are built on each element of the CCD array. Each element contains only one of these colors. As a consequence, the effective resolution of a color camera is lower than that of a black and white camera. Some expensive cameras use three CCD detectors to improve the image characteristics.

TV cameras for scientific applications may have TV systems different from NTSC or any other commercial system. Their resolution may be generally higher. There are two types of TV cameras: analog and digital. Analog cameras work in a similar manner to NTSC cameras, but they may have more scanning lines and a larger bandwidth to increase their resolution. Digital cameras, on the other hand, do not transform the signals from each row in the CCD detector into an analog signal. Instead, the signal from each element (pixel) in the detector is directly read and transmitted to the receiver or computer.

3.7.2 Frame Grabbers

When an analog camera is used to sample the image to be digitized, an electronic circuit has to be used to convert the analog signals from each line in the image into digital signals for each pixel image. This analog-to-digital converter is called a frame grabber. Frame grabbers are usually inserted inside the computer, but some models are external modules connected to an existing computer port. A typical frame grabber has one or more of the following components as shown in Fig. 3.29.

The *input multiplexer* selects from several available inputs, some with different specifications (RGB, composite video, S-video), into a single input channel.

The *signal conditioner* adjusts the input signal to a level compatible with the analog-to-digital converter. For monochrome frame grabbers, the chroma signal is removed to avoid having the chrominance signal treated as a luminance signal. In color grabbers, three separate video signals are obtained for each color to be digitized independently.

The *analog-to-digital converter* is a key component. The precision and resolution of the whole grabber depends on it. All grabbers use the so-called flash converter, the fastest digital-to-analog converter available and the most expensive. Flash converters are available with lower resolution (6 to 8 bits) than other kinds of converters, since their main characteristic is the speed of conversion.

Image memory is a random access memory used for storing a digitized frame. Some frame grabbers have enough memory to store several original frames as well as frames resulting from processing other frames. Most frame grabbers have a double-port memory that enables simultaneous read and write at different

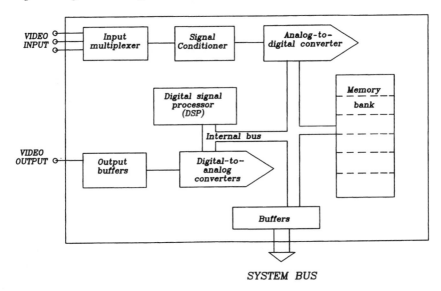

Figure 3.29 Block diagram of a typical frame grabber.

memory locations. The data can be written while they are being displayed. Color and high resolution grabbers require a large amount of memory.

Some grabbers include a *digital signal processor* (DSP) to do some high speed dedicated calculations. In other cases, a connector is provided for an external array or high speed processor board.

A *digital-to-analog converter* translates the digital image to an analog signal for display. The rate at which the data are being converted defines the output format. By selecting a window from the original data, and by adjusting the reading rate, a grabber may be used for format conversion.

The least expensive grabbers usually work at standard TV rates. Some of the more expensive handle nonstandard rates, including slow scan, line scan, high resolution, or custom-defined format. There are commercially available grabbers for several computer architectures, including PC bus, EISA, VMEbus, microVAX, and some others.

The selection of a frame grabber depends not only on the hardware compatibility but also on the software. Many grabbers are sold with bundled software such as drivers and demos. Manufacturers also sell a complete version of image processing software. Many programs are also available from third party providers.

REFERENCES

Gerchberg R. W., "Super-Resolution Through Error Energy Reduction," *Opt. Acta*, **21**, 709–720 (1974).

Gonzales R. C., and P. Wintz, *Digital Image Processing*, 2nd Edition, Addison-Wesley, Reading, MA, 1987.

Hariharan P., "Quasi-Heterodyne Hologram Interferometry," *Opt. Eng.*, **24**, 632–638 (1985).

Jain A. K., *Fundamentals of Digital Image Processing*, Prentice-Hall, Englewood Cliffs, NJ, 1989.

Kani L. M. and J. C. Dainty, "Super-resolution Using the Gerchberg Algorithm," *Opt. Commun.*, **68**, 11–15 (1988).

Kreis T. and H. Kreitlow, "Quantitative Evaluation of Holographic Interference Patterns Under Image Processing Aspects," *Proc. SPIE*, **210**, 196–202 (1979).

Kuan D. T., A. A. Sawchuk, T. C. Strand, and P. Chavel, "Adaptive Restoration of Images with Speckle," *Proc. SPIE*, **359**, 28–38 (1982).

Marroquin J. L., "Deterministic Interactive Particle Models for Image Processing and Computer Graphics," *Comput. Vision Graphics Image Process.*, **55**, 408–417 (1993).

Morimoto Y., "Digital Image Processing," in *Handbook of Experimental Mechanics*, A. S. Kobayashi, Ed., VCH Publishers, New York, 1993.

Oreb B. F., N. Brown, and P. Hariharan, "Microcomputer System for Acquisition and Processing of Video Data," *Rev. Sci. Instrum.*, **53**, 697–699 (1982).

Pratt W. K., *Digital Image Processing*, John Wiley and Sons, New York, 1978.

Prettyjohns K. N., "Charge-Coupled Device Image Acquisition for Digital Phase Measurement Interferometry," *Opt. Eng.*, **23**, 371–378 (1984).

Roddier C. and F. Roddier, "Interferogram Analysis Using Fourier Transform Techniques," *Appl. Opt.*, **26**, 1668–1673 (1987).

Stahl H. P. and C. L. Koliopoulos, "Interferometric Phase Measurement Using Pyroelectric Vidicons," *Appl. Opt.*, **26**, 1127–1136 (1987).

Tredwell T. J., "Visible Array Detectors," in *Handbook of Optics*, 2nd Edition, Vol. I, M. Bass, Ed., Optical Society of America, Washington, DC, 1995.

Womack K. H., "A Frequency Domain Description of Interferogram Analysis," *Proc. SPIE*, **429**, 166–173 (1983).

Womack K. H., "Frequency Domain Description of Interferogram Analysis," *Opt. Eng.*, **23**, 396–400 (1984).

4

Fringe Contouring and Polynomial Fitting

4.1 FRINGE DETECTION USING MANUAL DIGITIZERS

If a large tilt is introduced in a Twyman-Green interferogram of a perfectly flat wavefront interfering with a reference flat wavefront, the fringes will look straight, parallel, and equidistant. If the wavefront under test is not flat, the fringes are not straight but curved. These fringes are called equal-thickness fringes because they represent the locus of the points with constant wavefront separation. The wavefront deformations may be easily estimated from a visual examination of their deviation from straightness. If the maximum deviation of a fringe from its ideal straight shape is Δx, and the average separation between the fringes is equal to s, its wavefront deviation, in wavelengths, from the flat is equal to $\Delta x/s$.

This visual method can give us a precision that greatly depends on the skills of the person making the measurements. In the best case, one can probably approximate $\lambda/20$. Norms have been established to define and classify visually detected errors (Boutellier and Zumbrunn 1986). Even image quality can be determined from manual measurements in an interferogram (Platt *et al.* 1978). Some measuring devices have been proposed to aid in this fringe measurement (Dew 1964; Dyson 1963; Zanoni 1978). This procedure is still used in many manufacturing facilities, using test plates as the reference.

The simplest interferometric quantitative analysis method involves visually identifying and then tracking fringes in an interferogram. In this method a photograph of the interferogram is taken and then a digitizing tablet is used to enter into the computer the (x, y) coordinates of some selected points on the interferogram, placed over the peak of the fringes. Kingslake (1926–1927) computed the primary aberration coefficients by measuring a few points on the fringe peaks in an interferogram.

Alternatively, to avoid the need for a photograph, the image of the interferogram may be captured with a television camera and displayed on the computer screen, where the peaks of the fringes are manually sampled with the computer mouse (Augustyn *et al.* 1978; Augustyn 1979a, 1979b). When the image is digitized with a television camera, mechanical vibrations may introduce errors, but there are some methods that may reduce these errors (Crescentini and Fioco 1988; Crescentini 1988).

For manual sampling, the fringes are assigned consecutive numbers that increase by one from one fringe to the next. This is the interference order number m. If the tilt is large enough, so that there are no closed fringes, this presents no problem. Every time a point on top of a fringe is selected, the coordinates x and y are read by the graphic tablet or computer and an order number m is assigned. This number is typed on the computer keyboard by the operator every time a new fringe is beginning to be measured.

The wavefront deformation $W(x, y)$ at the sampled points on top of the fringes is

$$W(x, y) = m\lambda \tag{4.1}$$

The value of m may differ from the real number m by a constant quantity at all measurements, but this is not important. It is more important to know the direction in which the number m must increase. Otherwise, the sign of the wavefront deformations will be undetermined. It is impossible to determine the direction in which the fringe order number increases from a single picture of the interference pattern unless the sign of at least one of the component aberrations is known. For example, it would be sufficient if the sign of the tilt is known. This sign has to be determined when adjusting the interferometer to take the interferogram picture.

If some of the fringes form closed loops, the order number assignment is a little more difficult, but not impossible if carefully done, as illustrated in Fig. 4.1.

Many commercial and research systems had been developed to perform semi-automatic analysis of fixed interferograms of pictures or in real time (Jones and Kadakia 1968; Moore 1979; Augustyn 1979a, 1979b; Womack *et al.* 1979; Trolinger 1985; Truax and Selberg 1986/87; Truax 1986; Cline *et al.* 1982; Vrooman and Maas 1989). Reviews on the problems associated with the automatic analysis of fringes have been published by several authors, for example Choudry (1987) and Reid (1986/87, 1988).

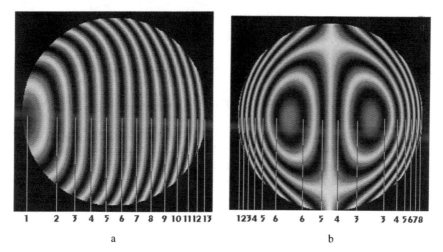

Figure 4.1 Sampling fringe positions at some points and assigning order numbers in an interferogram with (a) open fringes and (b) closed loop fringes.

4.2 FRINGE TRACKING AND FRINGE SKELETONIZING

The next stage in the automation process is to detect the fringes and assign order number by reading the interferogram image with a two-dimensional light detector or television camera and analyzing the image with a computer. The main problem is to locate the fringe maxima or minima by searching with algorithms based on line tracking, threshold comparison, or adaptive binarization. This automatic location of the fringe maxima has been done since the end of the 1970s (Hot and Durou 1979). Once the maxima are located, a subsequent fringe thinning or skeletonization is performed (Tichenor and Madsen 1978; Schluter 1980; Becker *et al.* 1982; Yatagai *et al.* 1982a and 1982b; Robinson 1983a, 1983b; Nakadate *et al.* 1983; Button *et al.* 1985; Becker and Yung 1985; Osten *et al.* 1987; Gillies 1988; Eichhorn and Osten 1988; Liu and Yang 1989; Hunter *et al.* 1989a, 1989b; Matczak and Budziński 1990; Yan *et al.* 1992; Huang 1993). Skeletonizing is based on the search of local irradiance peaks by segmentation algorithms based on adaptive thresholds, gradient operators, piecewise approximations, thinning procedures, or spatial frequency filtering. The result is a skeleton of the interferogram formed by lines one pixel wide.

Servin *et al.* (1990) described a technique they call rubber band to find the shape of a fringe. The method is based on a set of points linked together in a way similar to a rubber band, which attracts the point to a local maximum of the fringe.

Before sampling the fringes it is useful to add a tilt to the interferogram. This tilt straightens the fringes and reduces the fringe spacing, making it more uniform. Another benefit of the tilt is that it makes the fringe measurement and order identification easier. A wide spacing between the fringes increases the accuracy with which the top of the fringe is located. On the other hand, a large number of fringes increases the number of fringes to be sampled and hence the amount of measured information. So, an optimum intermediate tilt is desirable. For the case of digital sampling, using a two-dimensional light detector array, Macy (1983) and Hatsuzawa (1985) found that the optimum fringe spacing is that which produces a fringe separation of about four pixels.

The complete fringe analysis procedure may be summarized as follows (Reid 1986/87, 1988):

1. Spatial filtering of the image
2. Identification of fringe maxima
3. Assignment of order number to fringes
4. Interpolation of results between fringes

In the next few sections these steps are studied in some detail.

4.2.1 Spatial Filtering of the Image

The first step, spatial filtering of the image, is used to reduce the noise. This noise reduction can be performed in several different ways (Varman and Wykes 1982). If the spatial frequency of the noise is higher than that of the fringes, a low pass filtering is appropriate. When the spatial frequency of the noise is much lower than that of the fringes, for example, due to an uneven illumination, a high pass filtering improves the fringe contrast. A more difficult situation appears when the spatial frequency of the noise is similar to that of the fringes. Sometimes the noise is fixed to the aperture, for example, due to diffracting particles in the interferometer components. In this case (Kreis and Kreitlow 1983), if we take a second interferogram, moving the fringes by a change in the OPD by $\lambda/2$, the two interferograms will be complementary. That is, a dark fringe in one pattern will correspond to a clear fringe in the other. If we subtract one fringe pattern from the other, the fixed noise will be greatly reduced.

4.2.2 Identification of Fringe Maxima

Skeletonizing techniques detect the fringe peaks within the entire area of the digitized interferogram. Many different methods may be used to detect the fringe peaks. Schemm and Vest (1983) reduced the noise and located the fringe peaks using a nonlinear regression analysis, with a least squares fit of the irradiance

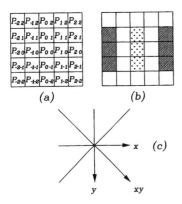

Figure 4.2 Yatagai *et al.* matrix to find fringe maxima. See text for details.

measurements in a small region to a sinusoidal function. Snyder (1980) plotted the fringe profiles in a direction perpendicular to the fringes by first smoothing and reducing the data by processing with an adaptive digital filter that locates the symmetry points of the fringe pattern. Mastin and Ghiglia (1985) skeletonized fringe patterns by using the fast Fourier transform and then locating the dominant spatial frequency in the vicinity of each fringe and also by performing a set of logical transformations in the neighborhood of a fringe peak. Zero crossing algorithms have also been used (Gåsvik 1989).

These peaks can also be detected by a matrix of 5×5 pixels, illustrated in Fig. 4.2, as proposed by Yatagai *et al.* (1982a). Let us assume that the matrix in Fig. 4.2a is centered on top of a vertical fringe. Then the average values of the irradiance of the shaded pixels in Fig. 4.2b will have a smaller value than the average values of the irradiance of the pixels with dots. The same principle may be applied to horizontal fringes and to inclined fringes, as shown in Fig. 4.2c.

Thus, the conditions for detecting fringe maxima in the x direction are

$$P_{00} + P_{0-1} + P_{01} = P_{-21} + P_{-20} + P_{-2-1} \tag{4.2}$$

and

$$P_{00} + P_{0-1} + P_{01} = P_{21} + P_{20} + P_{2-1} \tag{4.3}$$

In the y direction they are

$$P_{00} + P_{-10} + P_{10} = P_{-1-2} + P_{0-2} + P_{1-2} \tag{4.4}$$

and

$$P_{00} + P_{-10} + P_{10} = P_{-12} + P_{02} + P_{12} \qquad (4.5)$$

in the xy direction they are

$$P_{00} + P_{-1-1} + P_{11} = P_{-2+2} + P_{-21} + P_{-12} \qquad (4.6)$$

and

$$P_{00} + P_{-1-1} + P_{11} = P_{2-2} + P_{2-1} + P_{1-2} \qquad (4.7)$$

and finally in the $-xy$ direction they are

$$P_{00} + P_{-11} + P_{1-1} = P_{22} + P_{21} + P_{12} \qquad (4.8)$$

and

$$P_{00} + P_{-11} + P_{1-1} = P_{-2-2} + P_{-2-1} + P_{-1-2} \qquad (4.9)$$

When at least two of these conditions are satisfied, the point is assumed to be on top of a fringe. Figure 4.3 shows an example of fringe skeletonizing using this method. Yu *et al.* (1994) showed that if the interferogram illumination has a strong modulation, for example, if a large-aperture Gaussian beam is used, the central peak of the fringes will shift laterally a small amount. This shift is greater where a larger slope of the interferogram illumination exists.

The extracted skeletons may contain many disconnections. Then the next step is to localize them and make some corrections. Many sophisticated methods have been devised to perform this operation (Becker *et al.* 1982).

For simple interferograms with low noise and good contrast, the matrix operators described in Chap. 3 may be used.

4.2.3 Assignment of Order Numbers to Fringes

The assignment of order numbers to the fringes is an extremely important step. A mistake in just one of the fringes can lead to coarse errors in the calculated wavefront deformation. This step can be made quite simple if a large amount of tilt is introduced to avoid closed fringes (Hovanesian and Hung 1990). In this case the order number increases monotonically from one fringe to the next.

Sometimes, however, such a large tilt is not possible or practical. In this case (Livnat *et al.* 1980) suggested a method using two interferograms taken with different colors or with slightly different optical path differences. This is

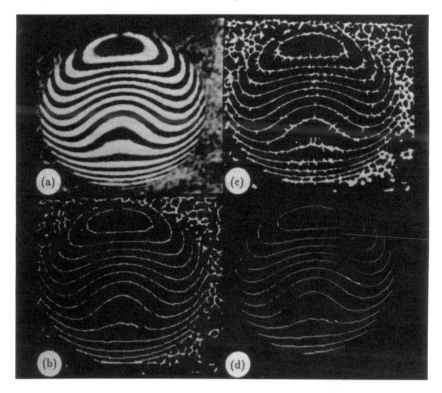

Figure 4.3 Skeletonizing and thinning of interferometric fringes. (a) Original interferogram; (b) result after low pass filtering; (c) result after detecting peaks in two orthogonal directions; (d) thinned skeletons with noise outside of pupil being removed. (After Yatagai 1993.)

equivalent to the methods used in the optical shop with test plates to determine if the surface under test is concave or convex with respect to the test plate (Mantravadi 1992). Hovanesian and Hung (1990) studied three similar methods to identify the fringe order number.

Frequently, if an automatic method is difficult, the order number has to be determined by visual observation of the fringes. Trolinger (1985) pointed out the problems of a completely automatic fringe analysis. Then interactive procedures are convenient. These are semiautomatic algorithms, which allow for the operator to direct the computer during the interferogram processing. Yatagai *et al.* (1982b) reported an interactive interferogram analyzing system in which the operator uses a light pen to take some decisions. Funell (1981) made another interactive system in which the operator helps the machine in the fringe identification through commands on the computer keyboard. Still another interactive system was made by Yatagai *et al.* (1984) to test the flatness of very large integrated circuit wafers. Finally, Parthiban and Sirohi (1989) constructed an interactive system in which the operator helps the machine to identify the fringe order number using a gray scale coding, using different gray levels according to the order of the fringes.

The problem of fringe number identification may be simplified if some *a priori* information is known (Robinson 1983a). A clear example occurs when we know in advance that the fringes are circular.

4.3 GLOBAL POLYNOMIAL INTERPOLATION

Once the values of the wavefront deformations have been found for many points over the interferogram, an interpolation between the points has to be made to estimate the complete wavefront shape. This interpolation is accomplished by employing a two-dimensional function. This is a global interpolation, because a single analytical function is used to represent the wavefront for the whole interferogram.

To perform a global interpolation the polynomials most frequently used are the Zernike polynomials (Loomis 1978; Plight 1980; Swantner and Lowre 1980; Wang and Silva 1980; Kim 1982; Hariharan *et al.* 1984; Kim and Shannon 1987; Mahajan 1981, 1984; Malacara 1983; Malacara *et al.* 1976, 1987, 1990; Prata and Rusch 1989; Malacara and De Vore 1992). Zernike polynomials are now briefly described.

Since the pupil of an optical system is frequently circular, it seems logical to express this two-dimensional function in polar coordinates as

$$x = \rho \sin \theta \tag{4.10}$$

and

$$y = \rho \cos \theta \tag{4.11}$$

where the angle θ is measured with respect to the y axis, as in Fig. 4.4.

The wavefront deformations can be represented by means of many types of two-dimensional analytical functions, but the most common are the Zernike polynomials. If the fit is not perfect we define the fit variance σ_f^2 as the difference between the actual sampled wavefront W' and the analytical wavefront $W(\rho, \theta)$ as follows:

$$\sigma_f^2 = = \frac{1}{\pi} \int_0^1 \int_0^{2\pi} [W' - W(\rho, \theta)]^2 \rho \, d\rho \, d\theta \tag{4.12}$$

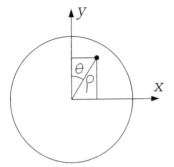

Figure 4.4 Polar coordinates used for two-dimensional polynomials.

The normalizing factor in front of the integral is $1/\pi$. If the fit variance is zero, the analytic function is an exact representation of the wavefront.

Sometimes it is also important to specify the mean wavefront deformation W_{av} including the normalizing factor, which is defined by

$$W_{av} = \frac{1}{\pi} \int_0^1 \int_0^{2\pi} W(\rho, \theta) \rho \, d\rho \, d\theta \tag{4.13}$$

Wavefront deformations are nearly always measured with respect to a close spherical reference. This spherical reference is defined by the position of the center of curvature and the radius of curvature.

The average wavefront deviations with respect to the spherical reference is the variance σ_w^2, defined as

$$\sigma_w^2 = \frac{1}{\pi} \int_0^1 \int_0^{2\pi} [W(\rho, \theta) - W_{av}]^2 \rho \, d\rho \, d\theta$$

$$= \frac{1}{\pi} \int_0^1 \int_0^{2\pi} W^2(\rho, \theta) \rho \, d\rho \, d\theta - W_{av}^2 \tag{4.14}$$

which is frequently called the root mean square (rms) value of the wavefront deformations. The reference spherical wavefront may be defined with any value of the radius of curvature (piston term) without modifying the position of the center of curvature. Nevertheless, the value of the wavefront variance may be affected by this selection, because the average wavefront is also affected. A convenient way to eliminate this problem is to select the reference sphere in the definition of the wavefront variance as one with the same position as the mean wavefront deformation. This is the reason for subtracting W_{av}^2 in this expression.

4.3.1 Zernike Polynomials

The Zernike polynomials are used because of their unique and desirable properties derived from their orthogonality. These polynomials have been described in many places in the literature (Zernike 1934, 1954; Born and Wolf 1964; Barakat 1980; Bathia and Wolf 1952, 1954; Malacara and De Vore, 1992; Wyant and Creath 1992). A brief review is given here. The Zernike polynomials $U(\rho, \theta)$, written in polar coordinates, are orthogonal in the unit circle in a continuous fashion (exit pupil with radius 1), with the condition

$$\int_0^1 \int_0^{2\pi} U_n^l(\rho, \theta) U_{n'}^{l'}(\rho, \theta) \rho \, d\rho \, d\theta = \frac{\pi}{2(n+1)} \delta_{nn'} \delta_{ll'} \tag{4.15}$$

where $\rho = S/S_{\max}$ is the normalized radial coordinate, with S being the non-normalized radial coordinate. The Kronecker $\delta_{nn'}$ is zero if n is different from n'. The Zernike polynomials are represented with two indices n and l, since they are dependent on two coordinates. The index n is the degree of the radial polynomial, and l is the angular dependence index. The numbers n and l are both even or both odd, making $n - l$ always even. There are $(1/2)(n+1)(n+2)$ linearly independent polynomials $U_n^l(\rho, \theta)$ of degree $\leq n$, one for each pair of numbers n and l. Then the polynomials can be separated into two functions, one depending only on the radius ρ and the other being dependent only on the angle θ, obtaining

$$U_n^l(\rho, \theta) = R_n^l(\rho) \begin{bmatrix} \sin \\ \cos \end{bmatrix} l\theta$$

$$= U_n^{n-2m}(\rho, \theta) = R_n^{n-2m}(\rho) \begin{bmatrix} \sin \\ \cos \end{bmatrix} (n - 2m)\theta \tag{4.16}$$

where the sine function is used when $n - 2m > 0$ (antisymmetrical functions) and the cosine function is used when $n - 2m \leq 0$ (symmetrical functions). The degree of the radial polynomial $R_n^l(\rho)$ is n, and $0 \leq m \leq n$. It may be shown that $|l|$ is the minimum exponent of these polynomials R_n^l. The radial polynomial is given by

$$R_n^{n-2m}(\rho) = R_n^{-(n-2m)}(\rho)$$

$$= \sum_{s=0}^{m} (-1)^s \frac{(n-s)!}{s!(m-s)!(n-m-s)!} \rho^{n-2s} \tag{4.17}$$

All Zernike polynomials $U_n(\rho)$ may be ordered with a single index r, defined by

$$r = \frac{n(n+1)}{2} + m + 1 \tag{4.18}$$

Table 4.1 shows the first 15 Zernike polynomials. Kim and Shannon (1987) showed isometric plots for the first 37 Zernike polynomials. Figure 4.5 shows plots for some of these polynomials.

The triangular and ashtray astigmatisms may be visualized as the shape that a flexible disc adopts when supported on top of three or four points equally distributed around the edge. It should be pointed out that these polynomials are orthogonal only if the pupil is circular, without any central obscurations.

Any continuous wavefront shape $W(x, y)$ may be represented by a linear combination of the Zernike polynomials, as follows:

$$W(\rho, \theta) = \sum_{n=0}^{k} \sum_{m=0}^{n} A_{nm} U_{nm}(\rho, \theta) = \sum_{r=0}^{L} A_r U_r(\rho, \theta) \tag{4.19}$$

If the maximum power is L, the coefficients A_r are found by any of several possible procedures, for example, by requiring that the fit variance defined be minimized.

4.3.2 Properties of Zernike Polynomials

The advantage of expressing the wavefront by a linear combination of orthogonal polynomials is that the wavefront deviation represented by each term is a best fit (minimum fit variance) with respect to the actual wavefront. Then any combination of these terms must also be a best fit. Each Zernike polynomial is obtained by adding to each type of aberration the proper amount of piston, tilt, and defocusing, so that the rms value σ_w^2 for each Zernike polynomial is minimized. To illustrate this with an example, let us consider the spherical aberration polynomial, where we may see that a term $+1$ (piston term) and a term $-6\rho^2$ (defocusing) have been added to the spherical aberration term $6\rho^4$. These additional terms minimize the rms deviation of spherical aberration with respect to a flat wavefront.

The practical consequence of the orthogonality of the Zernike polynomials is that any aberration terms, like defocusing, tilt, or any other, may be added or subtracted from the wavefront function $W(x, y)$ without losing the best fit to the data points.

Using the orthogonality condition, the mean wavefront deformation for each Zernike polynomial may be shown to be

Table 4.1   First 15 Zernike Polynomials

n	m	r	Zernike polynomial	Meaning
0	0	1	1	Piston term
1	0	2	$\rho \sin \theta$	Tilt about x axis
1	1	3	$\rho \cos \theta$	Tilt about y axis
2	0	4	$\rho^2 \sin 2\theta$	Astigmatism with axis at $\pm 45°$
2	1	5	$2\rho^2 - 1$	Defocusing
2	2	6	$\rho^2 \cos 2\theta$	Astigmatism, axis at $0°$ or $90°$
3	0	7	$\rho^3 \sin 3\theta$	Triangular astig., base on x axis
3	1	8	$(3\rho^3 - 2\rho) \sin \theta$	Primary coma along x axis
3	2	9	$(3\rho^3 - 2) \cos \theta$	Primary coma along y axis
3	3	10	$\rho^3 \cos 3\theta$	Triangular astig., base on y axis
4	0	11	$\rho^4 \sin 4\theta$	Ashtray astig., nodes on axes
4	1	12	$(4\rho^4 - 3\rho^2) \sin 2\theta$	
4	2	13	$64\rho^4 - 6\rho^2 + 1$	Primary spherical aberration
4	3	14	$4(\rho^4 - 3\rho^2) \cos 2\theta$	
4	4	15	$\rho^4 \cos 4\theta$	Ashtray astig., crests on axis

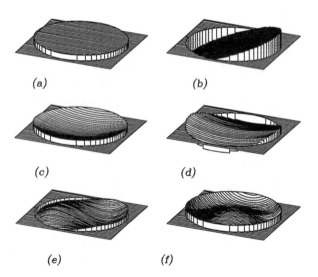

(a)   (b)

(c)   (d)

(e)   (f)

Figure 4.5   Isometric plots for some Zernike polynomials. (a) Piston term; (b) tilt; (c) defocusing; (d) astigmatism; (e) coma; (f) spherical aberration.

$$W_{av} = \frac{1}{\pi} \int_0^1 \int_0^{2\pi} U_r(\rho, \theta) \rho \, d\rho \, d\theta$$

$$= \frac{1}{2} \quad \text{if } r = 1 \tag{4.20}$$

$$= 0 \quad \text{if } r > 1$$

This means that the mean wavefront deformation is zero for all Zernike polynomials with the exception of the piston term. Thus, the wavefront variance is given by

$$\sigma_W^2 = \frac{1}{2} \sum_{r=1}^{L} \frac{A_r^2}{n+1} - W_{av}^2 = \frac{1}{2} \sum_{r=2}^{L} \frac{A_r^2}{n+1} \tag{4.21}$$

where n is related to r by

$$n = \text{next integer greater than } \frac{-3 + [1 + 8r]^{1/2}}{2} \tag{4.22}$$

4.3.3 Least Squares Fit to Zernike Polynomials

The analytical wavefront in terms of Zernike polynomials may be obtained by a two-dimensional least squares fit as shown by Malacara *et al.* (1990). The details of this procedure may be found in Malacara and De Vore (1992). A brief description is given here.

If we have N measured points with coordinates (ρ_n, θ_n) and values W_n' measured with respect to a closed analytical function $W(\rho, \theta)$, the discrete variance v^2 is defined by

$$v^2 = \frac{1}{N} \sum_{n=1}^{N} [W_n - W(\rho_n, \theta_n)] \tag{4.23}$$

Then the best least squares fit to the function $W(\rho, \theta)$ is defined when the analytical function is chosen so that this variance is a minimum with respect to the parameters of this function. We can see that this discrete variance S^2 and the variance σ_f^2 are the same if the number of points is infinity and they are uniformly distributed on the sampling region (aperture of the interferogram).

Let us now consider that the analytical function $W(\rho, \theta)$ is a linear combination of some predefined polynomials $V(\rho, \theta)$ as follows:

$$W(\rho_n, \theta) = \sum_{r=1}^{L} B_r V_r(\rho_n, \theta_n) \tag{4.24}$$

Then, in order to have the best fit we require that

$$\frac{\partial v}{\partial B_p} = 0 \tag{4.25}$$

with $p = 1, 2, 3, \ldots, L$. Then the following system of L linear equations is obtained:

$$\sum_{r=1}^{L} B_r \sum_{n=1}^{N} V_r(\rho_n, \theta_n) V_p(\rho_n, \theta_n) = \sum_{n=1}^{N} W'_n V_p(\rho_n, \theta_n) = 0 \tag{4.26}$$

The matrix of this linear system of equations becomes diagonal if the polynomials V_r satisfy the condition that

$$\sum_{n=1}^{N} V_r(\rho_n, \theta_n) V_p(\rho_n, \theta_n) = \left(\sum_{n=1}^{N} V_n^2(\rho_n, \theta_n) \right) \delta_{rp} \tag{4.27}$$

This expression means that the polynomials V_r are orthogonal on the discrete base of the measured data points, as opposed to the Zernike polynomials, which are orthogonal in a continuous manner. That is, they are not orthogonal in the unitary circle like the Zernike polynomials. Then, the solution to the system of equations becomes

$$B_p = \frac{\displaystyle\sum_{n=1}^{N} W'_n V_p(\rho_n, \theta_n)}{\displaystyle\sum_{n=1}^{N} V_n^2(\rho_n, \theta_n)} \tag{4.28}$$

The polynomials V_p are not the Zernike polynomials U_p, but they approach them when the number of sampling points is extremely large and they are uniformly distributed on the unitary circle. The most important and useful property of orthogonal polynomials, as was pointed out before, is that once a least squares fit is made, any polynomial in the linear combination can be taken out without losing the best fit. Hence, it is more convenient to use these polynomials V_p instead of the Zernike polynomials U_p to make the wavefront representation. If desired, they can later be transformed into Zernike polynomials.

A small problem, however, is that since the locations of sampling points are different for different interferograms, the polynomials V_p are not universally defined. Therefore, they have to be found for every particular case by a process called Gram-Schmidt orthogonalization.

4.3.4 Gram-Schmidt Orthogonalization

The desired polynomials, orthogonal in the data points base, can be found as a linear combination of the Zernike polynomials as

$$
V_r(\rho, \theta) = U_r + \sum_{s=1}^{r-1} D_{rs} V_s(\rho, \theta) \tag{4.29}
$$

where $r = 1, 2, 3, \ldots, L$. Now, using the orthogonality property and summing for all data points, we obtain, for all values of r different from p,

$$
\sum_{n=1}^{N} V_r(\rho_n, \theta_n) V_p(\rho_n, \theta_n) = \sum_{n=1}^{N} U_r(\rho_n, \theta_n) V_p(\rho_n, \theta_n)
$$

$$
+ D_{rp} \sum_{n=1}^{N} V_p^2(\rho_n, \theta_n) \tag{4.30}
$$

Thus, D_{rp} can be written

$$
D_{rp} = \frac{\displaystyle\sum_{n=1}^{N} U_r(\rho_n, \theta_n) V_p(\rho_n, \theta_n)}{\displaystyle\sum_{n=1}^{N} V_p^2(\rho_n, \theta_n)} \tag{4.31}
$$

with $r = 2, 3, 4, \ldots, L$ and $p = 1, 2, \ldots, r - 1$. These coefficients give us the desired orthogonal polynomials.

The factors affecting the accuracy of global interpolation using Zernike polynomials were studied by Wang and Ling (1989).

4.4 LOCAL INTERPOLATION BY SEGMENTS

A set of data points may be fitted to a polynomial, as we saw in the preceding section. This approach, however, has some problems. Perhaps the most important is that when the number of sampling points is large, the fit tends to have many oscillations and to deviate strongly at the edges, as illustrated in Fig. 4.6. Global and local fitting of interferograms have been studied and compared by several researchers, for example, by Roblin and Prévost (1978), Hayslett and Swantner (1978, 1980), and Freniere *et al.* (1979, 1981).

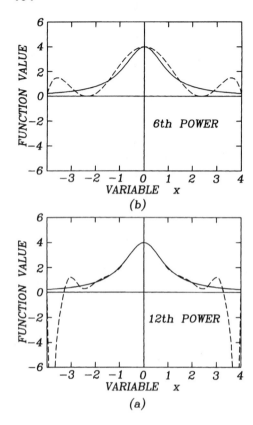

Figure 4.6 Errors in curve fitting for several polynomial degrees.

Local interpolation can be performed with several possible methods. The simplest one is the Newtonian trapezoidal interpolation. However, better approximations are frequently necessary. The three most common procedures are (Mieth and Osten 1990)

1. One-dimensional spline interpolation
2. Two-dimensional bilinear interpolation
3. Triangular interpolation

A brief review of splines follows. A spline is a mechanical device made with a flexible material that is used in drafting to draw curves. In mathematics, a spline is an extension-limited piece of curve that may be used to represent a small interval in the set of points to be interpolated. This method has the great advantage that more control is gained over the quality of the interpolation, since we proceed segment by segment, constructing the whole curve with many segments. The problem is that there is not a single analytical representation for the whole curve.

The theory of splines has been treated in several books (e.g., Lancaster and Šalkauskas 1986). The points to be joined by splines are called *knots*. Two knots may be connected with a straight line; this is a linear spline.

Additionally, at two consecutive knots joined by a spline, we may require that one or both of the following conditions be satisfied:

1. To have the same slope (first derivative) at the common knot. This condition can be satisfied with a third-degree polynomial. This is called a cubic spline.
2. To have the same curvature (second derivative) at the common knot. Under certain conditions this condition can also be satisfied with a cubic spline.

In interferometric data fitting, the cubic spline is the most popular and useful. To construct a cubic spline the first derivative (slope) at the knots must be continuous. However, we have two different possible ways to construct this spline:

1. The slope at each knot is first calculated. The final result critically depends on the choice of these slopes. A possible method is to choose the slope of the second-degree curve (parabola) that passes through the point being considered and the two points on each side. The slopes at the extremes are those of the straight lines joining the first two and the last two points. When the slopes at all the knots are defined, the cubic spline may be calculated.
2. Another possibility is not to previously define the slope values at each knot but require only that they be continuous. We use this extra degree of freedom to require that the curvatures (second derivatives) also be continuous at the knots. This is the *classic cubic spline*. We only have to define the slopes or the curvatures at the first and last knots. If we define these curvatures as zero, we have the *natural cubic spline*. Figure 4.7 shows an example of spline fitting.

An algorithm in C to calculate the classic spline is found in the book by Press *et al.* (1988), where we find the algebraic expressions to calculate the splines for interpolation of an array of points (y_i, x_i) with $x_1 < x_2 < \cdots < x_N$.

Besides the point coordinates we have to supply the program with the values of the slopes at the beginning and end of the array.

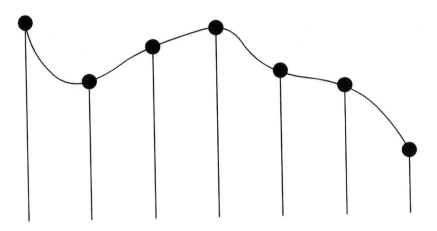

Figure 4.7 An example of spline fitting.

The procedure begins with the solution of a system of N linear equations with N unknowns. The first $N-2$ equations are

$$\frac{x_j - x_{j-1}}{6} y''_{j-1} + \frac{x_{j+1} - x_{j-1}}{3} y''_j + \frac{x_{j+1} - x_j}{6} y''_{j+1}$$

$$= \frac{y_{j+1} - y_j}{x_{j+1} - x_j} - \frac{y_j - y_{j-1}}{x_j - x_{j-1}}, \qquad j = 2, \ldots, N-1 \qquad (4.32)$$

where the unknowns y'' are second derivatives at each of the knots. The two other necessary equations to solve this system are

$$y''_1 = 0 \qquad \text{and} \qquad y''_N = 0 \qquad (4.33)$$

if the natural cubic spline is desired, or alternatively we may set both of the first derivatives at the beginning and end of the array of points to the desired values and use the following two equations.

$$y'_1 = \frac{y_2 - y_1}{x_2 - x_1} - \frac{3A_1^2 - 1}{6}(x_2 - x_1)y''_1 - \frac{3B_1^2 - 1}{6}(x_2 - x_1)y''_2 \qquad (4.34)$$

with

$$A_1 = \frac{y_2 - x_1}{x_2 - x_1}; B_1 = 1 - A_1 \qquad (4.35)$$

and

$$y'_N = \frac{y_N - y_{N-1}}{x_N - x_{N-1}} - \frac{3A_N^2 - 1}{6}(x_N - x_{N-1})y''_{N-1}$$

$$- \frac{3B_N^2 - 1}{6}(x_N - x_{N-1})y''_N \tag{4.36}$$

with

$$A_N = \frac{y_N - x_N}{x_N - x_{N-1}}; \; B_N = 1 - A_N \tag{4.37}$$

In two dimensions, a similar approach may be used, with bicubic splines.

REFERENCES

Augustyn W. H., "Automatic Data Reduction of Both Simple and Complex Interference Patterns," *Proc. SPIE*, **171**, 22–31 (1979a).

Augustyn W. H., "Versatility of a Microprocessor-Based Interferometric Data Reduction System," *Proc. SPIE*, **192**, 128–133 (1979b).

Augustyn W. H., A. H. Rosenfeld, and C. A. Zanoni, "An Automatic Interference Pattern Processor with Interactive Capability," *Proc. SPIE*, **153**, 146–155 (1978).

Barakat R., "Optimum Balanced Wave-Front Aberrations for Radially Symmetric Amplitude Distributions: Generalizations of Zernike Polynomials," *J. Opt. Soc. Am.*, **70**, 739–742 (1980).

Bathia A. B. and E. Wolf, "The Zernike Circle Polynomials Occurring in Diffraction Theory," *Proc. Phys. Soc.*, **B65**, 909–910 (1952).

Bathia A. B. and E. Wolf, "On the Circle Polynomials of Zernike and Related Orthogonal Tests," *Proc. Cambr. Phil. Soc.*, **50**, 40–48 (1954).

Becker F., "Zur Automatischen Auswertung von Interferogrammen," *Mitte. Max-Planck-Inst. Stroemungsforsch.*, Nr. 74 (1982).

Becker F. and Y. H. Yung, "Digital Fringe Reduction Techniques Applied to the Measurement of Three Dimensional Transonic Flow Fields," *Opt. Eng.*, **24**, 429–434 (1985).

Becker F., G. E. A. Maier, and H. Wegner, "Automatic Evaluation of Interferograms," *Proc. SPIE*, **359**, 386–393 (1982).

Born M. and E. Wolf, *Principles of Optics*, Pergamon Press, New York, 1964.

Boutellier R. and R. Zumbrunn, "Digital Interferogram Analysis and DIN Norms," *Proc. SPIE*, **656**, 128–134 (1986).

Button B. L., J. Cutts, B. N. Dobbins, J. C. Moxon, and C. Wykes, "The Identification of Fringe Positions in Speckle Patterns," *Opt. and Laser Technol.*, **17**, 189–192 (1985).

Choudri A., "Automated Fringe Reduction Analysis," *Proc. SPIE*, **816**, 49–55 (1987).

Cline H. E., A. S. Holik, and W. E. Lorensen, "Computer-Aided Surface Reconstruction of Interference Contours," *Appl. Opt.*, **21**, 4481–4488 (1982).

Crescentini L., "Fringe Pattern Analysis in Low-Quality Interferograms," *Appl. Opt.*, **28**, 1231–1234 (1988).

Crescentini L. and G. Fiocco, "Automatic Fringe Recognition and Detection of Sub-wavelength Phase Perturbations with a Michelson Interferometer," *Appl. Opt.*, **27**, 118–123 (1988).

Dew G. D., "A Method for the Precise Evaluation of Interferograms," *J. Sci. Instrum.*, **41**, 160–162 (1964).

Dyson J., "The Rapid Measurement of Photographic Records of Interference Fringes," *Appl. Opt.*, **2**, 487–489 (1963).

Eichhorn N. and W. Osten, "An Algorithm for the Fast Derivation of the Line Structures from Interferograms," *J. Mod. Opt.*, **35**, 1717–1725 (1988).

Freniere E. R., O. E. Toler, and R. Race, "Interferogram Evaluation Program for the HP-9825A Calculator," *Proc. SPIE*, **171**, 39-42 (1979).

Freniere E. R., O. E. Toler, and R. Race, "Interferogram Evaluation Program for the HP-9825A Calculator," *Opt. Eng.*, **20**, 253–255 (1981).

Funnell W. R. J., "Image Processing Applied to the Interactive Analysis of Interferometric Fringes," *Appl. Opt.*, **20**, 3245-3249 (1981).

Gåsvik K. J., "Fringe Location by Means of a Zero Crossing Algorithm," *Proc. SPIE*, **1163**, 64–70 (1989).

Gillies A. C., "Image Processing Approach to Fringe Patterns," *Opt. Eng.*, **27**, 861–866 (1988).

Hariharan P., B. F. Oreb, and Z. Wanzhi, "Measurement of Aspheric Surfaces Using a Microcomputer Controlled Digital Radial-Shear Interferometer," *Opt. Acta*, **31**, 989–999 (1984).

Hatsuzawa T., "Optimization of Fringe Spacing in a Digital Flatness Test," *Appl. Opt.*, **24**, 2456–2459 (1985).

Hayslett C. R. and W. H. Swantner, "Mathematical Methods for Deriving Wavefronts from Interferograms," in *Optical Interferograms—Reduction and Interpretation*, ASTM Symposium, Tech. Publ. 666, A. H. Guenther and D. H. Liedbergh, Eds., American Society for Testing and Materials, West Consohoiken, PA, 1978.

Hayslett C. R. and W. H. Swantner, "Wave-Front Derivation from Interferograms by Three Computer Programs," *Appl. Opt.*, **19**, 3401–3406 (1980).

Hot J. P. and C. Durou, "System for the Automatic Analysis of Interferograms Obtained by Holographic Interferometry," *Proc. SPIE*, **210**, 144–151 (1979).

Hovanesian J. Der and Y. Y. Hung, "Fringe Analysis and Interpretation," *Proc SPIE*, **1121**, 64–71 (1990).

Huang Z., "Fringe Skeleton Extraction Using Adaptive Refining," *Opt. and Lasers in Eng.*, **18**, 281–295 (1993).

Hunter J. C., M. W. Collins, and B. A. Tozer, "An Assessment of Some Image Enhancement Routines for Use with an Automatic Fringe Tracking Programme," *Proc. SPIE*, **1163**, 83–94 (1989a).

Hunter J. C., M. W. Collins, and B. A. Tozer, "A Scheme for the Analysis of Infinite Fringe Systems," *Proc. SPIE*, **1163**, 206–219 (1989b).

Jones R. A. and P. L. Kadakia, "An Automated Interferogram Analysis," *Appl. Opt.*, **7**, 1477–1481 (1968).

Kim C.-J., "Polynomial Fit of Interferograms," *Appl. Opt.*, **21**, 4521–4525 (1982).

Kim C.-J. and R. Shannon, "Catalog of Zernike Polynomials," in *Applied Optics and Optical Engineering*, Vol. 10, R. Shannon and J. C. Wyant, Eds., Academic Press, New York, 1987.

Kingslake R., "The Analysis of an Interferogram," *Trans. Opt. Soc.*, **28**, 1 (1926-1927).

Kreis T. M. and H. Kreitlow, "Quantitative Evaluation of Holographic Interferograms Under Image Processing Aspects," *Proc. SPIE*, **210**, 2850–2853 (1983).

Lancaster P. and K. Šalkauskas, *Curve and Surface Fitting: An Introduction*, Academic Press, San Deigo, 1986.

Liu K. and J. Y. Yang, "New Method of Extracting Fringe Curves from Images," *Proc. SPIE*, **1163**, 71–76 (1989).

Livnat A., O. Kafri, and G. Erez, "Hills and Valleys Analysis in Optical Mapping and Its Application to Moiré Contouring," *Appl. Opt.*, **19**, 3396–3400 (1980).

Loomis J. S., "A Computer Program for Analysis of Interferometric Data," in *Optical Interferograms—Reduction and Interpretation*, p. 193, ASTM Symposium, Tech. Publ. 666, A. H. Guenther and D. H. Liedbergh, Eds., American Society for Testing and Materials, West Consohocken, PA, 1978.

Macy W. W., Jr., "Two-Dimensional Fringe Pattern Analysis," *Appl. Opt.*, **22**, 3898–3901 (1983).

Mahajan V. N., "Zernike Annular Polynomials for Imaging Systems with Annular Pupils," *J. Opt. Soc. Am.*, **71**, 75–85 (1981).

Mahajan V. N., "Zernike Annular Polynomials for Imaging Systems with Annular Pupils," *J. Opt. Soc. Am. A*, **1**, 685–685 (1984).

Malacara D., "Set of Orthogonal Aberration Coefficients," *Appl. Opt.*, **22**, 1273–1274 (1983).

Malacara D. and S. L. De Vore, in *Optical Shop Testing*, 2nd Edition, p. 455, D. Malacara, Ed., John Wiley and Sons, New York, 1992.

Malacara D., A. Cornejo, and A. Morales, "Computation of Zernike Polynomials in Optical Testing," *Bol. Inst. Tonantzintla*, **2**, 121–126 (1976).

Malacara D., J. M. Carpio-Valadéz, and J. Sánchez-Mondragón, "Interferometric Data Fitting on Zernike Like Orthogonal Basis," *Proc. SPIE*, **813**, 35–36 (1987).

Malacara D., J. M. Carpio, and J. J. Sánchez, "Wavefront Fitting with Discrete Orthogonal Polynomials in a Unit Radius Circle," *Opt. Eng.*, **29**, 672–675 (1990).

Mantravadi M. V., "Newton, Fizeau and Haidinger Interferometers," in *Optical Shop Testing*, 2nd Edition, p. 1, D. Malacara, Ed., John Wiley and Sons, New York, 1992.

Mastin G. A. and D. C. Ghiglia, "Digital Extraction of Interference Fringe Contours," *Appl. Opt.*, **24**, 1727–1728 (1985).

Matczac M. J. and J. Budziński, "A Software System for Skeletonization of Interference Fringes," *Proc. SPIE*, **1121**, 136– 141 (1990).

Mieth U. and Osten W., "Three Methods for the Interpolation of Phase Values Between Fringe Pattern Skeletons," *Proc. SPIE*, **1121**, 151–153 (1990).

Moore R. C., "Automatic Method of Real-Time Wavefront Analysis," *Opt. Eng.*, **18**, 461–463 (1979).

Nakadate S., T. Yatagai, and H. Saito, "Computer Aided Speckle Pattern Interferometry," *Appl. Opt.*, **22**, 237–243 (1983).

Osten W., R. Höfling, and J. Saedler, "Two Computer Methods for Data Reduction from Interferograms," *Proc. SPIE*, **863**, 105–113 (1987).

Parthiban V. and R. J. Sirohi, "Use of Gray Scale Coding in Labeling Closed Fringe Patterns," *Proc. SPIE*, **1163**, 77–82 (1989).

Platt B. C., S. G. Reynolds, and T. R. Holt, "Determining Image Quality and Wavefront Profiles from Interferograms," in *Optical Interferograms—Reduction and Interpretation*, ASTM Symposium, Tech. Publ. 666, A. H. Guenther and D. H. Liedbergh, Eds., American Society for Testing and Materials, 1978.

Plight A. M., "The Calculation of the Wavefront Aberration Polynomial," *Opt. Acta*, **27**, 717–721 (1980).

Prata A., Jr., and W.V. T. Rusch, "Algorithm for Computation of Zernike Polynomials Expansion Coefficients," *Appl. Opt.*, **28**, 749–754 (1989).

Press W. H., B. P. Flannery, S. A. Teukolsky, and W. T. Vetterling, *Numerical Recipes in C*, Cambridge University Press, Cambridge, 1988.

Reid G. T., "Automatic Fringe Pattern Analysis: A Review," *Opt. and Lasers in Eng.*, **7**, 37–68 (1986/87).

Reid G. T., "Image Processing Techniques for Fringe Pattern Analysis," *Proc. SPIE*, **954**, 468–477 (1988).

Robinson D. W., "Automatic Fringe Analysis with a Computer Image-Processing System," *Appl. Opt.*, **22**, 2169–2176 (1983a).

Robinson D. W., "Role for Automatic Fringe Analysis in Optical Metrology," *Proc. SPIE*, **376**, 20–25 (1983b).

Roblin G. and M. Prévost, "A Method to Interpolate Between Two-Beam Interference Fringes," *Proc. ICO-11, Madrid*, 667–670 (1978).

Schemm J. B. and C. M. Vest, "Fringe Pattern Recognition and Interpolation Using Nonlinear Regression Analysis," *Appl. Opt.*, **22**, 2850–2853 (1983).

Schluter M., "Analysis of Holographic Interferogram with a TV Picture System," *Opt. and Laser Technol.*, **12**, 93–95 (1980).

Servin M., R. Rodríguez-Vera, M. Carpio, and A. Morales, "Automatic Fringe Detection Algorithm Used for Moiré Deflectometry," *Appl. Opt.*, **29**, 3266–3270 (1990).

Snyder J. J., "Algorithm for Fast Digital Analysis of Interference Fringes," *Appl. Opt.*, **19**,1223–1225 (1980).

Swantner W. H. and W. H. Lowrey, "Zernike-Tatian Polynomials for Interferogram Reduction," *Appl. Opt.*, **19**, 161–163 (1980).

Tichenor D. A. and V. P. Madsen, "Computer Analysis of Holographic Interferograms for Non Destructive Testing," *Proc. SPIE*, **155**, 222–227 (1978).

Trolinger J. D., "Automated Data Reduction in Holographic Interferometry," *Opt. Eng.*, **24**, 840–842 (1985).

Truax B. E., "Programmable Interferometry," *Proc. SPIE*, **680**, 10–18 (1986).

Truax B. E. and L. A. Selberg, "Programmable Interferometry," *Opt. and Lasers in Eng.*, **7**, 195–220 (1986/7).

Varman C. and C. Wykes, "Smoothing of Speckle and Moiré Fringes by Computer Processing," *Opt. and Lasers in Eng.*, **3**, 87–100 (1982).

Vrooman H. A. and A. Maas, "Interferogram Analysis Using Image Processing Techniques," *Proc. SPIE*, **1121**, 655–659 (1989).

Wang G.-Y. and X.-P. Ling, "Accuracy of Fringe Pattern Analysis," *Proc. SPIE*, **1163**, 251–257 (1989).

Wang J. Y. and D. E. Silva, "Wave-Front Interpretation with Zernike Polynomials," *Appl. Opt.*, **19**, 1510–1518 (1980).

Womack K. H., J. A. Jonas, C. L. Koliopoulos, K. L. Underwood, J. C. Wyant, J. S. Loomis, and C. R. Hayslett, "Microprocessor- Based Instrument for Analysis of Video Interferograms," *Proc. SPIE*, **192**, 134–139 (1979).

Wyant, J. C. and K. Creath, "Basic Wavefront Aberration Theory for Optical Metrology," in *Applied Optics and Optical Engineering*, Vol. XI, R. R. Shannon and J. C. Wyant, Eds., Academic Press, Boston, 1992, Chap. 1.

Yan D.-P., A He, and P. C. Miao, "Method of Rapid Fringe Thinning for Flow-Field Interferograms," *Proc. SPIE*, **1755**, 190–193 (1992).

Yatagai T., "Intensity Based Analysis Methods," in *Interferogram Analysis, Digital Fringe Pattern Measurement Techniques*, p. 72, D. W. Robinson and G. T. Reid, Eds., Institute of Physics Publ., Philadelphia, 1993.

Yatagai T., S. Nakadate, M. Idesawa, and H. Saito, "Automatic Fringe Analysis Using Digital Image Processing Techniques," *Opt. Eng.*, **21**, 432–435 (1982a).

Yatagai T., M. Idesawa, Y. Yamaashi, and M. Suzuki, "Interactive Fringe Analysis System: Applications to Moiré Contourgram and Interferogram," *Opt. Eng.*, **21**, 901–906 (1982b).

Yatagai T., S. Inabu, H. Nakano, and M. Suzuki, "Automatic Flatness Tester for Very Large Scale Integrated Circuit Wafers," *Opt. Eng.*, **23**, 401–405 (1984).

Yu Q., K. Andersen, W. Osten, and W. P. O. Juptner, "Analysis and Removal of the Systematic Phase Error in Interferograms," *Opt. Eng.*, **33**, 1630–1637 (1994).

Zanoni C. A., "A New, Semiautomatic Interferogram Evaluation Technique," in *Optical Interferograms—Reduction and Interpretation*, ASTM Symposium, Tech. Publ. 666, A. H. Guenther and D. H. Liedbergh, Eds., American Society for Testing and Materials, West Consohocken, PA, 1978.

Zernike F., "Begunstheorie des Schneidenver-Fahrens und Seiner Verbesserten Form der Phasenkontrastmethode," *Physica*, **1**, 689 (1934).

Zernike F., "The Diffraction Theory of Aberrations," in *Optical Image Evaluation*, Circular 526, Natl. Bureau of Standards, Washington, DC, 1954.

Zhi H. and R. B. Johansson, "Adaptive Filter for Enhancement of Fringe Patterns," *Opt. and Lasers in Eng.*, **15**, 241–251 (1991).

5

Signal Phase Detection

5.1 LEAST SQUARES PHASE DETECTION OF A SINUSOIDAL SIGNAL

The detection (or measurement) of the real phase of a real sinusoidal signal whose frequency is known, by means of a sampling procedure, is an important problem that has been solved with several different procedures. Let us begin by studying the least squares method. From Eq. 1.4 the $s(x)$ may be written in a very general manner as

$$s(x) = a + b\cos(\omega x + \phi) \tag{5.1}$$

where x is the coordinate (spatial or temporal) at which the irradiance is to be measured, ω is the angular spatial (or temporal) frequency, and ϕ is the phase at the origin ($x = 0$). If we want to make a least squares fit of these data measurements of the irradiance to a sinusoidal function as in Eq. 5.1 and Fig. 5.1, we see that we have to determine four unknown constants, i.e., a, b, ϕ, and ω. However, the analysis is simpler if we assume that the frequency of the sinusoidal function, ω, is known, as is normally the case.

To make the least squares analysis following Greivenkamp (1984) it is better to write this expression in the equivalent manner

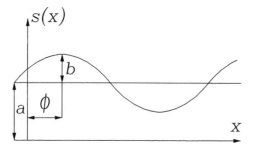

Figure 5.1 Unknowns when sampling a sinusoidal function. The frequency ω is assumed to be known.

$$s(x) = D_1 + D_2 \cos \omega x + D_3 \sin \omega x \tag{5.2}$$

where

$$D_1 = a$$
$$D_2 = b \cos \phi \tag{5.3}$$
$$D_3 = -b \sin \phi$$

Now, the following N measurements of the signal are taken:

$$s_n = D_1 + D_2 \cos \omega x_n + D_3 \sin \omega x_n, \qquad n = 1, \ldots, N \tag{5.4}$$

with $N \geq 3$ since there are three constants to be determined. The best fit of these measurements to the sinusoidal analytical function is obtained if the coefficients D_1, D_2, and D_3 are chosen so that the variance ε, defined by

$$\varepsilon = \frac{1}{N} \sum_{n=1}^{N} (D_1 + D_2 \cos \omega x_n + D_3 \sin \omega x_n - s_n)^2 \tag{5.5}$$

is minimized. Thus, taking the partial derivatives of this variance ε with respect to the three unknown constants D_1, D_2, and D_3, we find a set of simultaneous equations, which may be written in matrix form as

$$\begin{pmatrix} N & \sum \cos \omega x_n & \sum \sin \omega x_n \\ \sum \cos \omega x_n & \sum \cos^2 \omega x_n & \sum \cos \omega x_n \sin \omega x_n \\ \sum \sin \omega x_n & \sum \cos \omega x_n \sin \omega x_n & \sum \sin^2 \omega x_n \end{pmatrix} \begin{pmatrix} D_1 \\ D_2 \\ D_3 \end{pmatrix}$$

$$= \begin{pmatrix} \sum s_n \\ \sum s_n \cos \omega x_n \\ \sum s_n \sin \omega x_n \end{pmatrix} \tag{5.6}$$

This matrix is evaluated with the values of the phases at which the signal is measured, but it does not depend on the values of the signal. Thus, if necessary, the signal may be measured as many times as desired without the need to calculate the matrix elements every time. It is only necessary that the same phase values be used. This is the case in phase shifting interferometry, as will be seen in Chap. 6.

As shown by Greivenkamp (1984), this is a general least squares procedure, for any separation between the measurements, only assuming that the frequency ω is known.

The system expressed by Eq. 5.6 may also be written as

$$\begin{pmatrix} a_{11} & a_{12} & a_{13} \\ a_{12} & a_{22} & a_{23} \\ a_{13} & a_{23} & a_{33} \end{pmatrix} \begin{pmatrix} D_1 \\ D_2 \\ D_3 \end{pmatrix} = \begin{pmatrix} \sum s_n \\ \sum s_n \cos \omega x_n \\ \sum s_n \sin \omega s_n \end{pmatrix} \tag{5.7}$$

Then, from Eqs. 5.3 the phase may be found to be given by

$$\tan \phi = - \left(\frac{D_3}{D_2} \right)$$

$$= - \frac{\sum\limits_{n=1}^{N} s_n \left[A_{11} + A_{12} \cos \left(\dfrac{2\pi n}{N} \right) + A_{13} \sin \left(\dfrac{2\pi n}{N} \right) \right]}{\sum\limits_{n=1}^{N} s_n \left[A_{21} + A_{22} \cos \left(\dfrac{2\pi n}{N} \right) + A_{23} \sin \left(\dfrac{2\pi n}{N} \right) \right]} \tag{5.8}$$

where

$$A_{11} = (a_{12}a_{23} - a_{13}a_{22})$$
$$A_{12} = (a_{12}a_{13} - a_{11}a_{23})$$
$$A_{13} = (a_{11}a_{22} - a_{12}^2)$$
$$A_{21} = (a_{12}a_{33} - a_{13}a_{23})$$
$$A_{22} = (a_{13}^2 - a_{11}a_{33})$$
$$A_{23} = (a_{11}a_{23} - a_{12}a_{13})$$

$$(5.9)$$

A particular least squares sampling procedure was previously analyzed by Morgan (1982), who assumed that the measurements were taken at equally spaced intervals, uniformly spaced in k signal periods, defined by

$$\omega x_n = \frac{2\pi(n-1)}{N} + \omega x_1 \tag{5.10}$$

where x_1 is the location of the first sampling point and $n = 1, 2, \ldots, kN$. In the most frequent case the sampling points are distributed in only one signal period $(k = 1)$. To understand this angular distribution we plot these sampling points with unit vectors from the origin, each vector with an angle $2\pi(n-1)/N$ with respect to the x axis, as shown in Fig. 5.2. Then we may see that this sampling distribution for $N \geq 3$ requires that the vector sum of all the vectors from the origin to each point be zero. This condition may be expressed by

$$\sum_{n=1}^{N} \sin \omega x_n = 0; \qquad \sum_{n=1}^{N} \cos \omega x_n = 0 \tag{5.11}$$

This condition is necessary but not sufficient to guarantee the equally spaced and uniform distribution in Eq. 5.10. As shown in the lower row in Fig. 5.2, we also need the following conditions for twice the phase angle:

$$\sum_{n=1}^{N} \sin 2\omega x_n = 0; \qquad \sum_{n=1}^{N} \cos 2\omega x_n = 0 \tag{5.12}$$

From the first expression in Eqs. 5.12 we may see that

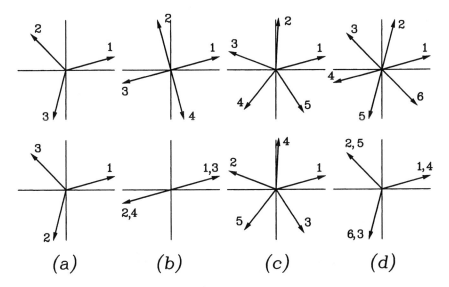

Figure 5.2 Polar representation of sampling points uniformly spaced in a signal period. (a) Three points; (b) four points; (c) five points; (d) six points. The upper row plots the phase for Eq. 5.11, and the lower row plots twice the phase angle for Eq. 5.12.

$$\sum_{n=1}^{N} \cos \omega x_n \sin \omega x_n = \frac{1}{2} \sum_{n=1}^{N} \sin 2\omega x_n = 0 \qquad (5.13)$$

and from the second expression and a well-known trigonometric relation we find

$$\sum_{n=1}^{N} \cos^2 \omega x_n = \sum_{n=1}^{N} \sin^2 \omega x_n = \frac{N}{2} \qquad (5.14)$$

With these relations we may see that the system matrix becomes diagonal

$$\begin{pmatrix} N & 0 & 0 \\ 0 & \frac{N}{2} & 0 \\ 0 & 0 & \frac{N}{2} \end{pmatrix} \begin{pmatrix} \alpha_1 \\ \alpha_2 \\ \alpha_3 \end{pmatrix} = \begin{pmatrix} \sum s_n \\ \sum s_n \cos\left(\frac{2\pi n}{N}\right) \\ \sum s_n \sin\left(\frac{2\pi n}{N}\right) \end{pmatrix} \qquad (5.15)$$

with the solutions

$$\alpha_1 = \frac{1}{N} \sum_{n=1}^{N} s_n \tag{5.16}$$

$$\alpha_2 = \frac{2}{N} \sum_{n=1}^{N} s_n \cos\left(\frac{2\pi n}{N}\right) \tag{5.17}$$

and

$$\alpha_3 = \frac{2}{N} \sum_{n=1}^{N} s_n \sin\left(\frac{2\pi n}{N}\right) \tag{5.18}$$

Substituting Eqs. 5.17 and 5.18 into Eq. 5.8, the phase ϕ at the origin may be obtained from

$$\tan\phi = -\left(\frac{\alpha_3}{\alpha_2}\right) = -\left(\frac{\displaystyle\sum_{n=1}^{N} s_n \sin\left(\frac{2\pi n}{N}\right)}{\displaystyle\sum_{n=1}^{N} s_n \cos\left(\frac{2\pi n}{N}\right)}\right) \tag{5.19}$$

This algorithm is so interesting that it deserves a name. We will call it the *diagonal least squares algorithm*. The minimum acceptable number of sampling points is $N = 3$. Then the sampling spacing as given by Eq. 5.10 is

$$\Delta x = \frac{2\pi}{3\omega} = \frac{1}{3f} \tag{5.20}$$

and the phase ϕ becomes

$$\tan\phi = \frac{\sqrt{3}(s_1 - s_2)}{s_1 + s_2 - 2s_3} \tag{5.21}$$

If the sampling points are not properly spaced as required by Eq. 5.20, the phase value obtained with Eq. 5.19 or 5.21 will not be correct, as will be shown later.

5.2 QUADRATURE PHASE DETECTION OF A SINUSOIDAL SIGNAL

Let us consider a sinusoidal signal $s(x)$ as in Eq. 5.1, now written as follows

$$s(x) = a + b\cos(2\pi f x + \phi) \tag{5.22}$$

where f is the frequency of this signal. Let us now take the Fourier transform $S(f)$ of this signal, at a reference frequency f_r as

$$S(f_r) = \int_{-\infty}^{\infty} \exp(-i2\pi f_r x)\, dx \tag{5.23}$$

obtaining

$$S(f_r) = a\delta(f_r) + \frac{b}{2}\delta(f_r - f)\exp(i\phi) + \frac{b}{2}\delta(f_r + f)\exp(-i\phi) \tag{5.24}$$

If the reference frequency f_r is equal to the frequency of the signal ($f = f_r$), this function has the value

$$S(f_r) = \frac{b}{2}\exp(i\phi) = \frac{b}{2}(\cos\phi + \sin\phi) \tag{5.25}$$

Then, as pointed out in Chapter 2, the phase ϕ of the real periodic signal in Eq. 5.1 evaluated at the origin ($x = 0$) is equal to the phase of its Fourier transform at the frequency of the signal ($f = f_r$). Thus, using Eq. 5.23 we may write

$$\tan\phi = \left(\frac{\mathrm{Im}\{S(f_r)\}}{\mathrm{Re}\{S(f_r)\}}\right) = -\left(\frac{\int_{-\infty}^{\infty} s(x)\sin(2\pi f_r x)\, dx}{\int_{-\infty}^{\infty} s(x)\cos(2\pi f_r x)\, dx}\right) \tag{5.26}$$

To gain some insight into the nature of these integrals, we may perform the multiplication of the signal with a frequency f by sine and cosine functions with frequency f_r, obtaining

$$
\begin{aligned}
z_S(x) &= s(x)\sin\omega_r x \\
&= -\frac{b}{2}\sin(\omega x - \omega_r x + \phi) + a\sin(\omega_r x) \\
&\quad + \frac{b}{2}\sin(\omega x + \omega_r x + \phi)
\end{aligned} \tag{5.27}
$$

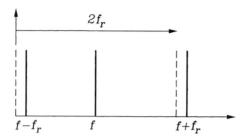

Figure 5.3 Spectrum of functions resulting from the multiplication of the sinusoidal signal by two reference sinusoidal functions, sine and cosine.

and

$$z_C(x) = s(x)\cos\omega_r x$$
$$= \frac{b}{2}\cos(\omega x - \omega_r x + \phi) + a\cos(\omega_r x)$$
$$+ \frac{b}{2}\cos(\omega x + \omega_r x + \phi) \tag{5.28}$$

where $\omega = 2\pi f$ and $\omega_r = 2\pi f_r$. These two functions $z_S(x)$ and $z_C(x)$ are periodic, but they contain three harmonic components: (1) the first term, with a very low frequency, equal to the difference between the signal and the reference frequencies; (2) the second term with the reference frequency; and (3) the last term with a frequency equal to the sum of the signal and reference frequencies. The spectrum of these functions is illustrated in Fig. 5.3.

If the terms with frequencies ω_r and $\omega + \omega_r$ are properly eliminated by a suitable low pass filter that also preserves the ratio of the amplitudes of the low frequency terms, we obtain the filtered versions of these functions as

$$\bar{z}_S(x) = -\frac{b}{2}\sin(\omega x - \omega_r x + \phi) \tag{5.29}$$

and

$$\bar{z}_C(x) = \frac{b}{2}\cos(\omega x - \omega_r x + \phi) \tag{5.30}$$

Thus, we may obtain

$$\tan(\omega x - \omega_r x + \phi) = -\frac{\bar{z}_S(x)}{\bar{z}_C(x)} \tag{5.31}$$

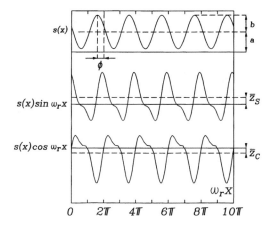

Figure 5.4 Functions resulting from the multiplication of the sinusoidal signal by two reference sinusoidal functions, sine and cosine, with the same frequency as the signal.

When the signal and reference frequencies are equal, functions 5.29 and 5.30 are constants. Figure 5.4 plots functions 5.27 and 5.28 for this case, where, since the signal is not phase-modulated, the filtered functions $\bar{z}_S(x)$ and $\bar{z}_C(x)$ become constants.

The phase ϕ at the origin $(x = 0)$ may be calculated with

$$\tan \phi = -\frac{\bar{z}_S(0)}{\bar{z}_C(0)} \tag{5.32}$$

The conditions for this method to produce accurate results and the effect of several possible sources of error were studied by Nakadate (1988a, 1988b). In the next section we will study how the low pass filtering has to be performed to obtain the phase ϕ at the origin or the phase $(\omega x - \omega_r x + \phi)$ at any point x.

5.2.1 Low Pass Filtering in Phase Detection

The simplest case for phase detection is when there is no detuning, that is, when the signal frequency and the reference frequency are equal. In this case, evaluating the integrals in Eq. 5.26, we may obtain the graphs in Fig. 5.5. We see that the values of both integrals tend to infinity, although the ratio of the two integrals has a finite value which is equal to the ratio of their average slopes.

There are several ways to find this finite ratio of the integrals. One method, since the signal is periodic, performs the integration only in the finite size interval $-1/2f < x < 1/2f$, $(-\pi < \omega x < \pi)$ or integer multiples of this value, as

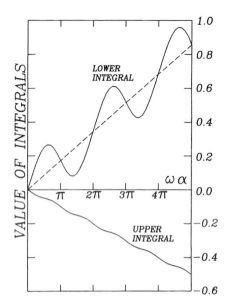

Figure 5.5 Plots of the values of the integrals in Eq. 5.23 for a signal phase equal to 30°, and signal constants $a = 1.3$ and $b = 1$.

shown in Fig. 5.6. Two disadvantages of this method are that a large number of sampling points are needed to emulate a continuous measurement and also that the signal frequency must be accurately known in order to correctly fix the sampling interval.

Another equivalent method is a discrete sampling low pass filtering process that may be performed by means of a convolution, as described in Chap. 2, with a pair of suitable filtering functions $h_S(x)$ and $h_C(x)$. Let us now consider this method but remove the restriction for zero detuning. The whole process of multiplication by the sinusoidal reference and low pass filtering to obtain the filtered functions $\bar{z}_S(x)$ and $\bar{z}_C(x)$ may be expressed by

$$\bar{z}_S(x) = \int_{-\infty}^{\infty} z_S(\alpha) h_S(x - \alpha) \, d\alpha \tag{5.33}$$

and in an analogous manner, with a filtering function $h_C(x)$,

$$\bar{z}_C(x) = \int_{-\infty}^{\infty} z_C(\alpha) h_C(x - \alpha) \, d\alpha \tag{5.34}$$

To obtain the correct value of the phase $(\omega x - \omega_r x + \phi)$ at any point x in the presence of detuning by using Eq. 5.31 we need to satisfy three conditions:

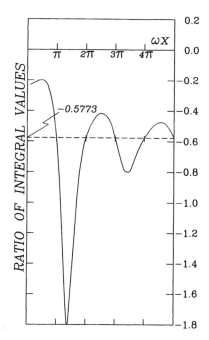

Figure 5.6 Plot of the values of the ratio of the integrals in Eq. 5.26 for a signal phase equal to 30°, and signal constants $a = 1.3$ and $b = 1$.

1. The low pass filtering must be performed using the convolution operation, as expressed by Eqs. 5.33 and 5.34.
2. The terms with frequencies ω_r and $\omega + \omega_r$ must be completely eliminated, so that their value is zero at any value of x.
3. The ratio of the amplitudes of the low frequency terms with frequency $\omega - \omega_r$ must be preserved by the filtering process

In general, the filtering functions for $z_S(x)$ and $z_C(x)$ can be different, although sometimes they are the same, as we will see later. If the filtering function is the same for both functions, the third condition is automatically satisfied, but not if they are different.

Let us now consider the case when we are interested, not in the phase at any value of x, but only in the phase ϕ at the origin. In this case we need to satisfy slightly different conditions, as will now be described. In order to obtain the correct phase with expression 5.32, the contribution of the high frequency components of $\bar{z}_S(x)$ or $\bar{z}_C(x)$ to the value of the filtered signals $\bar{z}_S(0)$ or $\bar{z}_C(0)$, respectively, must be zero. In other words, we do not require that the high frequency components be completely eliminated, only that their value at $x = 0$ be zero. The conditions to be satisfied in this case are

1. The low pass filtering must be complete only for the point at the origin, using the convolution with $x = 0$.
2. The contributions to $\bar{z}_S(0)$ and $\bar{z}_C(0)$ of the terms with frequencies ω_r and $\omega + \omega_r$ evaluated at the origin must be zero.
3. The ratio of the amplitudes of the low frequency terms with frequency $\omega - \omega_r$ must be preserved by the filtering process.

To better understand the second condition, let us assume that we need to avoid any effect on the phase in Eq. 5.32 of a certain high frequency component present in $z_S(x)$ or $z_C(x)$ which is sinusoidal and real. We require that the value of this sinusoidal component be zero at the origin. The value at the origin of this sinusoidal component is zero not only if its amplitude is zero but also if it is antisymmetrical (sine function). Then, its Fourier transform at this frequency must be imaginary and antisymmetrical, as shown in Table 2.3.

We have seen in Chap. 2 that the convolution of two functions is equal to the inverse Fourier transform of the product of the Fourier transforms of the two functions. Hence, we may write

$$F\{\bar{z}_S(x)\} = Z_S(f)H_S(f) \tag{5.35}$$

and similarly for $z_C(x)$. Thus, the right-hand side of this expression at the frequency to be filtered, like the left-hand side, must be imaginary and antisymmetrical.

On the other hand, the sinusoidal component of $z_S(x)$ that we want to filter out is real, and thus its Fourier transform $Z_S(f)$, according to Table 2.3, can be (1) real and symmetrical, (2) imaginary and antisymmetrical, or (3) complex and Hermitian. For these cases we can see that $H(f)$ must be (1) imaginary and antisymmetrical, (2) real and symmetrical, or (3) complex and Hermitian, respectively. The results are summarized in Table 5.1.

The second term in Eq. 5.27 is real and antisymmetrical. Thus, we need a filter function such that its Fourier transform is real and symmetrical at this frequency, satisfying the condition

$$H_S(f_r) = H_S(-f_r) \tag{5.36}$$

Similarly, the second term in Eq. 5.28 is real and symmetrical. Thus, we need a filter function such that its Fourier transform is imaginary and antisymmetrical at this frequency, satisfying the condition

$$H_C(f_r) = -H_C(-f_r) \tag{5.37}$$

Table 5.1 Necessary Properties of the Fourier Transform of the Filtering Function, for Each Kind of Function to be Filtered, in order to Make the Right-Hand Side of Eq. 5.35 Imaginary and Antisymmetrical

Sinusoidal component of $z(x)$	Fourier transform $Z_S(f_r)$ or $Z_C(f_r)$	Function $H(f_r)$
Real and symmetrical	Real and symmetrical	Imaginary and anti-symmetrical
Real and antisymmetrical	Imaginary and anti-symmetrical	Real and symmetrical
Real and asymmetrical	Complex and Hermitian	Complex and Hermitian

The terms with frequency $2f_r$ (assuming $f = f_r$) are asymmetrical, that is, they are neither symmetrical nor antisymmetrical. Moreover, the degree of asymmetry is not predictable, since it depends on the phase of the signal. So the only solution is that the Fourier transforms of the filtering functions have zeros at this frequency, as follows:

$$H_S(2f_r) = H_S(-2f_r) = 0$$
$$H_C(2f_r) = H_C(-2f_r) = 0 \tag{5.38}$$

Besides these conditions, the filtering function $h(x)$ must not modify the ratio between the constant (zero frequency) terms in the functions in Eqs. 5.27 and 5.28, thus also requiring that

$$H_S(0) = H_C(0) \tag{5.39}$$

These conditions in Eqs. 5.36 to 5.39 are quite general. There are an infinite number of possible filter functions, continuous and discrete, that satisfy these conditions. Each pair of possible filter functions leads to a different algorithm with different properties.

A particular case of the conditions in Eqs. 5.36 and 5.37 is the stronger condition

$$H_S(f_r) = H_S(-f_r) = H_C(f_r) = H_C(-f_r) = 0 \tag{5.40}$$

which occurs when the sampling points distribution satisfies Eq. 5.10. In this case the two filter functions become identical at all frequencies

A continuous filtering function with a continuous sampling of points, satisfying Eq. 5.10, is the square function

$$h(\alpha) = 1 \quad \text{for} \quad -\frac{1}{2f_r} \le \alpha \le \frac{1}{2f_r}$$
$$= 0 \quad \text{for} \quad |\alpha| > \frac{1}{2f_r} \tag{5.41}$$

whose Fourier transform has zeros at nf_r, with n being any nonzero integer. We then see that this filtering process is equivalent to performing the integration in a finite limited interval, as suggested before.

5.3 DISCRETE LOW PASS FILTERING FUNCTIONS

Now some discrete sampling low pass filtering functions will be described. We write the filtering functions $h_S(x)$ and $h_C(x)$ for the sampled signal process as

$$h_S(x) = \sum_{n=1}^{N} w_{Sn}\delta(x - \alpha_n) \tag{5.42}$$

and

$$h_C(x) = \sum_{n=1}^{N} w_{Cn}\delta(x - \alpha_n) \tag{5.43}$$

where α_n are the positions of the sampling points. The Fourier transforms of these functions are given by

$$H_S(f) = \sum_{n=1}^{N} w_{Sn}\exp(-i2\pi f\alpha_n) \tag{5.44}$$

and

$$H_C(f) = \sum_{n=1}^{N} w_{Cn}\exp(-i2\pi f\alpha_n) \tag{5.45}$$

where w_{Sn} and w_{Cn} are the filtering weights.

Filtering functions of special interest are discrete functions with equally spaced and uniformly distributed sampling points in a signal interval, as stated by

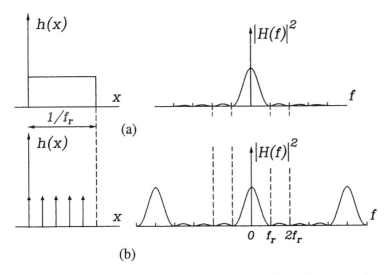

Figure 5.7 Spectra of the filtering function when five points are used to sample a sinusoidal function. (a) Rectangular filter; (b) discrete sampling filter.

Eq. 5.10. The filtering functions $h_S(x)$ and $h_C(x)$ satisfy Eq. 5.39, and thus they are identical, equal to $h(x)$ with all the filtering weights equal to 1. With these filtering functions, synchronous detection as expressed by Eq. 5.26 may become identical to the *diagonal least squares algorithm* as expressed by Eq. 5.15.

To consider this case, we impose the condition that the sampling points have a constant separation $\Delta\alpha$, with the first point at the position $\alpha = 0$, as in Eq. 5.10. Then this expression becomes

$$H(f) = \frac{1 - \exp(-i2\pi f N \Delta\alpha)}{1 - \exp(-i2\pi f \Delta\alpha)}$$

$$= \frac{\sin(\pi f N \Delta\alpha)}{\sin(\pi f \Delta\alpha)} \exp[-i2\pi(N - 1)f \Delta\alpha] \qquad (5.46)$$

Hence, the power spectrum of this filtering function is

$$|H(f)|^2 = \frac{\sin^2(\pi f N \Delta\alpha)}{\sin^2(\pi f \Delta\alpha)} \qquad (5.47)$$

and it is illustrated in Fig. 5.7a for the case of an infinite number of points and in Fig. 5.7b for the discrete case of five sampling points.

We see that the zeros and peaks of this function occur at frequencies $n/(N\Delta\alpha)$, where n is any integer and the zeros when n/N is not an integer. Thus, we have $N - 1$ minima (zeros) between two consecutive lobes. There is a lobe at zero frequency ($n = 0$). Since we want zeros at the signal frequency f_S and at twice this frequency, we need at least three sampling points ($N \geq 3$). In order to locate the first two zeros at these frequencies we require equally and uniformly spaced sampling points on the signal period, as

$$\Delta\alpha = \frac{1}{Nf_S} \tag{5.48}$$

This condition is the same as that in Eq. 5.10, used to make the least squares matrix diagonal. Thus, if we use this filtering function $h(x)$ for equally spaced sampling points, we obtain Eq. 5.19.

We may see that the zeros of this function occur at frequencies nf, with the exception of Nf and integer multiples of Nf, where n is any integer and N is the number of sampling points. Since we must filter out frequencies f and $2f$, we see that there must be at least three sampling points ($N = 3$), so that there are at least two minima (zeros) between two consecutive peaks of the filtering function.

Filtering functions were studied by de Groot (1955), who referred to them as data sampling windows.

5.3.1 Examples of Discrete Filtering Functions

To better illustrate the concepts of discrete filtering functions, let us describe three interesting algorithms that will be studied in more detail from another point of view in Chap. 6.

Wyant's Three-step Algorithm

The first example is Wyant's three-step algorithm to be described in Sec. 6.2.3, which uses three sampling points at $-45°$, $45°$, and $135°$. This algorithm is obtained if we use the filtering functions

$$h_S(x) = \delta\left(x + \frac{X_r}{8}\right) + \delta\left(x - \frac{X_r}{8}\right) \tag{5.49}$$

and

$$h_C(x) = \delta\left(x - \frac{X_r}{8}\right) + \delta\left(x - 3X_r 8\right) \tag{5.50}$$

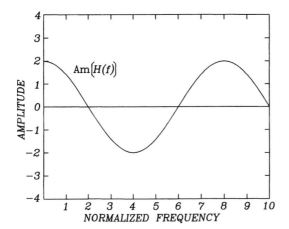

Figure 5.8 Amplitudes of the Fourier transforms of the filtering function for Wyant's algorithm.

where $X_r = 1/f_r$. These two filtering functions are different. Their Fourier transforms are

$$H_S(f) = \cos\left(\frac{\pi}{4}\frac{f}{f_r}\right) \tag{5.51}$$

and

$$H_C(f) = \cos\left(\frac{\pi}{4}\frac{f}{f_r}\right)\exp\left(-i\frac{\pi}{2}\frac{f}{f_r}\right)$$
$$= \cos\left(\frac{\pi}{4}\frac{f}{f_r}\right)\cos\left(\frac{\pi}{s}\frac{f}{f_r}\right) + i\cos\left(\frac{\pi}{4}\frac{f}{f_r}\right)\sin\left(\frac{\pi}{s}\frac{f}{f_r}\right) \tag{5.52}$$

We can see that although the two filtering functions are different, the amplitudes of the two Fourier transforms are equal, as shown in Fig. 5.8. There is a zero of this amplitude at $2f_r$, as required by Eq. 5.38. The conditions in Eqs. 5.36 and 5.39 are also satisfied.

Four Steps in Cross Algorithm

The second example is the four steps in cross algorithm described in Sec. 6.3.1, which uses four sampling points at $0°$, $90°$, $180°$, and $270°$. This is a diagonal least squares algorithm. It may be obtained if we use the filtering function

$$h_S(x) = h_C(x)$$

$$= \delta(x) + \delta\left(x - \frac{X_r}{4}\right) + \delta\left(x - \frac{X_r}{2}\right) + \delta\left(x - \frac{3X_r}{4}\right) \tag{5.53}$$

The Fourier transform of this function is

$$H_S(f) = 2\left[\cos\left(\frac{3\pi}{4}\frac{f}{f_r}\right) + \cos\left(\frac{\pi}{4}\frac{f}{f_r}\right)\right]\exp\left(-i\frac{3\pi}{4}\frac{f}{f_r}\right) \tag{5.54}$$

and its amplitude is illustrated in Fig. 5.9.

We can see that the amplitude has zeros at the reference frequency f_r and at twice this frequency. Conditions in Eqs. 5.38 to 5.40 are satisfied.

Schwider-Hariharan Five (4 + 1) Step Algorithm

The third example is the Schwider-Hariharan five (4+1) step algorithm described in Sec. 6.5.2, which uses five sampling points at $0°$, $90°$, $180°$, $270°$, and $360°$. This algorithm is obtained when we use the filtering function

$$h_S(x) = h_C(x)$$

$$= \frac{1}{2}\delta(x) + \delta\left(x - \frac{X_r}{4}\right) + \delta\left(x - \frac{x_r}{2}\right)$$

$$+ \delta\left(x - \frac{3X_r}{4}\right) + \frac{1}{2}\delta(x - X_r) \tag{5.55}$$

The Fourier transform of this function is

$$H_S(f) = H_C(f)$$

$$= \left[\cos\left(\pi\frac{f}{f_r}\right) + 2\cos\left(\frac{\pi}{2}\frac{f}{f_r}\right) + 1\right]$$

$$\times \exp\left(-i\pi\frac{f}{f_r}\right) \tag{5.56}$$

and its amplitude is illustrated in Fig. 5.10.

We can see that the amplitude of this Fourier transform of the filtering functions has zeros at the reference frequency and at twice the reference frequency, satisfying Eqs. 5.38, 5.39, and 5.40.

It is interesting to note that in Eqs. 5.27 and 5.28 as well as in Fig. 5.3 the term with frequency f_r is fixed and its position is independent of any possible

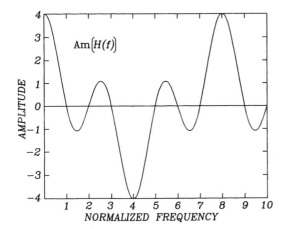

Figure 5.9 Amplitude of the Fourier transform of the filtering function for the four steps in cross algorithm.

difference between the reference frequency f_r and the signal frequency f (detuning). On the other hand, the Fourier components with the lowest frequency and the one with frequency $f + f_r$ may have slight frequency variations with these frequency deviations. The slope of the amplitude in these two regions is nearly zero, making this algorithm insensitive to small detuning.

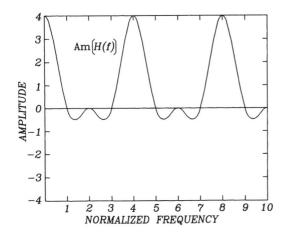

Figure 5.10 Amplitude of the Fourier transform of the filtering function for the Schwider-Hariharan algorithm.

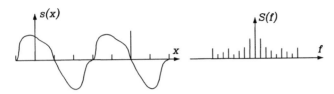

Figure 5.11 A periodic distorted signal and its spectrum.

5.4 FOURIER DESCRIPTION OF SYNCHRONOUS PHASE DETECTION

In this section we will study the synchronous detection in a more general manner, from a Fourier domain point of view, as developed by Freischlad and Koliopoulos (1990) and by Parker (1991) and later reviewed by Larkin and Oreb (1992).

If we want to remove this restriction of equally and uniformly spaced sampling points, the product of the sine function and the low pass filtering function $h(-x)$ has to be more generally considered as a function $g_1(x)$. This function does not necessarily need to be the product of a sine function and a filtering function. In an analogous manner, a function $g_2(x)$ replaces the product of the cosine and filtering functions. The two functions $g_1(x)$ and $g_2(x)$ will be called the sampling reference functions.

This treatment considers synchronous detection with the following two general assumptions:

1. The signal to be detected is periodic but not necessarily sinusoidal. In other words, it may contain harmonics.
2. The two reference functions $g_1(x)$ and $g_2(x)$ just mentioned are used instead of the products of the sine and cosine functions and the low pass filtering function.

The theory will permit us to analyze many possible sources of errors. It will also permit the study of the detection of a sinusoidal signal with a different frequency than that of the reference functions.

A real periodic distorted signal $s(x)$ as shown in Fig. 5.11 has several harmonic frequencies, i.e., frequencies that are an integer multiples of the fundamental frequency f, and may be written as

$$s(x) = S_0 + 2 \sum_{m=1}^{\infty} S_m \cos(2\pi m f x + \phi_m) \tag{5.57}$$

or equivalently

$$s(x) = \sum_{m=-\infty}^{\infty} S_m \exp(i(2\pi mfx + \phi_m)) \tag{5.58}$$

where we have defined $S_{-m} = -S_m$, $\phi_{-m} = -\phi_m$, and $\phi_0 = 0$.

Thus, the Fourier transform of this signal may be represented by

$$S(f) = \sum_{m=-\infty}^{\infty} S_m \delta(f - mf) \exp(i\phi_m) \tag{5.59}$$

In this expression m is the harmonic component number, S_m and ϕ_m are the amplitude and phase at the origin, respectively, of this harmonic component m, and f is the fundamental frequency of the signal.

The two sampling reference functions $g_i(x)$ are real and not necessarily periodic but have continuous Fourier transforms with many sinusoidal components with different frequencies. Also, the sinusoidal element components of the two functions do not necessarily have the same amplitude, nor are they necessarily orthogonal at any frequency, only at certain selected frequencies. To be able to use the sampling functions as references we need the Fourier elements at the desired reference frequency (normally the fundamental or lowest frequency) of these functions to be orthogonal, to have the same amplitude, and not to have any DC bias. In other words, the elemental reference components $\delta g_i(x)$ at the reference frequency ideally should be the typical sine and cosine functions

$$\delta g_1(x) = \pm A \sin(2\pi f_r x - \psi(f_r))\delta f$$
$$= A \cos(2\pi f_r x - \psi(f_r) \mp \pi/2)\delta f \tag{5.60}$$

and

$$\delta g_2(x) = A \cos(2\pi f_r x - \psi(f_r))\delta f \tag{5.61}$$

where $\psi(f_r)$ is the displacement in the positive direction of the Fourier element $\delta g_i(x)$ with frequency f_r, of the reference function $g_i(x)$, with respect to the origin of the phase. The frequency interval δf is formed by two symmetrically placed intervals, to cover positive as well as negative frequencies of value $|f_r|$. The first maxima of the Fourier transform $G_i(f)$ is frequently located near the reference frequency f_r.

We have seen before that the phase is the ratio of the two convolutions in Eqs. 5.33 and 5.34, using the proper filtering function. On the other hand, we also have seen that if the goal is to find the phase ϕ at the origin, we need to evaluate the convolution only at the origin. So it is reasonable to expect that the phase will be given by the ratio $r(f)$ of the correlations,

$$r(f) = \frac{C_1}{C_2} = \frac{\displaystyle\int_{-\infty}^{\infty} s(x)g_1(x)\,dx}{\displaystyle\int_{-\infty}^{\infty} s(x)g_2(x)\,dx} \tag{5.62}$$

if the functions $g_1(x)$ and $g_2(x)$ are properly selected. This correlation ratio is a function of the signal frequency f as well as of the signal phase ϕ. If the two reference functions $g_1(x)$ and $g_2(x)$ satisfy the conditions intuitively stated before, by analogy with Eq. 5.28, we expect the phase ϕ of the signal harmonic with frequency f being detected to be given by

$$\tan(\phi - \psi(f_r)) = \mp r(f_r) \tag{5.63}$$

We will prove this expression to be correct if these conditions are satisfied. Otherwise, the phase ϕ cannot be found with this expression. Let us now study in some detail what happens when these conditions are satisfied. The quantity C_j has been defined as

$$C_j = \int_{-\infty}^{\infty} s(x)g_j(x)\,dx, \qquad j = 1,2 \tag{5.64}$$

which is the cross-correlation of the two functions $s(x)$ and $g_i(x)$ evaluated at the origin. For simplicity we will simply call these quantities correlations.

We may see that the ratio of the correlations $r(f)$ is a function of the reference and signal frequencies and that it is directly related to the phase of the real signal only if the proper conditions for the functions $g_j(x)$ are met. From the central ordinate theorem expressed by Eqs. 3.10 we find

$$C_j = (\mathrm{F}\{s(x)g_j(x)\})_{f=0}, \qquad j = 1,2 \tag{5.65}$$

evaluated at the origin ($f = 0$), since the quantity to be determined is the phase of the fundamental frequency of the signal with respect to the phase of the reference functions. Using now the convolution theorem in Eq. 2.18, we find

$$C_j = (S(f) * G_j(f))_{f=0}, \qquad j = 1,2 \tag{5.66}$$

where $S(f)$ and $G_j(f)$ are the Fourier transforms of $s(x)$ and $g_j(x)$, respectively. Hence, we write the convolution at $f = 0$ as

$$C_j = \int_{-\infty}^{\infty} S(\nu) G_j(-\nu) \, d\nu, \qquad j = 1, 2 \tag{5.67}$$

where ν is the dummy variable used in the convolution. Since $s(x)$ and $g_j(x)$, are real, $S(f)$ and $G_j(f)$ are Hermitian, and we obtain

$$C_j = 2\text{Re} \int_{-\infty}^{\infty} S(f) G_j^*(f) \, df, \qquad j = 1, 2 \tag{5.68}$$

where Re stands for the real part and the asterisk (*) denotes the complex conjugate. For clarity the dummy variable ν has been changed to the frequency variable f.

If we substitute here the value of $S(f)$ from Eq. 5.59, we obtain

$$C_j = 2\text{Re} \sum_{m=-\infty}^{\infty} S_m G_j^*(mf) \exp(i\phi_m), \qquad j = 1, 2 \tag{5.69}$$

The reference functions $g_1(x)$ and $g_2(x)$ are real; hence their Fourier transforms are complex and Hermitian. Quite generally, we may express these functions $G_j(f)$ as

$$G_j(f) = |G_j(f)| \exp(i\gamma_j(f)), \qquad j = 1, 2 \tag{5.70}$$

where $\gamma_i(f)$ is the phase of the Fourier element with frequency f of the reference functions $g_j(x)$. Also, $\gamma_j(-mf) = -\gamma_j(-mf)$ since $G_j(f)$ is Hermitian. Hence

$$C_j = 2\text{Re} \sum_{m=-\infty}^{\infty} S_m |G_j(mf)| \exp[i(\phi_m - \gamma_j(mf))], \qquad j = 1, 2 \tag{5.71}$$

Since the argument of the exponential function is antisymmetric with respect to m, this equation may also be written as

$$C_j = 2S_0 |G_j(0)| + 4 \sum_{m=1}^{\infty} S_m |G_j(mf)| \cos(\phi_m - \gamma_j(mf)),$$

$$j = 1, 2 \tag{5.72}$$

This expression is valid for C_1 as well as for C_2 and for any harmonic component of the signal with frequency mf. The correlation ratio $r(f)$ is then given by

$$r(f) = \frac{S_0|G_1(0)| + 2\sum_{m=1}^{\infty} S_m|G_1(mf)|\cos(\phi_m - \gamma_1(mf))}{S_0|G_2(0)| + 2\sum_{m=1}^{\infty} S_m|G_2(mf)|\cos(\phi_m - \gamma_2(mf))} \tag{5.73}$$

This is a completely general expression for the value of $r(f)$ but, as pointed out before, does not produce correct results for the signal phase unless certain conditions are met, as will be seen next. The elemental Fourier components of these functions at the frequency of the signal being selected must satisfy the following conditions, briefly mentioned before:

1. The Fourier elements of the reference functions $g_1(x)$ and $g_2(x)$ must have a zero DC term. Then the Fourier transforms $G_1(f)$ and $G_2(f)$ of the two reference functions at zero frequency must be equal to zero.
2. All interference (cross talk) between undesired harmonics in the signal and in the reference functions must be avoided.
3. The Fourier elements of the reference functions $g_1(x)$ and $g_2(x)$ at the frequency f_r must be orthogonal to each other. This means that the Fourier transforms $G_1(f)$ and $G_2(f)$ of the two reference functions at the frequency f_r must have a phase difference equal to $\pm\pi/2$. (The plus sign corresponds to the upper sign in Eq. 5.57, and then the phase of $G_2(f)$ is $\pi/2$ greater than the phase of $G_1(f)$.)
4. The Fourier transforms $G_1(f)$ and $G_2(f)$ of the two reference functions, at the frequency f_r must have the same magnitude.

Given a reference frequency, these four conditions can in general be satisfied only at certain signal frequencies. To illustrate these conditions, in Fig. 5.12 we have the Fourier spectra of two reference functions plotted together with the Fourier spectra of a periodic signal. Here we notice the following for the functions $G_1(f)$ and $G_2(f)$: (1) They pass through the origin, indicating that their DC bias is zero; (2) the harmonics of the signal are located at zeros of these functions; (3) they have the same magnitude at the fundamental frequency of the signal f. If also these functions are orthogonal to each other, all conditions are satisfied at the fundamental frequency of the signal.

Let us now consider the four conditions and apply them to expression 5.71. The first condition of a zero DC term may be easily satisfied if from the central theorem studied in Chap. 2 we write

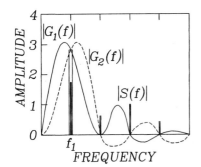

Figure 5.12 Fourier spectra of the two reference functions and a signal.

$$G_1(0) = G_2(0) = 0 \tag{5.74}$$

Then Eq. 5.73 becomes

$$r(f) = \frac{\displaystyle\sum_{m=1}^{\infty} S_m |G_1(mf)| \cos(\phi_m - \gamma_1(mf))}{\displaystyle\sum_{m=1}^{\infty} S_m |G_2(mf)| \cos(\phi_m - \gamma_2(mf))} \tag{5.75}$$

The second condition, that there is no interference from undesired harmonics, is satisfied if for all harmonics m, with the exception of the fundamental frequency which is being measured, we have

$$S_m G_i(mf) = 0 \qquad \text{for } m > 1 \tag{5.76}$$

This means that the harmonic components $m > 1$ should not be present either in the signal or in the reference functions. Obviously, if the signal is perfectly sinusoidal, this condition is always satisfied.

Applying these two conditions to a sinusoidal signal with frequency f, Eq. 5.73 becomes

$$r(f) = \frac{|G_1(f)| \cos(\phi - \gamma_1(f))}{|G_2(f)| \cos(\phi - \gamma_2(f))} = \frac{\text{Re}\{G_1(f) \exp(-i\phi)\}}{\text{Re}\{G_2(f) \exp(-i\phi)\}} \tag{5.77}$$

During the phase detection process the frequency of the signal has to be estimated, so that the reference frequency f_r is as close as possible to this value. We say that there is a detuning error if the reference frequency f_r is different from the signal frequency f.

Now, we need to satisfy only two more conditions. For the two elements of the two reference functions to be orthogonal to each other at the reference frequency f_r, we need

$$G_1(f_r) = \mp iz(f_r)G_2(f_r) = z(f_r)G_2(f_r)\exp\left(\mp i\frac{\pi}{2}\right) \qquad (5.78)$$

at the harmonic m being considered, where $z(f_r)$ is any positive-valued real function or a positive constant. The upper (minus) sign indicates that the phase of $G_2(f_r)$ is $\pi/2$ greater than the phase of $G_1(f_r)$. This case corresponds to the upper sign in Eq. 5.60. This is what we have implicitly considered in the preceding sections in this chapter, although both cases are possible. Thus, the phases $\gamma_1(f_r)$ and $\gamma_2(f_r)$ in Eq. 5.70 are related by

$$\gamma_1(f_r) = \gamma_2(f_r) \mp \frac{\pi}{2} \qquad (5.79)$$

The values of these angles depend on the location of the point being selected as the origin of coordinates ($x = 0$). If, as defined before, $\psi(f_r)$ is the phase displacement in the positive direction, of the zero phase point of the Fourier elements of the reference functions with frequency f_r, with respect to the origin of coordinates, we may write

$$\gamma_1(f_r) = \psi(f_r) \mp \frac{\pi}{2}$$
$$\gamma_2(f_r) = \psi(f_r) \qquad (5.80)$$

Thus, we see that when $\psi(f_r)$ is equal to zero, the function $G_2(f)$ becomes real at the reference frequency. Then we may conclude that the function element $\delta g_1(x)$ is antisymmetrical. In other words, the origin of coordinates is located at the zero phase point of this sine function.

The condition that the magnitudes of the Fourier components at the frequency being detected be equal requires that

$$|G_1(f_r)| = |G_2(f_r)| \qquad (5.81)$$

Thus, applying these last two conditions, we finally obtain

$$r(f_r) = \mp\frac{\sin(\phi - \psi(f_r))}{\cos(\phi - \psi(f_r))} = \mp\tan(\phi - \psi(f_r)) \qquad (5.82)$$

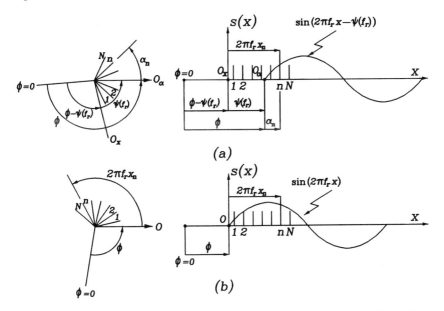

Figure 5.13 Sampling a signal with equally spaced points; (a) with different locations for origins O_x and O_α and (b) with both origins at the same location O.

where, as we explained before, the upper sign is taken when the phase of $G_1(f_r)$ is $\pi/2$ ahead of the phase of $G_2(f_r)$. Hence the phase is

$$\tan(\phi - \psi(f_r)) = \mp r(f_r) \tag{5.83}$$

as was intuitively expected.

5.5 SYNCHRONOUS DETECTION USING A FEW SAMPLING POINTS

Let us now apply the general theory of synchronous detection just developed to the particular case of a discrete sampling procedure, with only a few sampling points. As illustrated in Fig. 5.13, let us take $N \geq 3$ points with their relative phases α_n, referred to the origin O_α. The phases of the sampling points are measured with respect to the origin of the reference function, which may be located at any arbitrary position, not necessarily the origin of coordinates or any sampling point in particular. Thus, we obtain N equations from which the signal phase ϕ at the origin of the reference function may be calculated.

The location of the phase origin $\mathbf{O}_\alpha$ for the sampling points is the same as the zero phase point for the sampling reference functions at the reference frequency, but not necessarily at any other frequency. According to the translation property in Fourier theory, since the two reference functions are orthogonal to each other at the reference frequency f_r, the location of the zero phase point with respect to the sampling points may be selected so that the Fourier transform $G_1(f_r)$ is real and the Fourier transform $G_2(f_r)$ is imaginary, or vice versa.

Given a phase detecting sampling algorithm with the positions of the sampling points with respect to the origin of coordinates ($x = 0$) and their associated sampling weights already defined, the value of $\psi(f_r)$ is already determined and can be found after the Fourier transforms $G_1(f)$ and $G_2(f)$ have been calculated. Then the value of $\psi(f_r)$ is equal to minus the value of the phase of $G_1(f)$ at the reference frequency at the origin ($x = 0$). Thus, we have

$$\alpha_n(x) = 2\pi f_r x_n - \psi(f_r) \tag{5.84}$$

When the sampling weights have not yet been determined, a common approach is to place the zero phase origin $\mathbf{O}_\alpha$, that is, the origin of the reference functions $\cos(2\pi f_r x)$ and $\sin(2\pi f_r x)$, at the same place as the origin $\mathbf{O}_x$, thus making $\psi(f_r) = 0$, as in Fig. 5.13b. Then the values of the sampling weights can be calculated.

The sampling points may also be shifted and located with respect to the phase origin at a position such that $G_1(f)$ becomes imaginary and $G_2(f)$ becomes real at the reference frequency. There are two interesting particular cases when this occur:

1. If $g_1(x)$ is symmetrical and $g_2(x)$ is antisymmetrical about the point with phase $m\pi$, where m is any integer.
2. If $g_1(x)$ is antisymmetrical and $g_2(x)$ is symmetrical about the point with phase $(m + 1/2)\pi$, where m is any integer.

If desired, the first sampling point may be placed at the origin of coordinates, but frequently this is not the case.

5.5.1 General Discrete Sampling

If we sample any N points, with any arbitrary separation between them, we may see that the sampling reference functions are then given by

$$g_1(x) = \sum_{n=1}^{N} W_{1n}\delta(x - x_n) \tag{5.85}$$

and

$$g_2(x) = \sum_{n=1}^{N} W_{2n}\delta(x - x_n) \tag{5.86}$$

where W_{in} is the sampling weight for sampling point i and N is the number of sampling points with coordinates $x = x_n$. The Fourier transforms of these sampling reference functions are

$$G_1(f) = \sum_{n=1}^{N} W_{1n} \exp(-i2\pi f x_n) \tag{5.87}$$

and

$$G_2(f) = \sum_{n=1}^{N} W_{2n} \exp(-i2\pi f x_n) \tag{5.88}$$

but from 5.84 we may write

$$2\pi f x_n = [\alpha_n + \psi(f_r)]\frac{f}{f_r} \tag{5.89}$$

hence, these Fourier transforms become

$$G_1(f) = \exp\left(-i\psi(f_r)\frac{f}{f_r}\right) \sum_{n=1}^{N} W_{1n} \exp\left(-i\alpha_n \frac{f}{f_r}\right) \tag{5.90}$$

and

$$G_2(f) = \exp\left(-i\psi(f_r)\frac{f}{f_r}\right) \sum_{n=1}^{N} W_{2n} \exp\left(-i\alpha_n \frac{f}{f_r}\right) \tag{5.91}$$

Now, since the reference functions are to be orthogonal to each other and have the same magnitude at the frequency $f = f_r$, we need, as in Eq. 5.74,

$$G_1(f_r) = \mp i G_2(f_r) \tag{5.92}$$

where the upper (plus) sign indicates that the phase of $G_2(f_r)$ is $\pi/2$ ahead of the phase of $G_1(f_r)$. Using this expression with Eqs. 5.87 and 5.88, we find

$$\sum_{n=1}^{N}(W_{2n} \mp i W_{1n}) \exp(-i2\pi f_r x_n) = 0 \tag{5.93}$$

Thus, we have

$$\sum_{n=1}^{N}(W_{2n} \mp i W_{1n}) \cos(2\pi f_r x_n) - i \sum_{n=1}^{N}(W_{2n} \mp i W_{1n}) \sin(2\pi f_r x_n) = 0 \tag{5.94}$$

or

$$\sum_{n=1}^{N}[W_{2n} \cos(2\pi f_r x_n) \mp W_{1n} \sin(2\pi f_r x_n)]$$

$$- i \sum_{n=1}^{N}[W_{2n} \sin(2\pi f_r x_n) \pm W_{1n} \cos(2\pi f_r x_n)] = 0 \tag{5.95}$$

which can be true only if

$$\sum_{n=1}^{N}[W_{2n} \cos(2\pi f_r x_n) \mp W_{1n} \sin(2\pi f_r x_n)] = 0 \tag{5.96}$$

and

$$\sum_{n=1}^{N}[W_{2n} \sin(2\pi f_r x_n) \pm W_{1n} \cos(2\pi f_r x_n)] = 0 \tag{5.97}$$

We now may define the Fourier transform vectors $\mathbf{G}_1$ and $\mathbf{G}_2$ as

$$\mathbf{G}_1 = \left(\sum_{n=1}^{N} W_{1n} \cos(2\pi f_r x_n), \sum_{n=1}^{N} W_{1n} \sin(2\pi f_r x_n) \right) \tag{5.98}$$

and

$$\mathbf{G}_2 = \left(\sum_{n=1}^{N} W_{2n} \cos(2\pi f_r x_n), \sum_{n=1}^{N} W_{2n} \sin(2\pi f_r x_n) \right) \tag{5.99}$$

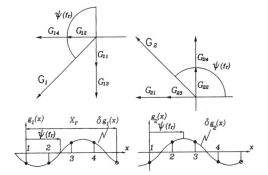

Figure 5.14 Sampling reference vectors for a sampling algorithm.

where, from Eqs. 5.87 and 5.88, we see that the x and y components of the vector are the real and imaginary parts of the Fourier transforms of the reference functions. These Fourier transform vectors may also be written as

$$\mathbf{G}_1 = \mathbf{G}_{11} + \mathbf{G}_{12} + \mathbf{G}_{13} + \cdots + \mathbf{G}_{1N} \tag{5.100}$$

and

$$\mathbf{G}_2 = \mathbf{G}_{21} + \mathbf{G}_{22} + \mathbf{G}_{23} + \cdots + \mathbf{G}_{2N} \tag{5.101}$$

where this is a vector sum of the vectors $\mathbf{G}_{in}$ defined by

$$\mathbf{G}_{in} = (W_{in} \cos(2\pi f_r x_n), W_{in} \sin(2\pi f_r x_n)) \tag{5.102}$$

If we use these vectors in Eqs. 5.96 and 5.97 we see that the vectors are orthonormal, that is, they are mutually perpendicular and have the same magnitude at the frequency f_r. Thus, we may say that the two reference sampling functions are orthogonal and have the same amplitude if the two Fourier transform vectors are mutually perpendicular and have the same magnitude, as illustrated in Fig. 5.14. The angle of $\mathbf{G}_1$ is $\pi/2$ greater than that of $\mathbf{G}_2$ for the upper sign. The angle of $\mathbf{G}_1$ with respect to the horizontal axis is equal to $\psi_1(f_r)$. In the same manner, the angle of $\mathbf{G}_2$ with respect to the horizontal axis is equal to $\psi_2(f_r)$.

Quite frequently, the phase origin in algorithms is located at a point such that $G_1(f)$ is imaginary and $G_2(f)$ is real at the reference frequency. Under these conditions the vector $\mathbf{G}_1$ is vertical, the vector $\mathbf{G}_2$ is horizontal, and Eqs. 5.97 and 5.96 may be written as

$$\sum_{n=1}^{N}[W_{1n}\cos(2\pi f_r x_n)] = 0 \tag{5.103}$$

$$\sum_{n=1}^{N}[W_{2n}\sin(2\pi f_r x_n)] = 0 \tag{5.104}$$

and

$$\sum_{n=1}^{N}[W_{1n}\sin(2\pi f_r x_n)] = \sum_{n=1}^{N}[W_{2n}\cos(2\pi f_r x_n)] \tag{5.105}$$

Additionally, we must require that there be no bias in the reference functions, which is true if

$$\sum_{n=1}^{N}W_{1n} = 0 \tag{5.106}$$

and

$$\sum_{n=1}^{N}W_{2n} = 0 \tag{5.107}$$

The value of the phase ϕ may be calculated by using Eqs. 5.85 and 5.86 in Eq. 5.62 and then with 5.82 obtaining

$$\tan\phi = \mp\frac{\sum_{n=1}^{N}s(x_n)W_{1n}}{\sum_{n=1}^{N}s(x_n)W_{2n}} \tag{5.108}$$

The phase $\psi(f_r)$ at the origin ($x = 0$) of the Fourier elements $\delta g_1(x)$ and $\delta g_2(x)$ with frequency f_r is assumed to be equal to zero.

This is a completely general treatment, with any number of sampling functions, with any separation, and with different weights.

5.5.2 Equally Spaced and Uniform Sampling

A frequent particular case is when the sampling points are equally separated and uniformly distributed in a signal period X_r, with the positions defined as in Eq. 5.10, by

$$x_n = \frac{(n-1)X_r}{N} + x_1 = \frac{n-1}{Nf_r} + x_1; \qquad n = 1, \ldots, N \tag{5.109}$$

In this expression the origin $\mathbf{O}_\alpha$ for the reference function and the first sampling point had been taken at the same location of the origin of coordinates $\mathbf{O}_x$, as shown in Fig. 5.13b. The reference frequency f_r is defined as $1/X_r$, and it is usually equal to the signal frequency, but it may be different.

As described in Sec. 5.1, with this sampling distribution we have

$$\sum_{n=1}^{N} \sin(2\pi f_r x_n - \psi(f_r)) = \sum_{n=1}^{N} \sin(2\pi f_r x_n) = 0 \tag{5.110}$$

$$\sum_{n=1}^{N} \cos(2\pi f_r x_n - \psi(f_r)) = \sum_{n=1}^{N} \cos(2\pi f_r x_n) = 0 \tag{5.111}$$

$$\sum_{n=1}^{N} \cos(4\pi_r x_n - \psi(f_r)) = \sum_{n=1}^{N} \cos(4\pi f_r x_n) = 0 \tag{5.112}$$

and

$$\sum_{n=1}^{N} \sin(4\pi f_r x_n - \psi(f_r)) = \sum_{n=1}^{N} \sin(4\pi f_r x_n) = 0 \tag{5.113}$$

These results are independent of the location of the origin for the phases, that is, for any value of $\psi(f_r)$. The reason for this becomes clear if we notice that the vector diagram in Fig. 5.2 remains in equilibrium when all vectors are rotated by an angle $\psi(f_r)$.

The condition that there should be no DC term (bias) on the reference functions is expressed by Eqs. 5.106 and 5.107. From Eq. 5.112 we may see that

$$\sum_{n=1}^{N} \cos(2\pi f_r x_n - \psi(f_r)) \cos(2\pi f_r x_n)$$

$$-\sum_{n=1}^{N} \sin(2\pi f_r x_n - \psi(f_r)) \sin(2\pi f_r x_n) = 0 \tag{5.114}$$

and from Eq. 5.113

$$\sum_{n=1}^{N} \cos(2\pi_r x_n - \psi(f_r)) \sin(2\pi f_r x_n)$$

$$+ \sum_{n=1}^{N} \sin(2\pi f_r x_n - \psi(f_r)) \cos(2\pi f_r x_n) = 0 \qquad (5.115)$$

Now we may see that these last two expressions become identical to Eqs. 5.96 and 5.97 if the sampling weights are defined by

$$W_{1n} = \pm \sin(2\pi_r x_n - \psi(f_r)) \qquad (5.116)$$

and

$$W_{2n} = \cos(2\pi f_r x_n - \psi f_r) \qquad (5.117)$$

Expressions 5.110, 5.111, 5.114, and 5.115 when $\psi(f_r) = 0$ are the same as those used in Sec. 5.1 to make the least squares matrix diagonal.

Now we may obtain the phase value with the ratio of the correlations by using these sampling weights in Eq. 5.108, assuming a value $\psi(f_r) = 0$.

$$\tan \phi = \mp \left(\frac{\displaystyle\sum_{n=1}^{N} s(x_n) \sin(2\pi f_r x_n)}{\displaystyle\sum_{n=1}^{N} s(x_n) \cos(2\pi f_r x_n)} \right) \qquad (5.118)$$

and the signal may be calculated with Eq. 5.83. A plus sign here corresponds to the upper sign in Eq. 5.60. As we might expect, this result is the one obtained with the diagonal least squares algorithm.

We have pointed out before that the location of the origin of coordinates is important because it affects the algebraic appearance (phase) of the result. However, for any selected origin location, the relative phase for all points is the same. There are two typical locations for the origin, (1) the first sampling point or (2) the zero phase point for the Fourier elements.

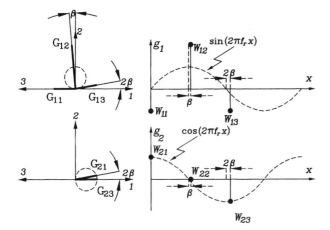

Figure 5.15 Effect of detuning in a three-point algorithm; upper part shows effects on g_1 and $\mathbf{G}_{1n}$ and lower part shows effects on g_2 and $\mathbf{G}_{2n}$.

5.5.3 Applications of the Graphical Vector Representation

We have three interesting properties of the graphical vector representation.

1. Examining the vectors of any two algorithms that satisfy the conditions for orthogonality and equal magnitudes of $G_1(f)$ and $G_2(f)$, we can see that a superposition of the two algorithms also satisfies the required conditions.
2. Any vector system with zero bias and in equilibrium may be added to the system without changing either the orthogonality or the equal-magnitudes condition at the reference frequency.
3. A detuning shifts the angular orientations of the vectors W_{ij} a small angle β in direct proportion to their phase α_n.

As examples of the application of the vector representation let us consider the effect of detuning using the vector representation in two algorithms with three sampling points. The first one to be considered is in Fig. 5.15. The three points have phases $0°$, $90°$, and $180°$. However, if there is a detuning, as shown in this figure, the sampling points have phases $0°$, $90° + \beta$, and $180° + 2\beta$. Examining the vector plots at the left of this figure we see that the vector sums $\mathbf{G}_1$ and $\mathbf{G}_2$ are both rotated an angle β, preserving their orthogonality. Since β

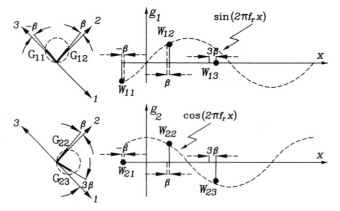

Figure 5.16 Effect of detuning in a three-point algorithm; upper part shows effects on g_1 and $\mathbf{G}_{1n}$ and lower part shows effects on g_2 and $\mathbf{G}_{2n}$.

is arbitrary, the orthogonality conditions are preserved at all frequencies, but not their magnitudes.

A second algorithm to analyze is in Fig. 5.16, where the sampling points are at $-45°$, $45°$, and $135°$. If there is detuning, the three phases will be $-(45° + \beta)$, $(45° + \beta)$ and $(135° + 3\beta)$. Then the vectors on the left side of the figure will be angularly displaced. We may easily observe that the angle between the vectors $\mathbf{G}_{1n}$ is preserved and also the angle between the vectors $\mathbf{G}_{2n}$. Thus, the magnitudes of $G_1(f)$ and $G_2(f)$ are preserved, but not their orthogonality.

5.5.4 Characteristic Polynomial of a Sampling Algorithm

A characteristic polynomial that can be associated with a discrete sampling algorithm was proposed by Surrel (1996). This polynomial may be used to derive all main properties of the algorithm, in a manner closely associated with the Fourier theory just described. To define this polynomial, let us use Eq. 5.108, considering that the phase ϕ is given by the phase of the complex function $V(\phi)$ defined by

$$V(\phi) = \sum_{n=1}^{N} (W_{1n} + i W_{2n}) s(x_n) \tag{5.119}$$

where $\psi(f_r) = 0$. Then using the Fourier expansion for the signal given by Eq. 5.58 in this expression we may find

$$V(\phi) = \sum_{m=-\infty}^{\infty} S_m \exp(i\phi_m) \sum_{n=1}^{N} (W_{1n} + i W_{2n}) \exp(i 2\pi m f x_n) \tag{5.120}$$

where $\phi = \phi_1$ is the phase of the signal at the fundamental frequency. Different harmonic components have different phases. Now, from Eq. 5.89 we have

$$V(\phi) = \sum_{m=-\infty}^{\infty} S_m \exp(i\phi_m) \sum_{n=1}^{N} (W_{1n} + i W_{2n}) \exp\left(im\alpha_n \frac{f}{f_r}\right) \tag{5.121}$$

where α_n is the phase for the sampling point n. This phase may be assumed to be equal to $\alpha_n = (n - 1)\Delta\alpha$, with $\Delta\alpha$ being the phase interval separation between the sampling points, transforming this expression into

$$V(\phi) = \sum_{m=-\infty}^{\infty} S_m$$

$$\times \exp(i\phi_m) \sum_{n=1}^{N} (W_{1n} + i W_{2n}) \exp\left(im(n-1)\frac{f}{f_r}\Delta\alpha\right) \tag{5.122}$$

If there is no detuning, so that $f = f_r$, this expression can be written as

$$V(\phi) = \sum_{m=-\infty}^{\infty} S_m \exp(i\phi_m) P[\exp(im\Delta\alpha)] \tag{5.123}$$

with a polynomial $P(z)$ defined by

$$P(z) = \sum_{n=1}^{N} (W_{1n} + i W_{2n})[\exp(im\Delta\alpha)]^{(n-1)} = \sum_{n=1}^{N} \sigma_n z^{(n-1)} \tag{5.124}$$

This is the characteristic polynomial proposed by Surrel (1996) which is associated with any sampling algorithm. It is quite simple to derive this polynomial from the sampling weights W_{in}. From this characteristic polynomial, many interesting properties of the sampling algorithm with which it is associated may be found.

Let us first consider the case when there is no detuning, so that $f = f_r$. We assume, however, that there is harmonic distortion in the signal. The signal

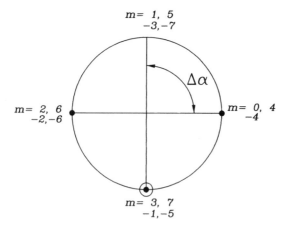

Figure 5.17 Points for each harmonic number for a sampling algorithm. If there is a polynomial root at any sampling point, the point is plotted with a large dot. If there is a double root, it is plotted with a circle around the dot.

harmonic component m with $m \neq 1$ will not influence the value of the complex function $V(\phi)$ if the polynomial $P(z)$ has a root (zero value) for the value of z that corresponds to that harmonic.

Each complex value of z is associated to a harmonic number m, by

$$\exp(im\Delta\alpha) = z \tag{5.125}$$

These values of z may be represented in a unit circle in the complex plane. Given a sampling algorithm, the value of the phase interval $\Delta\alpha$ between sampling points is fixed, i.e., there is a point for each possible value of the harmonic number, including positive as well as negative numbers, as illustrated in Fig. 5.17. This is a characteristic diagram of the sampling algorithm.

If there is detuning ($f \neq f_r$), we can obtain by expanding in a Taylor series

$$\sum_{n=1}^{N} \sigma_n \left[\exp\left(im\frac{f}{f_r}\Delta\alpha \right) \right]^{(n-1)}$$

$$= P(z) + im\left(\frac{f}{f_r} - 1\right)\exp(im\Delta\alpha)P'(z) \tag{5.126}$$

Then we see that there is insensitivity to harmonic m as well as to detuning of that harmonic only if both $P(z)$ and its derivative have roots at the corresponding value of z. In other words, a double root must be at that value of z.

The following are some of the important properties of this characteristic diagram:

1. An algorithm is insensitive to the harmonic component m if the characteristic polynomial has zeros for the values of z corresponding to $\pm m$. To state it in a different manner, the algorithm is insensitive to harmonic m with $m \neq 1$ if both $\exp(-im\Delta\alpha)$ and $\exp(im\Delta\alpha)$ are roots of the characteristic polynomial.

2. If only $\exp(-im\Delta\alpha)$ with $m > 0$ is a root and $\exp(im\Delta\alpha)$ is not a root of the characteristic polynomial, that harmonic component may be detected. If the fundamental frequency ($m = 1$) is to be detected, as is normally the case, $\exp(-i\Delta\alpha)$ should be a root and $\exp(i\Delta\alpha)$ should not be.

3. In an analogous manner, it is possible to prove that there is insensitivity, as well as detuning insensitivity to harmonic m with $m = 1$, if there is a double zero at the values of z corresponding to the $\pm m$ harmonic components. In other words, both $\exp(im\Delta\alpha)$ and $\exp(-im\Delta\alpha)$ are double roots of the characteristic polynomial.

4. If only $\exp(-im\Delta\alpha)$ with $m > 0$ is a double root and $\exp(im\Delta\alpha)$ is not a root of the characteristic polynomial, that harmonic component may be detected with detuning insensitivity. If the fundamental frequency ($m \neq 1$) is to be detected with detuning insensitivity, $\exp(-i\Delta\alpha)$ should be a double root and $\exp(i\Delta\alpha)$ should not be a root.

As an example, let us consider the Schwider-Hariharan algorithm with $\Delta\alpha = 90°$, to be studied in detail in Chap. 6. The phase equation is

$$\tan\phi = -\frac{2(s_2 - s_4)}{s_1 - 2s_3 + s_5} \tag{5.127}$$

and thus the corresponding characteristic polynomial is

$$V(\phi) = 1 - 2iz - 2z^2 + 2iz^3 + z^4$$
$$= (z - 1)(z + 1)(z + i)^2 \tag{5.128}$$

We can observe that the signal may be detected with detuning insensitivity at the fundamental frequency and also at the fifth harmonic. The characteristic diagram for this algorithm is presented in Fig. 5.18.

Many other properties can be derived from a detailed analysis of the characteristic diagram of a sampling algorithm. There is a close connection between this characteristic diagram and the Fourier theory studied before.

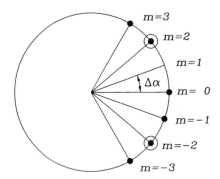

Figure 5.18   Characteristic diagram for a detuning-insensitive algorithm (Schwider-Hariharan).

The characteristic diagrams for many sampling algorithms are described by Surrel (1997).

5.6   ERROR ANALYSIS OF SYNCHRONOUS PHASE DETECTION ALGORITHMS

The theory developed in this chapter permits the error analysis of the sampling algorithms used for the synchronous detection of periodic signals. As examples, some possible sources of error will now be studied.

In the treatment by Freischlad and Koliopoulos (1990) we have seen that if the four conditions required in Sec. 5.4 are satisfied, the phase can be determined without any error. With a proper design of the algorithm, these conditions are satisfied when the reference frequency f_r is equal to the frequency of a harmonic component of the signal to be detected. If one or two of the four conditions are not satisfied, an error may appear in the calculated phase.

Considering the common case when $\psi(f_r) = 0$, in the absence of any error, when the four conditions are satisfied, the phase is calculated with

$$\tan \phi = \mp r(f_r) \tag{5.129}$$

but in the presence of an error, the calculated phase with the error introduced will be

$$\tan \phi_{\text{err}} = \mp r(f) \tag{5.130}$$

thus, the error in the tangent of the phase is

$$\delta \tan \phi = \tan \phi_{err} - \tan \phi$$
$$= \mp r(f) - \tan \phi \tag{5.131}$$

On the other hand, differentiating the tangent function we may obtain

$$\delta \phi = (\cos^2 \phi)(\delta \tan \phi) \tag{5.132}$$

hence substituting Eq. 5.130 into Eq. 5.131, the error in the calculated phase is

$$\delta \phi = [\mp r(f) - \tan \phi] \cos^2 \phi \tag{5.133}$$

This is a completely general expression for the phase errors if one or more of the four required conditions is not fulfilled. Depending on which condition is failing, the ratio of the two correlations $r(f)$ defined by Eq. 5.62 can be calculated

1. In the general case, with Eq. 5.73, when any or several of the four conditions fail.
2. If the zero bias condition is the only one being satisfied, with Eq. 5.75.
3. If besides satisfying the zero bias condition the signal is perfectly sinusoidal, or there is no cross talk between harmonic components in the signal and in the reference functions, Eq. 5.82 may be used. When the orthogonality condition has a small deviation, assuming again that $\psi(f_r) = 0$, the phases $\gamma_i(f)$ may be expressed as

$$\gamma_1(f) = \delta \gamma_1(f) \mp \frac{\pi}{2}$$
$$\gamma_2(f) = \delta \gamma_2(f) \tag{5.134}$$

In this case the phase error becomes

$$\delta \phi = \left[\frac{|G_1(f)| \sin(\phi - \delta \gamma_1(f))}{|G_2(f)| \cos(\phi - \delta \gamma_2(f))} - \tan \phi \right] \cos^2 \phi \tag{5.135}$$

which may be approximated by

$$\delta \phi = \frac{1}{2} \left[\frac{|G_1(f)|}{|G_2(f)|} - 1 \right] \sin 2\phi$$
$$+ \frac{1}{2} \frac{|G_1(f)|}{|G_2(f)|} (\delta \gamma_2(f) - \delta \gamma_1(f)) \cos 2\phi$$
$$- \frac{1}{2} \frac{|G_1(f)|}{|G_2(f)|} (\delta \gamma_2(f) + \delta \gamma_1(f)) \tag{5.136}$$

Given a detuning magnitude, the signal frequency is a constant; hence the last term in this expression is a constant phase shift, like a piston term. Thus, this term does not have any importance and can be ignored, obtaining

$$\delta\phi = \frac{1}{2}\left[\frac{|G_1(f)|}{|G_2(f)|} - 1\right]\sin 2\phi$$
$$+ \frac{1}{2}\frac{|G_1(f)|}{|G_2(f)|}(\delta\gamma_2(f) - \delta\gamma_1(f))\cos 2\phi \qquad (5.137)$$

Alternatively, if we define $\rho(f)$ as the ratio of the magnitudes of the Fourier transforms $G_1(f)$ and $G_2(f)$, which depends only on the signal frequency f, we may write

$$\delta\phi = \frac{1}{2}[\rho(f) - 1]\sin 2\phi$$
$$+ \rho(f)\left(\frac{\delta\gamma_2(f) - \delta\gamma_1(f)}{2}\right)\cos 2\phi \qquad (5.138)$$

This phase error $\delta\phi$ has a sinusoidal variation with the signal phase, with twice the frequency of the signal. This result is valid for any kind of error where the orthogonality and the equal-magnitudes conditions fail. When the cross talk between harmonics is present—for example, if the signal has harmonic distortion—this conclusion might not be true. As pointed out by Cheng and Wyant (1985), the phase error may be eliminated by averaging the results of two measurements with opposite errors, as will be seen in Chap. 6. The two measurements only need to have an offset of 90° with respect to each other.

4. When only the equal-magnitudes condition fails, we have

$$\delta\phi = \frac{1}{2}[\rho(f) - 1]\sin 2\phi \qquad (5.139)$$

As shown in Fig. 5.19, in this case the phase error becomes zero when the phase ϕ to be measured is an integer multiple of $\pi/2$. This error has a peak value equal to $[\rho(f) - 1]/2$.

5. Finally, if only the orthogonality condition fails, the phase error is

$$\delta\phi = \frac{1}{2}\rho(f)[\delta\gamma_2(f) - \delta\gamma_1(f)]\cos 2\phi \qquad (5.140)$$

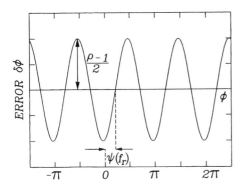

Figure 5.19 Phase error as a function of the measured phase, for an algorithm where the Fourier transforms $G_1(f)$ and $G_2(f)$ are orthogonal at all frequencies.

We can see that in this case the phase error again oscillates sinusoidally with the phase signal, between zero and a peak value equal to the difference between the derivatives of the phase difference $\gamma_2(f) - \gamma_1(f)$ with respect to the signal frequency, as shown in Fig. 5.20. This phase error becomes zero even if there is some detuning, when the phase ϕ to be measured is $\pi/4$ plus an integer multiple of $\pi/2$.

These expressions are the basis for the analysis of errors in phase shifting interferometry, as will be described in the next few sections.

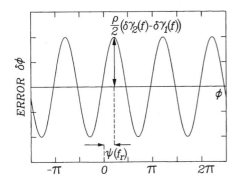

Figure 5.20 Phase error as a function of the measured phase, for an algorithm where the Fourier transforms $G_1(f)$ and $G_2(f)$ have equal magnitudes at all frequencies.

5.6.1 Detuning Error

This problem has been studied by several authors, for example, by Ransom and Kokal (1986). If there is a small detuning and the signal frequency deviates from the reference frequency, the zero bias condition is preserved. If the signal is assumed to be sinusoidal, the condition for no cross talk between the signal and reference functions harmonics is also preserved. However, the conditions for orthogonality and equal magnitudes of $G_1(f_r)$ and $G_2(f_r)$ may not be satisfied. Thus, the phase error in this case is given in general by Eq. 5.138, or by Eq. 5.139 or 5.140, according to the case.

An ideal algorithm in which the orthogonality condition is preserved for all signal frequencies is one for which $\rho(f)$ varies very little from the ideal value of 1, with detuning. If these magnitudes are to be the same in a small range of frequencies centered on the reference frequency, the two curves for the Fourier transforms should not only touch each other (same magnitudes), they should also be tangential to each other at that point. In other words, they should have the same magnitude and the same slope, as follows:

$$\left(\frac{d|G_2(f)|}{df}\right)_{f=f_r} = \left(\frac{d|G_1(f)|}{df}\right)_{f=f_r} \tag{5.141}$$

These magnitudes may be calculated as described in the appendix at the end of this chapter.

Another particular case is when the detuning affects the orthogonality condition but the ratio between the two magnitudes of the Fourier transforms is preserved. We may write the difference between the two phase increments in terms of the derivative of the phase difference with respect to the frequency. In this case $\rho(f)$ is equal to 1. Then the phase error, Eq. 5.140, is

$$\delta\phi = \frac{1}{2}[\delta\gamma_2(f) - \delta\gamma_1(f)]\cos 2\phi$$
$$= \frac{1}{2}\frac{d(\gamma_2(f) - \gamma_1(f))}{df}(f - f_r)\cos 2\phi \tag{5.142}$$

It is easy to see that the algorithm in this case is insensitive to small detuning if

$$\left(\frac{d\gamma_2(f)}{df}\right)_{f=f_r} = \left(\frac{d\gamma_1(f)}{df}\right)_{f=f_r} \tag{5.143}$$

When the signal is not sinusoidal, the treatment of detuning is more complicated, because any detuning affects not only the fundamental frequency of the signal but also its harmonics.

5.6.2 Phase Detection of a Harmonically Distorted Signal

A distorted periodic signal may be phase detected with a synchronous detection sampling method without any error only if the signal harmonic frequencies are located at places where there is a zero value of the Fourier transforms of the reference functions. Many sampling algorithms, like the two described in this chapter, have zeros at the reference function spectra at even harmonics.

Signal harmonics may appear for many reasons, for example:

1. When the signal is not sinusoidal, as in the measurement of aspherical wavefronts by means of spatial phase shifting analysis of interferograms.
2. When the signal is sinusoidal, but the phase shifting device has a nonlinear response in the phase scale. This is the case in temporal phase shifting interferometry with a nonlinear phase shifter.
3. When the signal is sinusoidal but the response of the light detector is not linear with the signal.
4. In multiple beam interferograms, or ronchigrams, as pointed out by Hariharan (1987).

To make the algorithm insensitive to the signal harmonic m, we need zeros of the Fourier transforms of the sampling reference functions at a given signal harmonic m, as follows

$$\sum_{n=1}^{N}[W_{1n}\cos(2\pi m f_r x_n)] = 0 \tag{5.144}$$

$$\sum_{n=1}^{N}[W_{2n}\sin(2\pi m f_r x_n)] = 0 \tag{5.145}$$

$$\sum_{n=1}^{N}[W_{1n}\sin(2\pi m f_r x_n) = 0 \tag{5.146}$$

and

$$\sum_{n=1}^{N}[W_{2n}\cos(2\pi m f_r x_n)] = 0 \tag{5.147}$$

which resemble Eqs. 5.93 to 5.95.

Table 5.2 Sensitivity to Signal Harmonics of Algorithms with Equally and Uniformly Spaced Points

Number of sampling points	Harmonics being suppressed									
	2	3	4	5	6	7	8	9	10	11
3		✓		✓		✓		✓		
4	✓		✓		✓		✓		✓	
5	✓	✓		✓		✓	✓		✓	
6	✓	✓	✓		✓		✓	✓	✓	

Source: Stetson and Brohinsky (1985).

Stetson and Brohinsky (1985) and Hibino et al. (1995) showed that to suppress all harmonics up to the mth order in algorithms with equally spaced points it is necessary that

1. The maximum phase spacing between sampling points be equal to $2\pi/(m+2)$.
2. The minimum number of sampling points be $m + 2$, when the phase interval is set to its maximum value. A smaller phase interval would require more sampling points.

To make this more clear, let us assume that we have N equally spaced sampling points with a phase separation equal to $2\pi/N$. Then all harmonic components up to the $m = N - 2$ order will be eliminated. Of course, some other higher harmonics may also be eliminated. Stetson and Brohinsky (1985) showed that an algorithm with equally and uniformly spaced sampling points, as given in Eq. 5.10, is sensitive to the harmonics m given by

$$m = N \pm 1 + pN \tag{5.148}$$

where p is an integer. The results are tabulated in Table 5.2.

If the phase detecting algorithm is sensitive to undesired harmonics, the response to these harmonics may be reduced by additional filtering by means of a bucket integration or with an additional filtering function, as described in Sec. 5.7.

In order to have insensitivity to a given harmonic order in the presence of detuning, we have two requirements on the Fourier transforms $G_1(f)$ and $G_2(f)$ of the reference sampling functions:

1. Both Fourier transforms must have zero magnitude at the harmonic frequency.
2. Both Fourier transforms must have a stationary magnitude with respect to the frequency (zero slope) at the harmonic frequency.

Hibino *et al.* (1995) showed that for an algorithm that is insensitive up to the mth harmonic order and is also insensitive to detuning of the fundamental frequency and its harmonics,

1. The maximum phase interval between sampling points is equal to $2\pi/(m + 2)$.
2. The minimum number of sampling points is equal to $2m + 3$ when the phase interval is set to its maximum value. However, later Surrel (1996) showed that this minimum number of sampling points should be equal to $2m + 2$. A phase interval smaller than its maximum value would require a greater number of sampling points. An exception is when the algorithm needs detuning insensitivity only at the fundamental frequency. Then the phase interval may be reduced from its maximum value of 120° to any smaller value without the need for more than five sampling points.

Given an unfiltered signal with harmonics, whose amplitude and phase are known, the phase error may be calculated by means of the general expression 5.130, with the ratio of the correlations $r(f)$ given by Eq. 5.75, where the only condition being satisfied is the zero bias. If we assume that (1) the orthogonality and equal-magnitudes conditions are fulfilled at the signal frequency and (2) that the algorithm has the relatively common property that the orthogonality of the reference sampling functions is preserved at all signal frequencies, we may write this expression as

$$r(f) = \mp \frac{S_1|G_1(f)|\sin\phi + \sum_{m=2}^{\infty} S_m|G_1(mf)|\sin\phi_n}{S_1|G_1(f)|\cos\phi + \sum_{m=2}^{\infty} S_m|G_2(mf)|\cos\phi_n} \tag{5.149}$$

Hence, using Eq. 5.132, the phase error may be shown to be given by

$$\phi = \sum_{m=2}^{\infty} \frac{S_m}{S_1} \left(\frac{|G_1(mf)|}{|G_1(f)|}\sin\phi_m\cos\phi - \frac{|G_2(mf)|}{|G_1(f)|}\cos\phi_m\sin\phi \right) \tag{5.150}$$

The values of the amplitudes S_m and of the phases ϕ_n of the harmonic components of the signal depend on the signal characteristics. The phase ϕ_m

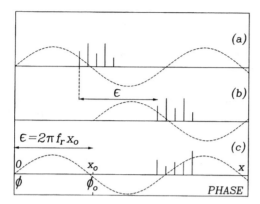

Figure 5.21 Shifting an algorithm.

may be written as $\phi_m = m\phi + \beta_m$. We may observe that the phase error does not change in a purely sinusoidal manner with the signal phase like the other phase errors considered before. The functional dependence on the signal phase ϕ is more complicated, but in a first approximation it oscillates with the same frequency as the signal.

5.7 SHIFTING ALGORITHMS WITH RESPECT TO THE PHASE ORIGIN

The sampling weights of an algorithm change if the sampling points are shifted with respect to the origin by a phase distance ϵ. Here we will study how the sampling weights change, modifying the algorithm structure. Let us first consider an algorithm in which the x origin $\mathbf{O}_x$ and the phase origin $\mathbf{O}_\alpha$ are at the same point, as in Fig. 5.21a. The phase of the signal at the origin, from Eqs. 5.62 and 5.63, is then given by

$$\tan \phi = \mp \frac{\displaystyle\int_{-\infty}^{\infty} s(x) g_1(x)\, dx}{\displaystyle\int_{-\infty}^{\infty} s(x) g_2(x)\, dx} \tag{5.151}$$

If the sampling points are shifted in the positive direction of x as in Fig. 5.21b, preserving the same reference sampling functions, the phase ϕ_0 at the position x_0 where $\epsilon = 2\pi f_r x_0$ is now given by

$$\tan \phi_0 = \mp \frac{\displaystyle\int_{-\infty}^{\infty} s(x)g_1(x - x_0)\, dx}{\displaystyle\int_{-\infty}^{\infty} \infty s(x)g_2(x - x_0)\, dx} \tag{5.152}$$

The phase at the origin with these shifted sampling points can be obtained only if the reference sampling functions are properly modified with the phase equation

$$\tan \phi = \mp \frac{\displaystyle\int_{-\infty}^{\infty} s(x)g_1'(x)\, dx}{\displaystyle\int_{-\infty}^{\infty} s(x)g_2'(x)\, dx} \tag{5.153}$$

or by means of the well-known trigonometric relation

$$\tan \phi = \tan(\phi_0 - \epsilon) = \frac{\tan \phi_0 - \tan \epsilon}{1 + \tan \epsilon \tan \phi_0} \tag{5.154}$$

Thus, from Eqs. 5.16, 5.17, and 5.18 we find

$$\frac{g_1'(x)}{g_2'(x)} = \frac{\cos(\epsilon g_1(x - x_0)) \pm \sin(\epsilon g_2(x - x_0))}{\cos(\epsilon g_2 x - x_0)) \mp \sin(\epsilon g_1(x - x_0))} \tag{5.155}$$

Thus, we may write

$$g_1'(x) = \cos(\epsilon g_1(x - x_0)) \pm \sin(\epsilon g_2(x - x_0)) \tag{5.156}$$

and

$$g_2'(x) = \cos(\epsilon g_2(x - x_0)) \mp \sin(\epsilon g_1(x - x_0)) \tag{5.157}$$

Hence, we may also write for the Fourier transforms of these reference sampling functions:

$$G_1'(f) = [\cos(\epsilon G_1(f)) \pm \sin(\epsilon G_2(f))] \exp(-i\epsilon) \tag{5.158}$$

and

$$G_2'(f) = [\cos(\epsilon G_2(f)) \mp \sin(\epsilon G_1(f))] \exp(-i\epsilon) \tag{5.159}$$

or in terms of the magnitudes and phases

$$G'_1(f) = [\cos(\epsilon|G_1(f)|)\exp(i\gamma_1(f))$$
$$\pm \ \sin(\epsilon|G_2(f)|)\exp(i\gamma_2(f))]\exp(-i\epsilon) \tag{5.160}$$

and

$$G'_2(f) = [\cos(\epsilon|G_2(f)|)\exp(i\gamma_2(f))$$
$$\mp \ \sin(\epsilon|G_1(f)|)\exp(i\gamma_1(f))]\exp(-i\epsilon) \tag{5.161}$$

5.7.1 Shifting the Algorithm by $\pi/2$

Of special interest is the case where the sampling points are shifted a phase ϵ equal to $\pi/2$. In this case, we may see from Eq. 5.156 that

$$g'_1(x) = g_2(x - x_0) \tag{5.162}$$

and from Eq. 5.157,

$$g'_2(x) = -g_1(x - x_0) \tag{5.163}$$

In other words, we can say that after shifting, the sampling reference functions are just interchanged with no change in sign. Thus, for this case we have

$$g'_1(x) = g_2\left(x - \frac{X_r}{4}\right) \tag{5.164}$$

and

$$g'_2(x) = -g_1\left(x - \frac{X_r}{4}\right) \tag{5.165}$$

where $X_r = 1/f_r$.

5.7.2 Shifting the Algorithm by $\pi/4$

This is another particular case of special interest. In this case, from Eq. 5.156 we can see that

$$g'_1(x) = \frac{1}{\sqrt{2}}[g_1(x - x_0) \pm g_2(x - x_0)] \tag{5.166}$$

and from Eq. 5.157,

$$g_2'(x) = \frac{1}{\sqrt{2}}[g_2(x - x_0) \mp g_1(x - x_0)] \tag{5.167}$$

Thus, if we ignore the unimportant constant factor, we have

$$g_1'(x) = g_1\left(x - \frac{X_r}{8}\right) \pm g_1\left(x - \frac{X_r}{8}\right) \tag{5.168}$$

and

$$g_2'(x) = g_2\left(x - \frac{X_r}{8}\right) \mp g_1\left(x \frac{X_r}{8}\right) \tag{5.169}$$

Let us now compare the sensitivity to detuning of the original and shifted algorithms. The Fourier transforms of these sampling reference functions from Eqs. 5.160 and 5.161 are

$$G_1'(f) = \frac{1}{\sqrt{2}}[|G_1(f)| \exp(i\gamma_1(f)) \pm |G_2(f)| \exp(i\gamma_2(f))]$$
$$\times \exp\left(-i\frac{\pi}{4}\right) \tag{5.170}$$

and

$$G_2'(f) = \frac{1}{\sqrt{2}}[|G_2(f)| \exp(i\gamma_2(f)) \mp |G_1(f)| \exp(i\gamma_1(f))]$$
$$\times \exp\left(-i\frac{\pi}{4}\right) \tag{5.171}$$

Let us now study two different particular cases of this algorithm shifted by $\pi/4$. The first case is when the original reference functions have the same magnitudes but are not orthogonal. In this case, from Eqs. 5.170 and 5.171 we have

$$G_1'(f) = \frac{1}{\sqrt{2}}|G_1(f)|[\exp(i\gamma_1(f)) \pm \exp(i\gamma_2(f))] \exp\left(-i\frac{\pi}{4}\right) \tag{5.172}$$

and

$$G_2'(f) = \frac{1}{\sqrt{2}}|G_2(f)|[\exp(i\gamma_2(f)) \mp \exp(i\gamma_1(f))] \exp\left(-i\frac{\pi}{4}\right) \tag{5.173}$$

which may be transformed into

$$G_1'(x) = \sqrt{2}|G_1(x)|\cos\left(\frac{\gamma_1(f) - \gamma_2(f)}{2}\right)$$

$$\times \exp\left(i\frac{\gamma_1(f) + \gamma_2(f) - \pi/2}{2}\right) \quad (5.174)$$

and

$$G_2'(x) = \sqrt{2}i|G_1(x)|\sin\left(\frac{\gamma_1(f) - \gamma_2(f)}{2}\right)$$

$$\times \exp\left(i\frac{\gamma_1(f) + \gamma_2(f) - \pi/2}{2}\right) \quad (5.175)$$

These values are for the upper signs. For the lower signs, the values are just interchanged. The important conclusion is that these Fourier transforms are orthogonal, but their amplitudes are not the same. The ratio of the magnitudes of these Fourier transforms is given by

$$\frac{|G_1'(x)|}{|G_2'(x)|} = \cot\left(\frac{\gamma_1(f) - \gamma_2(f)}{2}\right) \quad (5.176)$$

The second case to study is when the original reference sampling functions are orthogonal but their amplitudes are not the same. Then from Eqs. 5.174 and 5.175 and using the orthogonality condition in Eq. 5.80, we have

$$G_1'(f) = \frac{1}{\sqrt{2}}[|G_1(f)| + i|G_2(f)|]\exp\left[i\left(\gamma_1(f) - \frac{\pi}{4}\right)\right] \quad (5.177)$$

and

$$G_2'(f) = \frac{1}{\sqrt{2}}[|G_2(f)| + i|G_1(f)|]\exp\left[i\left(\gamma_2(f) - \frac{\pi}{4}\right)\right] \quad (5.178)$$

Thus, the shifted algorithm in this case has the same magnitude but is not orthogonal.

A consequence of these last two results is that an algorithm whose reference sampling functions are orthogonal to all frequencies but whose magnitudes are not equal at all frequencies will convert, after shifting by $\pi/4$, to an algorithm whose sampling reference functions have equal magnitudes at all frequencies but are orthogonal only at some frequencies.

Let us now consider the detuning properties of the shifted algorithm. Assuming a detuning from the reference frequency f_r, thus shifting the phases γ_1 and γ_2, and using here Eqs. 5.175 we find

$$\frac{|G'_1(x)|}{|G'_2(x)|} = \cot\left(\frac{\delta\gamma_1(f) - \delta\gamma_2(f) - \frac{\pi}{2}}{2}\right) \tag{5.179}$$

Then, if the detuning is relatively small, we can obtain

$$\frac{1}{2}\left(\frac{|G'_1(x)|}{|G'_2(x)|} - 1\right) = \frac{\delta\gamma_1(f) - \delta\gamma_2(f)}{2} \tag{5.180}$$

If we examine Eqs. 5.136 and 5.138 we may see that the magnitude of the detuning effect is the same for the original and shifted algorithms. So shifting the algorithm will not modify its detuning sensitivity.

5.8 CONCLUSIONS

In this chapter we have established the foundations for the analysis of phase detection algorithms. This theory permits us to analyze the properties of any algorithm and even allows us to design better algorithms.

APPENDIX DERIVATIVE OF THE MAGNITUDE OF THE FOURIER TRANSFORM OF THE REFERENCE SAMPLING FUNCTIONS

The derivative of the Fourier transform of the sampling functions is frequently needed. In this appendix we derive the expression for this derivative. Equation 5.54 may be written as

$$|G_j(f)|\exp(i\gamma(f)) = X(f) + iY(f) \tag{5A.1}$$

where $X(f)$ is the real part and $Y(f)$ is the imaginary part. Taking the derivative of this expression with respect to f we find

$$i|G_j(f)|\frac{d\gamma_j(f)}{df}\exp(i\gamma(f)) + \frac{d|G_j(f)|}{df}\exp(i\gamma_j(f))$$
$$= \frac{dX(f)}{df} + i\frac{dY(f)}{df} \tag{5A.2}$$

which can be transformed into

$$\frac{d|G_j(f)|}{df} = \left(\frac{dX(f)}{df} + i\frac{dY(f)}{df}\right) \exp(i\gamma_j(f)) - i|G_j(f)|\frac{d\gamma_j(f)}{df} \quad (5A.3)$$

Since the left-hand member of this expression is real, the right-hand member must also be real. Thus, we may obtain

$$\frac{d|G_j(f)|}{df} = \frac{dX(f)}{df}\cos(\gamma_j(f)) + \frac{dY(f)}{df}\sin(\gamma_j(f)) \quad (5A.4)$$

To apply this expression to an algorithm with N sampling points, we now use Eqs. 5.74 and 5.75 in this expression, with $\psi(f_r) = 0$. Then we may show that

$$\frac{d|G_j(f)|}{df} = -\frac{1}{f_r}\cos(\gamma_j(f))\sum_{n=1}^{N} W_{jn}\alpha_n \sin\left(\alpha_n\frac{f}{f_r}\right)$$

$$-\frac{1}{f_r}\sin(\gamma_j(f))\sum_{n=1}^{N} W_{jn}\alpha_n \cos\left(\alpha_n\frac{f}{f_r}\right) \quad (5A.5)$$

Thus, this derivative at the signal harmonic k (including the signal frequency f_r with $k = 1$) becomes

$$\left(\frac{d|G_j(f)|}{df}\right)_{f=kf_r} = -\frac{1}{f_r}\cos(\gamma_j(kf_r))\sum_{n=1}^{N} W_{jn}\alpha_n \sin(k\alpha_n)$$

$$-\frac{1}{f_r}\sin(\gamma_j(kf_r))\sum_{n=1}^{N} W_{jn}\alpha_n \cos(k\alpha_n) \quad (5A.6)$$

REFERENCES

Cheng Y.-Y. and J. C. Wyant, "Phase Shifter Calibration in Phase-Shifting Interferometry," *Appl. Opt.*, **24**, 30–49 (1985).

de Groot P., "Derivation of Algorithms for Phase Shifting Interferometry Using the Concept of a Data-Sampling Window," *Appl. Opt.*, **34**, 4723–4730 (1995).

Freischlad K. and C. L. Koliopoulos, "Fourier Description of Digital Phase Measuring Interferometry," *J. Opt. Soc. Am. A*, **7**, 542–551 (1990).

Greivenkamp J. E., "Generalized Data Reduction for Heterodyne Interferometry," *Opt. Eng.*, **23**, 350–352 (1984).

Hariharan P., "Digital Phase-Stepping Interferometry: Effects of Multiple Reflected Beams," *Appl. Opt.*, **26**, 2506–2507 (1987).

Hibino K., B. F. Oreb, and D. I. Farrant, "Phase Shifting for Non- Sinusoidal Waveforms with Phase Shift Errors," *J. Opt. Soc. Am. A*, **12**, 761–768 (1995).

Larkin K. G. and B. F. Oreb, "Design and Assessment of Symmetrical Phase-Shifting Algorithm," *J. Opt. Soc. Am.*, **9**, 1740–1748 (1992).

Morgan C. J., "Least Squares Estimation in Phase-Measurement Interferometry," *Opt. Lett.*, **7**, 368–370 (1982).

Nakadate S., "Phase Detection of Equidistant Fringes for Highly Sensitive Optical Sensing: I. Principle and Error Analysis," *J. Opt. Soc. Am. A*, **5**, 1258–1264 (1988a).

Nakadate S., "Phase Detection of Equidistant Fringes for Highly Sensitive Optical Sensing: II. Experiments," *J. Opt. Soc. Am. A*, **5**, 1265–1269 (1988b).

Parker D. H., "Moiré Patterns in Three-Dimensional Fourier Space," *Opt. Eng.*, **30**, 1534–1541 (1991).

Ransom P. L. and J. B. Kokal, "Interferogram Analysis by a Modified Sinusoid Fitting Technique," *Appl. Opt.*, **25**, 4199–4204 (1986).

Stetson K. A. and W. R. Brohinsky, "Electrooptic Holography and Its Applications to Hologram Interferometry," *Appl. Opt.*, **24**, 3631–3637 (1985).

Surrel I., "Design of Algorithms for Phase Measurements by the Use of Phase Stepping," *Appl. Opt.*, **35**, 51–60 (1996).

Surrel I., "Additive Noise Effect in Digital Phase Detection," *Appl. Opt.*, **36**, 271–276 (1997).

6

Phase Detection Algorithms

6.1 GENERAL PROPERTIES OF SYNCHRONOUS PHASE DETECTION ALGORITHMS

In this chapter we describe several phase detection algorithms, all of them with different properties, that have been developed by many authors. Different phase measuring algorithms have been reviewed by Creath (1986, 1991), Greivenkamp and Bruning (1992) and Schwider *et al.* (1983), among others. Here we will apply the Fourier theory developed in Chap. 5 to the analysis of some of these phase detection schemes.

Since we have three unknowns in Eq. 1.4, i.e., a, b, and ϕ, we need a minimum of three signal measurements to determine the phase ϕ. The measurements may have any phase, as long as they are known. We may assume the first measurement to be at a phase α_1, the second at α_2, the third at α_3, and so on. Here, the zero value position for these phases α_n will be considered to be at the origin of coordinates, thus making $\psi(f_r) = 0$. In this case the Fourier transforms of the sampling functions, from Eqs. 5.90 and 5.91, are

$$G_1(f) = \sum_{n=1}^{N} W_{1n} \exp\left(-i\alpha_n \frac{f}{f_r}\right) \tag{6.1}$$

and

$$G_2(f) = \sum_{n=1}^{N} W_{2n} \exp\left(-i\alpha_n \frac{f}{f_r}\right) \tag{6.2}$$

where the phase shift α_n is measured with respect to the reference frequency.

A sampling phase detecting algorithm is defined by the number of sampling points, their phase positions, and their associated sampling weights. The minimum number of sampling points is three. In this case their positions automatically define the values of the sampling weights. When the number of sampling points is greater than three, the phase positions of the sampling points does not completely define the algorithm, since there are an infinite number of sampling weight sets that satisfy the conditions studied in Chap. 5. However, only one of these possible solutions is a least squares fit.

We have studied in Chap. 5 that if there is a detuning, either the condition for equal magnitudes or the condition for orthogonality of the Fourier transforms of the sampling points, or both, are lost. Given a number of sampling points, these properties are defined by the sampling point phase locations, as will now be described.

If we consider only nonzero sampling weights, we may show that

1. If $g_1(x)$ is symmetric and $g_2(x)$ is antisymmetric, or vice versa, about the same phase point, the two functions are orthogonal at all frequencies.
2. If $g_1(x)$ and $g_2(x)$ are equal, and only one is shifted with respect to the other—for example, if they are both symmetric or both antisymmetric, about different points, separated by 90°—they will have the same magnitudes at all frequencies.

6.2 THREE-STEP ALGORITHMS TO MEASURE THE PHASE

We have seen before that to determine the phase without any ambiguity, a minimum of three sampling points are necessary.

Let us now consider the case of three sampling points, with any phases α_1, α_2, and α_3. Hence, we may write

$$
\begin{aligned}
s_1 &= a + b\cos(\phi + \alpha_1) \\
s_2 &= a + b\cos(\phi + \alpha_2) \\
s_3 &= a + b\cos(\phi + \alpha_3)
\end{aligned}
\tag{6.3}
$$

where the x, y dependence is implicit. These expressions may also be written as

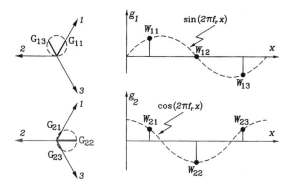

Figure 6.1 A three 120° step algorithm to measure the phase.

$$s_1 = a + b \cos \alpha_1 \cos - b \sin \alpha_1 \sin \phi$$
$$s_2 = a + b \cos \alpha_2 \cos - b \sin \alpha_2 \sin \phi \qquad (6.4)$$
$$s_3 = a + b \cos \alpha_3 \cos - b \sin \alpha_3 \sin \phi$$

hence we may find

$$\frac{s_2 - s_3}{2s_1 - s_2 - s_3}$$
$$= \frac{(\cos \alpha_2 - \cos \alpha_3) - (\sin \alpha_2 - \sin \alpha_3) \tan \phi}{(2 \cos \alpha_1 - \cos \alpha_2 - \cos \alpha_3) - (2 \sin \alpha_1 - \sin \alpha_2 - \sin \alpha_3) \tan \phi} \qquad (6.5)$$

This is a general expression for three-point sampling algorithms. Let us now consider some particular cases.

6.2.1 Three 120°-Step Algorithm

A particular case of the three-step method is obtained by taking $\alpha_1 = 60°$, $\alpha_2 = 180°$, and $\alpha_3 = 300°$, as shown in Fig. 6.1. Thus, we may obtain the following result for the phase

$$\tan \phi = -\sqrt{3} \frac{s_1 - s_3}{s_1 - 2s_2 + s_3} \qquad (6.6)$$

From this expression, by comparing with Eq. 5.85, we may see that the sampling reference weights have the values $W_{11} = \sqrt{3}/2$, $W_{12} = 0$, $W_{13} =$

$-\sqrt{3}/2$, $W_{21} = 1/2$, $W_{22} = -1$, $W_{23} = 1/2$. Thus, the sampling reference functions, also shown in Fig. 6.1, are

$$g_1(x) = \frac{\sqrt{3}}{2}\delta\left(x - \frac{X_r}{6}\right) - \frac{\sqrt{3}}{2}\delta\left(x - \frac{5X_r}{6}\right) \tag{6.7}$$

and

$$g_2(x) = \frac{1}{2}\delta\left(x - \frac{X_r}{6}\right) - \delta\left(x - \frac{3X_r}{6}\right) + \frac{1}{2}\delta\left(x - \frac{5X_r}{6}\right) \tag{6.8}$$

Since these three sampling points are equally spaced and uniformly distributed along the reference function period, as described by Eq. 5.10, the values of W_{1n} are equal to $\sin(2\pi f_r x_n)$ and the values of W_{2n} are equal to $\cos(2\pi f_r x_n)$. Thus, this is a diagonal least squares algorithm and expression 5.19 for the phase is valid. It may easily be shown that Eq. 5.19 reduces to Eq. 6.6 for these sampling points.

The sampling weights represented in a polar diagram are on the left side of the figure. We may see that the sampling vectors $\mathbf{G_1}$ and $\mathbf{G_2}$ are perpendicular to each other. We may also see on the right side of the figure that the sum of all sampling weights W_{1n} and similarly the sum of all sampling weights W_{2n} are equal to zero, since there is no DC term in functions $g_i(x)$.

The Fourier transforms of the sampling functions, using Eqs. 5.74 and 5.75, are

$$G_1(f) = \sqrt{3}\sin\left(\frac{2\pi}{3}\frac{f}{f_r}\right)\exp\left(-i\left(\pi\frac{f}{f_r} - \frac{\pi}{2}\right)\right) \tag{6.9}$$

and

$$G_2(f) = \left[1 - \cos\left(\frac{2\pi}{3}\frac{f}{f_r}\right)\right]\exp\left(-i\left(\pi\frac{f}{f_r} - \pi\right)\right) \tag{6.10}$$

The amplitudes of these functions are plotted in Fig. 6.2. Observing Eqs. 6.9 and 6.10, we see that these two functions are orthogonal at all frequencies. The normalized frequency is defined as the ratio of the frequency f to the reference frequency f_r. With a detuning, the condition for equal magnitudes is lost. It must be pointed out here that a phase π has been added, if necessary, to all expressions for the Fourier transforms $G_1(f)$ and $G_2(f)$ in this chapter, in order to change their sign and make their amplitudes positive at the reference frequency f_r.

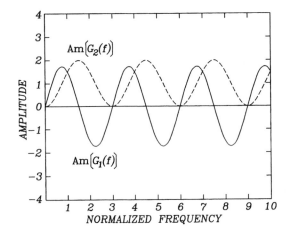

Figure 6.2 Amplitudes of the Fourier transforms of sampling functions for the three 120° step algorithm.

Given a reference frequency f_r, the value of $r(f)$ is a function of the signal phase and the signal frequency and is expressed by Eq. 5.75. This value of $r(f)$ is thus given by

$$r(f) = -\frac{\sqrt{3}\sin\left(\dfrac{2\pi}{3}\dfrac{f}{f_r}\right)\tan\left(\pi\dfrac{f}{f_r}+\phi\right)}{1-\cos\left(\dfrac{2\pi}{3}\dfrac{f}{f_r}\right)} \tag{6.11}$$

If both the reference and the signal frequencies are known, the phase may be obtained once the value of $r(f)$ is determined. If $f = f_r$, this expression reduces to Eq. 5.47. From Fig. 6.2 we may see that this algorithm has the following properties:

1. It is sensitive to detuning error, since the magnitudes of the Fourier transforms of the sampling functions become different after small detunings.
2. Signals with frequencies $f_r, 2f_r, 4f_r, 5f_r, 7f_r$, etc. can be detected, since the amplitudes of the Fourier transforms are the same (even if of different sign) at these frequencies.
3. Phase errors may be introduced by the presence in the signal of second, fourth, fifth, seventh, and eighth harmonics. However, it is insensitive to third, sixth, and ninth harmonics.

6.2.2 Three Steps in Inverted T Algorithm

Another particular case of the three-step method is obtained with $\alpha_1 = 0°$, $\alpha_2 = 90°$, and $\alpha_3 = 180°$, as shown in Fig. 6.3. Thus, we may obtain the following result for the phase:

$$\tan \phi = -\frac{-s_1 + 2s_2 - s_3}{s_1 - s_3} \tag{6.12}$$

These three points are equally but not uniformly spaced along the reference sampling function period. As a consequence, the sampling weights W_{1n} and W_{2n} are not equal to the functions $\sin(2\pi f_r \alpha_n)$ and $\cos(2\pi f_r \alpha_n)$, respectively, as in the case of uniformly spaced sampling points.

The sampling weights have the values $W_{11} = -1$, $W_{12} = 2$, $W_{13} = -1$, $W_{21} = 1$, $W_{22} = 0$, $W_{23} = -1$. Thus, the sampling reference functions are

$$g_1(x) = -\delta(x) + 2\delta\left(x - \frac{X_r}{4}\right) - \delta\left(x - \frac{2X_r}{4}\right) \tag{6.13}$$

and

$$g_2(x) = \delta(x) - \delta\left(x - \frac{X_r}{2}\right) \tag{6.14}$$

and the Fourier transforms of the sampling functions become

$$G_1(f) = 4\sin^2\left(\frac{\pi}{4}\frac{f}{f_r}\right)\exp\left(-i\left(\frac{\pi}{2}\frac{f}{f_r}\right)\right) \tag{6.15}$$

and

$$G_2(f) = 4\left[\sin\left(\frac{\pi}{4}\frac{f}{f_r}\right)\cos\left(\frac{\pi}{4}\frac{f}{f_r}\right)\right]\exp\left(-i\left(\frac{\pi}{2}\frac{f}{f_r} - \frac{\pi}{2}\right)\right) \tag{6.16}$$

We may see that these functions are orthogonal at all frequencies and their magnitudes are equal only at the reference frequency f_r and all of its harmonics. Their amplitudes are shown in Fig. 6.4.

The value of $r(f)$, from Eq. 5.75, is

$$r(f) = -\tan\left(\frac{\pi}{4}\frac{f}{f_r}\right)\tan\left(\phi + \frac{\pi}{2}\frac{f}{f_r} - \frac{\pi}{2}\right) \tag{6.17}$$

which, as expected, for $f = f_r$ becomes Eq. 5.47.

From Fig. 6.4 we may see that this algorithm has the following properties:

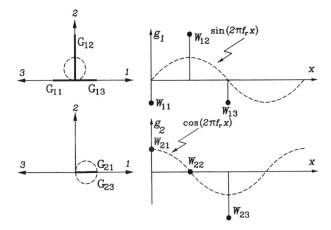

Figure 6.3 A three steps in inverted T algorithm to measure the phase.

1. It is quite sensitive to detuning error, since the magnitudes of the Fourier transforms of the sampling functions become very different after small detunings.
2. Signals with frequencies $f_r, 3f_r, 5f_r, 7f_r, 9f_r$, etc. can be detected, since the amplitudes of the Fourier transforms are the same (even if of different sign) at these frequencies.

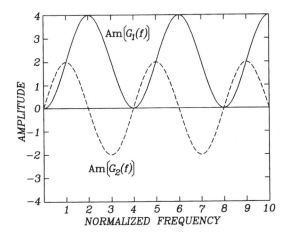

Figure 6.4 Amplitudes of the Fourier transforms of sampling functions for the three steps in inverted T algorithm.

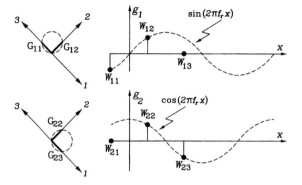

Figure 6.5 Wyant's three-step algorithm.

3. Phase errors may be introduced by the presence in the signal of second, third, fifth, sixth, seventh, and ninth harmonics. However, it is insensitive to fourth and eighth harmonics.

6.2.3 Wyant's Three Steps in Tilted T Algorithm

A particularly interesting case of three steps was proposed by Wyant *et al.* (1984) and later by Bhushan *et al.* (1985). In this case the expression for the phase is quite simple. The three sampling points are separated 90° as in the former algorithm, but with an offset of 45°, i.e., the first sampling point is taken at −45° with respect to the origin. It is interesting to note that a change in this offset changes the values of the sampling weights. They used $\alpha_1 = -45°$, $\alpha_2 = 45°$, and $\alpha_3 = 135°$, as shown in Fig. 6.5. Thus, we may obtain the following result for the phase:

$$\tan \phi = -\frac{-s_1 + s_2}{s_2 - s_3} \tag{6.18}$$

The sampling weights have the values $W_{11} = -1$, $W_{12} = 1$, $W_{13} = 0$, $W_{21} = 0$, $W_{22} = 1$, and $W_{23} = -1$. The sampling reference functions are

$$g_1(x) = -\delta\left(x + \frac{X_r}{8}\right) + \delta\left(x - \frac{X_r}{8}\right) \tag{6.19}$$

and

$$g_2(x) = \delta\left(x - \frac{X_r}{8}\right) - \delta\left(x - \frac{3X_r}{8}\right) \tag{6.20}$$

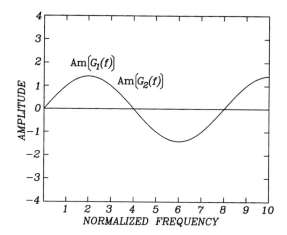

Figure 6.6 Amplitudes of Fourier transforms for sampling reference functions in Wyant's three-step algorithm.

Thus, the Fourier transform amplitudes of the sampling functions, illustrated in Fig. 6.6, are

$$G_1(f) = 2\sin\left(\frac{\pi}{4}\frac{f}{f_r}\right)\exp\left(-i\left(\frac{\pi}{2}\right)\right) \tag{6.21}$$

and

$$G_2(f) = 2\sin\left(\frac{\pi}{4}\frac{f}{f_r}\right)\exp\left(-i\left(\frac{\pi}{2}\frac{f}{f_r} - \frac{\pi}{2}\right)\right) \tag{6.22}$$

These functions have the same amplitudes at all frequencies, so their graphs are superimpose on each other. They are orthogonal only at the reference frequency f_r and at its odd harmonics.

From Eq. 5.75, the coefficient $r(f)$ is given by

$$r(f) = -\frac{\sin(\phi)}{\sin\left(\phi + \frac{\pi}{2}\frac{f}{f_r}\right)} \tag{6.23}$$

which may be used to find the phase in the presence of detuning if the magnitude of this detuning is known.

From Fig. 6.6 we may see that this algorithm has the following properties:

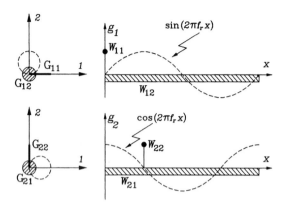

Figure 6.7 Sampling functions in two steps plus one algorithm.

1. It is quite sensitive to detuning error, since the orthogonality of the Fourier transforms of the sampling functions is lost after small detunings.
2. As in the preceding algorithm, signals with frequencies $f_r, 2f_r, 4f_r, 5f_r,$ $7f_r$, etc. can be detected, since the amplitudes of the Fourier transforms are the same (even if of different sign) at these frequencies.
3. Also as in the preceding algorithm, phase errors may be introduced by the presence in the signal of second, third, fifth, sixth, seventh, and ninth harmonics. As the former algorithm, it is also insensitive to fourth and eighth harmonics.

6.2.4 Two Steps Plus One Algorithm

If the constant term or bias is removed from the signal measurements, the phase can be determined with only two sampling points having a phase difference of 90°. The tangent of the phase is simply the ratio of the two measurements. Mendoza-Santoyo *et al.* (1988) determined the phase using this principle.

This principle has also been used in an interesting three-step method, illustrated in Fig. 6.7, suitable for systems with vibrations, as in the test of large astronomical mirrors (Angel and Wizinowich 1989, 1990; and Wizinowich 1990). The phase of one of the beams is rapidly switched between two values separated by 90°. This is done fast enough to reduce the effects of vibration. Further readings are taken any time later to obtain the sum of the irradiance of the beams independently of their relative phase. These later readings to find the irradiance sum may be performed in any of several ways. One possible way is to take two readings separated by 180°. An alternative is to use an integrating interval $\Delta = 360°$. The Fourier analysis of this algorithm thus depends on the approach used to find the irradiance. Here we will consider the second method, by integrating the signal over a period.

We write

$$s_1 = a + b\cos(\phi)$$

$$s_2 = a + b\cos(\phi + 90°) \tag{6.24}$$

$$s_3 = \frac{1}{X_r} \int_0^{X_r} s(x)\, dx = a$$

where $x = (X_r/2\pi)\phi$, which gives the following result for the phase:

$$\tan\phi = -\frac{s_2 - s_3}{s_1 - s_3} \tag{6.25}$$

The sampling reference functions are

$$g_1(x) = \delta(x) - f(x) \tag{6.26}$$

and

$$g_2(x) = \delta\left(x - \frac{X_r}{4}\right) - f(x) \tag{6.27}$$

with

$$f(x) = 0 \qquad \text{for } x \le 0$$

$$= \frac{1}{X_r} \qquad \text{for } 0 \le x \le X_r \tag{6.28}$$

$$= 0 \qquad \text{for } X_r \le x$$

Thus, the Fourier transforms of these sampling functions, shown in Fig. 6.8, are

$$G_1(f) = \left(1 - \frac{\sin\left(\pi \dfrac{f}{f_r}\right)}{\pi \dfrac{f}{f_r}} \exp\left(-i\frac{\pi}{2}\frac{f}{f_r}\right)\right) \exp\left(-i\frac{\pi}{2}\frac{f}{f_r}\right) \tag{6.29}$$

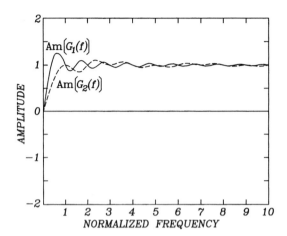

Figure 6.8 Amplitudes of Fourier transforms for sampling reference functions in two steps plus one algorithm.

and

$$G_2(f) = \left(1 - \frac{\sin\left(\pi\dfrac{f}{f_r}\right)}{\pi\dfrac{f}{f_r}} \exp\left(-i\frac{\pi}{2}\frac{f}{f_r}\right)\right) \exp\left(-i\frac{\pi}{2}\frac{f}{f_r}\right) \tag{6.30}$$

We may easily see that these two Fourier transforms are orthogonal to each other and have the same amplitude at the signal frequency and all of its harmonics. In other words, this algorithm is not insensitive to any of the signal harmonics. There is also sensitivity to detuning.

The value of $r(f)$, from Eq. 5.75, is given by

$$r(f) = \frac{\cos\left(\dfrac{\pi}{2}\dfrac{f}{f_r} + \phi\right) - \dfrac{\sin\left(\pi\dfrac{f}{f_r}\right)}{\pi\dfrac{f}{f_r}} \cos\left(\pi\dfrac{f}{f_r} + \phi\right)}{\cos\phi - \dfrac{\sin\left(\pi\dfrac{f}{f_r}\right)}{\pi\dfrac{f}{f_r}} \cos\left(\pi\dfrac{f}{f_r} + \phi\right)} \tag{6.31}$$

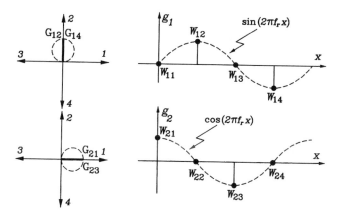

Figure 6.9 Four steps in cross algorithm.

6.3 FOUR-STEP ALGORITHMS TO MEASURE THE PHASE

In principle, three steps are enough to determine the three unknown constants; however, small measurement errors may have a large effect on the results. Four-step methods may be better in this respect.

With four steps, as we pointed out at the beginning of this chapter, given a sampling point distribution there are an infinite number of solutions for the phase, and some of them are diagonal least squares algorithm solutions. Some of these algorithms are described in this section.

6.3.1 Four Steps in Cross Algorithm

The values of the irradiance are measured using four different values of the phase, $\alpha_1 = 0°$, $\alpha_2 = 90°$, $\alpha_3 = 180°$, and $\alpha_4 = 270°$, are

$$s_1 = a + b\cos(\phi)$$

$$s_2 = a + b\cos(\phi + 90°)$$

$$s_3 = a + b\cos(\phi + 180°) \tag{6.32}$$

$$s_4 = a + b\cos(\phi + 270°)$$

These are shown in Fig. 6.9.

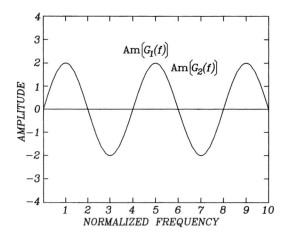

Figure 6.10 Amplitudes of Fourier transforms for sampling reference functions in the four steps in cross algorithm.

From these expressions, one possible solution for the phase is

$$\tan \phi = -\frac{s_2 - s_4}{s_1 - s_3} \tag{6.33}$$

The sampling weights have the values $W_{11} = 0$, $W_{12} = 1$, $W_{13} = 0$, $W_{14} = -1$, $W_{21} = 1$, $W_{22} = 0$, $W_{23} = -1$, and $W_{24} = 0$. We may see in Fig. 6.9 that these sampling weights are as described by Eq. 5.19. Hence this is a diagonal least squares solution, with the system matrix diagonal. The sampling reference functions are

$$g_1(x) = \delta\left(x - \frac{X_r}{4}\right) - \delta\left(x - \frac{3X_r}{4}\right) \tag{6.34}$$

and

$$g_2(x) = \delta(x) - \delta\left(x - \frac{2X_r}{4}\right) \tag{6.35}$$

Thus, the Fourier transforms of the sampling functions, illustrated in Fig. 6.10, are

$$G_1(f) = 2\sin\left(\frac{\pi}{2}\frac{f}{f_r}\right)\exp\left(-i\left(\pi\frac{f}{f_r} - \frac{\pi}{2}\right)\right) \tag{6.36}$$

and

$$G_2(f) = 2 \sin\left(\frac{\pi}{2}\frac{f}{f_r}\right) \exp\left(-i\left(\frac{\pi}{2}\frac{f}{f_r} - \frac{\pi}{2}\right)\right) \tag{6.37}$$

The amplitudes of these functions are the same at all frequencies and are orthogonal at the reference frequency f_r and all its odd harmonics.

Using Eq. 5.75, the value of $r(f)$ is given by

$$r(f) = -\frac{\sin\left(\phi + \pi\frac{f}{f_r}\right)}{\sin\left(\phi + \frac{\pi}{2}\frac{f}{f_r}\right)} \tag{6.38}$$

From Fig. 6.10 we may see that this algorithm has the following properties:

1. It is quite sensitive to detuning error, since as in Wyant's algorithm the orthogonality of the Fourier transforms of the sampling functions is lost after small detuning.
2. Phase errors may be introduced by the presence in the signal of all odd harmonics. However, it is insensitive to all even harmonics.

6.3.2 Four Steps in X Algorithm

The values of the irradiance are measured at four different values of the phase, equal to $\alpha_1 = 45°, \alpha_2 = 135°, \alpha_3 = 225°$, and $\alpha_4 = 315°$, as shown in Fig. 6.11, as follows

$$s_1 = a + b\cos(\phi + 45°)$$
$$s_2 = a + b\cos(\phi + 135°)$$
$$s_3 = a + b\cos(\phi + 225°) \tag{6.39}$$
$$s_4 = a + b\cos(\phi + 315°)$$

From these equations we may show that one solution for the phase is

$$\tan\phi = \frac{s_1 + s_2 - s_3 - s_4}{s_1 - s_2 - s_3 + s_4} \tag{6.40}$$

The sampling weights have the values $W_{11} = 1, W_{12} = 1, W_{13} = -1, W_{14} = -1, W_{21} = 1, W_{22} = -1, W_{23} = -1$, and $W_{24} = 1$. As in the preceding

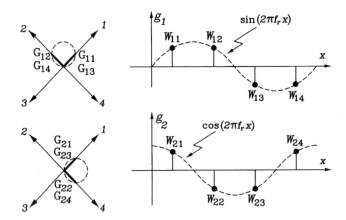

Figure 6.11   Four steps in X algorithm.

algorithm, we may see in Fig. 6.11 that these sampling weights are as described by Eq. 5.19 and hence that this is another diagonal least squares solution. Then the sampling reference functions are

$$g_1(x) = \delta\left(x - \frac{X_r}{8}\right) + \delta\left(x - \frac{3X_r}{8}\right) - \delta\left(x - \frac{5X_r}{8}\right) - \delta\left(x - \frac{7X_r}{8}\right)$$

(6.41)

and

$$g_2(x) = \delta\left(x - \frac{X_r}{8}\right) - \delta\left(x - \frac{3X_r}{8}\right) - \delta\left(x - \frac{5X_r}{8}\right) + \delta\left(x - \frac{7X_r}{8}\right)$$

(6.42)

The Fourier transforms of the sampling functions, illustrated in Fig. 6.12, are

$$G_1(f) = 4\sin\left(\frac{\pi}{2}\frac{f}{f_r}\right)\cos\left(\frac{\pi}{4}\frac{f}{f_r}\right)\exp\left(-i\left(\pi\frac{f}{f_r} - \frac{\pi}{2}\right)\right)$$

(6.43)

and

$$G_2(f) = 4\sin\left(\frac{\pi}{2}\frac{f}{f_r}\right)\sin\left(\frac{\pi}{4}\frac{f}{f_r}\right)\exp\left(-i\left(\pi\frac{f}{f_r} - \pi\right)\right)$$

(6.44)

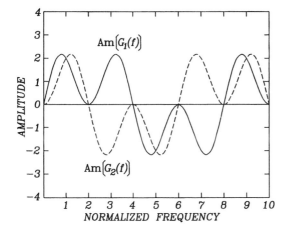

Figure 6.12 Amplitudes of Fourier transforms for sampling reference functions in four steps in X algorithm.

These functions are orthogonal at all frequencies and have the same amplitude only at the reference frequency f_r and all its odd harmonics.

From Eq. 5.75, the value of $r(f)$ may be shown to be given by

$$r(f) = -\frac{\tan\left(\pi \dfrac{f}{fr} + \phi\right)}{\tan\left(\dfrac{\pi}{4}\dfrac{f}{f_r}\right)} \tag{6.45}$$

Thus, any detuning may be compensated for, if the signal frequency is known, by dividing the calculated value of $r(f)$ by $\tan(\pi f/4 f_r)$.

From Fig. 6.12 we may see that this algorithm has the following properties:

1. It is quite sensitive to detuning error, since the amplitude of the Fourier transforms of the sampling functions become different after small detunings.
2. Signals with frequencies $f_r, 3f_r, 5f_r, 7f_r, 9f_r$, etc. can be detected because the amplitudes of the Fourier transforms are the same (even if of different sign) at these frequencies.
3. As in the preceding algorithm, phase errors may be introduced by the presence in the signal of all odd harmonics. Also, it is insensitive to all even harmonics.

6.4 FIVE-STEP ALGORITHM

In the five-step algorithm the values of the irradiance are measured at five different values of the phase, equal to $\alpha_1 = 36°$, $\alpha_2 = 108°$, $\alpha_3 = 180°$, $\alpha_4 = 252°$, and $\alpha_5 = 324°$, shown in Fig. 6.13, as follows:

$$s_1 = a + b\cos(\phi + 36°)$$
$$s_2 = a + b\cos(\phi + 108°)$$
$$s_3 = a + b\cos(\phi + 180°) \tag{6.46}$$
$$s_4 = a + b\cos(\phi + 252°)$$
$$s_5 = a + b\cos(\phi + 324°)$$

Then the diagonal least squares solution is

$$\tan\phi = -\frac{\displaystyle\sum_{n=1}^{5}\sin\left(\frac{2\pi n}{5}\right)s_n}{\displaystyle\sum_{n=1}^{5}\cos\left(\frac{2\pi n}{5}\right)s_n} \tag{6.47}$$

Thus, the sampling reference functions are

$$g_1(x) = \delta\left(x - \frac{X_r}{10}\right) + \delta\left(x - \frac{3X_r}{10}\right) - \delta\left(x - \frac{7X_r}{10}\right) - \delta\left(x - \frac{9X_r}{10}\right) \tag{6.48}$$

and

$$g_2(x) = \delta\left(x - \frac{X_r}{10}\right) - \delta\left(x - \frac{3X_r}{10}\right) - \delta\left(x - \frac{5X_r}{10}\right) - \delta\left(x - \frac{7X_r}{10}\right)$$
$$+ \delta\left(x - \frac{9X_r}{10}\right) \tag{6.49}$$

The Fourier transforms of the sampling functions, illustrated in Fig. 6.14, are

$$G_1(f) = 2\left[\sin\left(\frac{\pi}{5}\right)\sin\left(\frac{4\pi}{5}\frac{f}{f_r}\right) + \sin\left(\frac{3\pi}{5}\right)\sin\left(\frac{2\pi}{5}\frac{f}{f_r}\right)\right]$$
$$\times \exp\left(-i\left(\pi\frac{f}{f_r} + \frac{\pi}{2}\right)\right) \tag{6.50}$$

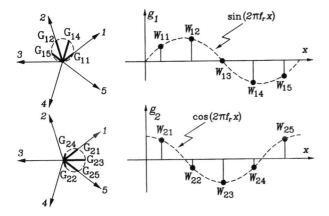

Figure 6.13 Five-step algorithm.

and

$$G_2(f) = 2\left[\frac{1}{2} - \cos\left(\frac{\pi}{5}\right)\cos\left(\frac{4\pi}{5}\frac{f}{f_r}\right) + \cos\left(\frac{2\pi}{5}\right)\cos\left(\frac{2\pi}{5}\frac{f}{f_r}\right)\right]$$
$$\times \exp\left(-i\left(\pi\frac{f}{f_r} + \pi\right)\right) \tag{6.51}$$

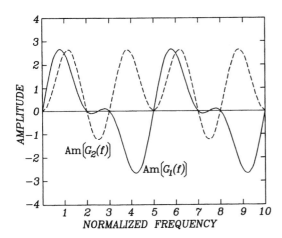

Figure 6.14 Amplitudes of the Fourier transforms for sampling reference functions in five-step algorithm.

These functions are orthogonal at all frequencies and have the same amplitude only at the reference frequency f_r and at the sixth harmonic.

From Eq. 5.75, the value of $r(f)$ may be shown to be given by

$$r(f) = \frac{\left[\sin\left(\frac{\pi}{5}\right)\cos\left(\frac{4\pi}{5}\frac{f}{f_r}\right) + \sin\left(\frac{3\pi}{5}\right)\cos\left(\frac{2\pi}{5}\frac{f}{f_r}\right)\right]\tan\left(\phi + \pi\frac{f}{f_r}\right)}{\frac{1}{2} - \cos\left(\frac{\pi}{5}\right)\cos\left(\frac{4\pi}{5}\frac{f}{f_r}\right) + \cos\left(\frac{2\pi}{5}\right)\cos\left(\frac{2\pi}{5}\frac{f}{f_r}\right)}$$

(6.52)

From Fig. 6.14 we may see that this algorithm has the following properties:

1. It is quite sensitive to detuning error, since the magnitudes of the Fourier transforms of the sampling functions become different after small detuning.
2. Signals with frequencies $f_r, 4f_r, 6f_r, 9f_r$, etc. can be detected, since the amplitudes of the Fourier transforms are the same (even if of different sign) at these frequencies.
3. Phase errors may be introduced by the presence in the signal of fourth, sixth, and ninth harmonics. It is insensitive to the second, third, fifth, seventh, eighth, and tenth harmonics.

6.5 SYMMETRICAL $N + 1$ PHASE STEPS ALGORITHMS

We have seen in Chapter 5 that any phase detection algorithm must satisfy the condition that the sampling reference vectors G_1 and G_2 must be orthogonal to each other and they must have the same magnitude. Also, the sums of their x and y components must be zero, as expressed by Eqs. 5.98 and 5.99. We have also seen in Chap. 5 that when there are N sampling points equally and uniformly spaced, as described by

$$x_n = \frac{n-1}{Nf_r}$$

(6.53)

these conditions are satisfied if the sampling weights are given by

$$W_{1n} = \sin\alpha_n$$

(6.54)

and

$$W_{2n} = \cos\alpha_n$$

(6.55)

where $\alpha_n = 2\pi f_r x_n$. Then the signal phase becomes

$$\tan \phi = - \left(\frac{\displaystyle\sum_{n=1}^{N} s(x_n) \sin \alpha_n}{\displaystyle\sum_{n=1}^{N} s(x_n) \cos \alpha_n} \right) \tag{6.56}$$

This expression is valid for all algorithms with N sampling points equally and uniformly spaced, according to Eq. 5.10. The first sampling point ($n = 1$) is located at a coordinate $x_n = 0$, and the last point is located at $x_N = (N-1)/Nf_r$. A point with $n = N + 1$ (which is not considered) would be located at $x_n = X_r = 1/f_r$, that is, at a phase 2π.

Let us now consider algorithms with $N + 1$ sampling points with the same separation as before. Then the last point has a phase equal to 2π. This modification disturbs the conditions for orthogonality and equal magnitudes required from the reference sampling weights, but they can be restored simply by splitting in half the magnitude of the first ($n = 1$) sampling weight W_{21} and setting the last ($n = N + 1$) sampling weight $W_{2(N+1)}$ equal to this value. Thus, the modified sampling weights W_{21} and $W_{2(N+1)}$ have the same value, equal to

$$W_{21} = W_{2(N+1)} = \frac{1}{2} \cos \alpha_1 = \frac{1}{2} \tag{6.57}$$

All other sampling weights remain the same. These algorithms, first described by Larkin and Oreb (1992) and Larkin (1992), are called symmetrical $N + 1$ sampling algorithms and have some interesting error-compensating properties.

The Fourier transforms of these sampling reference functions with $N + 1$ sampling points, from Eqs. 6.1 and 6.2, are given by

$$G_m(f) = \sum_{n=1}^{N+1} W_{mn} \exp(-i2\pi f x_n) \qquad m = 1, 2 \tag{6.58}$$

With the sampling point distribution just described for these algorithms, its Fourier transforms become, after adding together terms symmetrically placed in the sampling interval,

$$G_m(f) = \sum_{n=1}^{\frac{N+1}{2}} \left[W_{mn} \exp(-i2\pi f x_n) + W_{m(N+2-n)} \exp(-i2\pi f x_{(N+2-n)}) \right]$$

$$\tag{6.59}$$

for N odd, with no sampling point at the middle central position of the sampling interval, since the total number of points $(N + 1)$ is even; or

$$G_m(f) = \sum_{n=1}^{\frac{N}{2}} [W_{mn} \exp(-i2\pi f x_n) + W_{m(N+2-n)} \exp(-i2\pi f x_{(N+2-n)})]$$
$$+ W_{m(N/2+1)} \exp(-i2\pi f x_{N/2+1}) \tag{6.60}$$

for N even. Since the total number of sampling points is odd, there is a point at the middle. The weights defined by Eqs. 6.54 and 6.55 are antisymmetrical, while the terms defined by Eq. 6.57 are symmetrical. Then, we may show that $G_1(f)$ is given by

$$G_1(f) = 2i \sum_{n=1}^{\frac{N+1}{2}} W_{1n} \sin\left(\left(1 - \frac{2(n-1)}{N}\right) \pi \frac{f}{f_r}\right) \exp\left(-i\pi \frac{f}{f_r}\right) \tag{6.61}$$

for N odd, and that

$$G_1(f) = 2i \sum_{n=1}^{\frac{N}{2}} W_{1n} \sin\left(\left(1 - \frac{2(n-1)}{N}\right) \pi \frac{f}{f_r}\right) \exp\left(-i\pi \frac{f}{f_r}\right) \tag{6.62}$$

for N even. The last term has disappeared, since the weight $W_{1(N/2+1)}$ is equal to zero. In the same manner $G_2(f)$ is given by

$$G_2(f) = 2 \sum_{n=1}^{\frac{N+1}{2}} W_{2n} \cos\left(\left(1 - \frac{2(n-1)}{N}\right) \pi \frac{f}{f_r}\right) \exp\left(-i\pi \frac{f}{f_r}\right) \tag{6.63}$$

for N odd, and

$$G_2(f) = 2 \sum_{n=1}^{\frac{N}{2}} W_{2n} \cos\left(\left(1 - \frac{2(n-1)}{N}\right) \pi \frac{f}{f_r}\right) \exp\left(-i\pi \frac{f}{f_r}\right)$$
$$+ W_{2(N/2+1)} \exp\left(-i\pi \frac{f}{f_r}\right) \tag{6.64}$$

for N even.

From Eqs. 6.54, 6.55, and 6.57, since $\psi(f_r)$ is zero, the sampling weights, using the sampling point distribution in Eq. 6.53, are

$$W_{1n} = \sin\left(\frac{2\pi(n-1)}{N}\right) \tag{6.65}$$

for all values of n,

$$W_{2n} = \cos\left(\frac{2\pi(n-1)}{N}\right) \tag{6.66}$$

for $1 < n < N + 1$, and

$$W_{2n} = \frac{1}{2} \tag{6.67}$$

for $n = 1$ and $n = N + 1$.

We may see that due to their symmetry these two functions are orthogonal at all frequencies. This is an important result, since then we may conclude that with a detuning, the only condition that may fail is that requiring equal amplitudes of the Fourier transforms of the sampling functions.

Then the only requirement for insensitivity to detuning, as studied in Chap. 5, is that the amplitude of the Fourier transforms remain the same in a small frequency interval centered at f_r. As described in Chap. 4, this is so when the two plots for $G_1(f)$ and $G_2(f)$ touch tangentially at the frequency f_r.

One of the important properties of these symmetric $N + 1$ algorithms is that they can be made insensitive to small frequency detuning. The requirement that the slopes for $G_1(f)$ and $G_2(f)$ be equal, so that they touch tangentially, is satisfied in some of these algorithms for some values of N but not for all of them. When this happens, the algorithm may still be modified, so that insensitivity to detuning is obtained.

Let us assume, as described by Larkin and Oreb (1992), that an additional term $\Delta G_1(f)$ is added to the function $G_1(f)$, with the following conditions:

1. Its phase is equal to that of $G_1(f)$, so that the orthogonality condition is not disturbed at any frequency.
2. Its amplitude at the frequency f_r is zero, so that the equal-amplitudes condition is not disturbed at this frequency.
3. The sum of its sampling weights should be zero, so that the no DC bias condition remains.
4. Its amplitude is zero at the harmonics of the frequency f_r, so that the no harmonics cross talk status is not altered by the presence of this extra term.

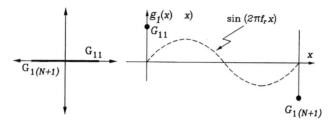

Figure 6.15 Sampling weights for extra term $\Delta G_1(f)$.

5. Its slope at the frequency f_r is different from zero, so that the final slope of the Fourier transform $G_1(f)$ may be changed as needed to make the algorithm insensitive to small detuning.

The sampling weights W_{11} and $W_{1(N+1)}$ have a zero value. Let us assume that the sampling weights for the additional term $\Delta G_1(f)$ are given a nonzero value with the same amplitude but with opposite sign at these locations, as shown in Fig. 6.15.

The necessary conditions are satisfied, and the slope of the amplitude of the Fourier transform $G_1(f)$ at the signal frequency may be modified. Thus, we see that $\Delta G_1(f)$ plotted in Fig. 6.16 is

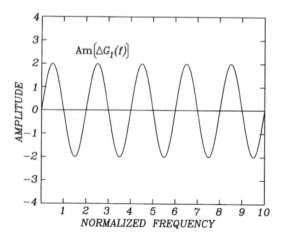

Figure 6.16 Amplitude of the Fourier transforms for extra term $\Delta G_1(f)$.

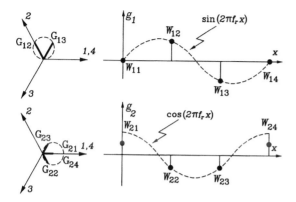

Figure 6.17 Symmetrical four $(3 + 1)$ step algorithm.

$$\Delta G_1(f) = 2i\, W_{11} \sin\left(\pi \frac{f}{f_r}\right) \exp\left(-i\pi \frac{f}{f_r}\right) \tag{6.68}$$

where $W_{11} = -W_{1(N+1)}$ is set to a value such that the two desired slopes become equal.

We will apply this extra term to some symmetrical algorithms later in this chapter to make them insensitive to detuning. Surrel (1993), independently from Larkin and Oreb (1992), developed the symmetrical detuning-insensitive algorithms. He showed that the sampling weights W_{11} and $W_{1(N+1)}$ must have the value

$$W_{11} = W_{1(N+1)} = \frac{1}{2 \tan\left(\dfrac{2\pi}{N}\right)} \tag{6.69}$$

6.5.1 Symmetrical Four (3 + 1) Step Algorithm

For this algorithm, with $N = 3$, illustrated in Fig. 6.17, the signal four measurements are written as follows

$$s_1 = a + b\cos(\phi)$$

$$s_2 = a + b\cos(\phi + 120°)$$

$$s_3 = a + b\cos(\phi + 240°)$$

$$s_4 = a + b\cos(\phi + 360°) \tag{6.70}$$

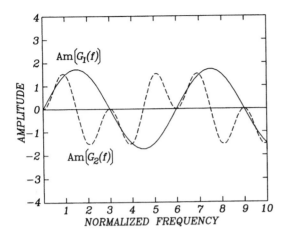

Figure 6.18 Amplitudes of the Fourier transforms for sampling reference functions in symmetrical four $(3 + 1)$ step algorithm.

The first and last points have the same phase. Thus, we may take the average of these points in order to reduce the number of equations to three. Then, from these equations we may find

$$\tan \phi = -\sqrt{3} \frac{s_2 - s_3}{s_1 - s_2 - s_3 + s_4} \tag{6.71}$$

It is interesting to note that this expression may be obtained from a three-point algorithm as the three $120°$ steps algorithm, with the first sampling point at $0°$ if s_1 is replaced by $(s_1 + s_4)/2$.

The sampling weights are $W_{11} = 0$, $W_{12} = \sqrt{3/2}$, $W_{13} = -\sqrt{3/2}$, $W_{14} = 0$, $W_{21} = 0.5$, $W_{22} = -0.5$, $W_{23} = -0.5$, $W_{24} = 0.5$. Then the sampling reference functions are

$$g_1(x) = \frac{\sqrt{3}}{2} \left(\delta \left(x - \frac{X_r}{3} \right) - \delta \left(x - \frac{2X_r}{3} \right) \right) \tag{6.72}$$

and

$$g_2(x) = \frac{1}{2} \left(\delta(x) - \delta \left(x - \frac{X_r}{3} \right) - \delta \left(x - \frac{2X_r}{3} \right) + \delta(x - X_r) \right) \tag{6.73}$$

The Fourier transforms of these sampling functions, plotted in Fig. 6.18, are

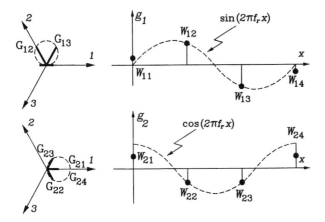

Figure 6.19 Symmetrical four $(3 + 1)$ step algorithm with extra term to obtain detuning insensitivity.

$$G_1(f) = \sqrt{3} \sin\left(\frac{\pi}{3}\frac{f}{f_r}\right) \exp\left(-i\left(\pi\frac{f}{f_r} - \frac{\pi}{2}\right)\right) \tag{6.74}$$

and

$$G_2(f) = 2\sin\left(\frac{2\pi}{3}\frac{f}{f_r}\right) \sin\left(\frac{\pi}{3}\frac{f}{f_r}\right) \exp\left(-i\left(\pi\frac{f}{f_r} - \pi\right)\right) \tag{6.75}$$

The value of $r(f)$, from Eq. 5.75, is given by

$$r(f) = \frac{\sqrt{3}\tan\left(\phi + \pi\dfrac{f}{f_r}\right)}{2\sin\left(\dfrac{2\pi}{3}\dfrac{f}{f_r}\right)} \tag{6.76}$$

These Fourier transforms are orthogonal at all frequencies. We see that the two curves do not touch each other tangentially at the reference frequency f_r. Then, in order to have detuning insensitivity we must add to the function $G_1(f)$ the additional term $\Delta G_1(f)$, with the proper amplitude σ. Then the value of W_{11} that makes the slope of $\Delta G_1(f)$ equal to minus this value is equal to $W_{11} = 1/(2\sqrt{3})$. The sampling weights for the final algorithm are in Fig. 6.19.

The plots of the amplitudes of the Fourier transforms are in Fig. 6.20.

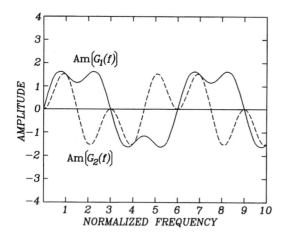

Figure 6.20   Amplitudes of the Fourier transforms for sampling reference functions in symmetrical four $(3 + 1)$ step algorithm with extra term.

From Fig. 6.20 we may see that this algorithm has the following properties:

1. It is insensitive to small detuning error, since the two plots for the Fourier transform magnitudes touch each other tangentially at the reference frequency.
2. Signals with frequencies $f_r, 2f_r, 4f_r, 5f_r, 7f_r$, etc. can be detected, since the amplitudes of the Fourier transforms are the same (even if of different sign) at these frequencies.
3. Phase errors may be introduced by the presence in the signal of second, fourth, fifth, seventh, and eighth harmonics. It is insensitive to third, sixth, and ninth harmonics.

6.5.2   Schwider-Hariharan Five (4 + 1) Step Algorithm

This algorithm was described by Schwider *et al.* (1983) and later by Hariharan *et al.* (1987). The irradiance measurements for the five sampling points are

$$s_1 = a + b\cos(\phi)$$

$$s_2 = a + b\cos(\phi + 90°)$$

$$s_3 = a + b\cos(\phi + 180°)$$ (6.77)

$$s_4 = a + b\cos(\phi + 270°)$$

$$s_5 = a + b\cos(\phi + 360°)$$

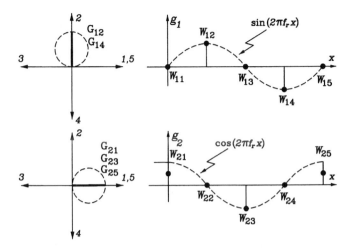

Figure 6.21 Symmetrical five $(4 + 1)$ step algorithm.

From these equations, the phase can be obtained as follows:

$$\tan \phi = -\frac{s_2 - s_4}{\frac{1}{2}s_1 - s_3 + \frac{1}{2}s_5} \tag{6.78}$$

This expression may be obtained from the four steps at $n\pi/2$ algorithm by substituting for the measurement s_1 the average of the measurements s_1 and s_5. The sampling weights, as shown in Fig. 6.21, have the values $W_{11} = 0$, $W_{12} = 1$, $W_{13} = 0$, $W_{14} = -1$, $W_{15} = 0$, $W_{21} = 1/2$, $W_{22} = 0$, $W_{23} = -1$, $W_{24} = 0$, and $W_{25} = 1/2$. Then the sampling reference functions are

$$g_1(x) = \delta\left(x - \frac{X_r}{4}\right) - \delta\left(x - \frac{3X_r}{4}\right) \tag{6.79}$$

and

$$g_2(x) = \frac{1}{2}\delta(x) - \delta\left(x - \frac{X_r}{2}\right) + \frac{1}{2}\delta(x - X_r) \tag{6.80}$$

The amplitudes of the Fourier transforms of the sampling functions, shown in Fig. 6.22, are

$$G_1(f) = 2\sin\left(\frac{\pi}{2}\frac{f}{f_r}\right)\exp\left(-i\left(\pi\frac{f}{f_r} - \frac{\pi}{2}\right)\right) \tag{6.81}$$

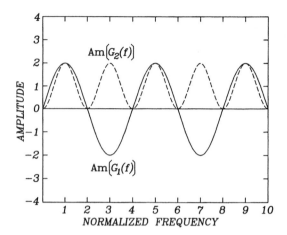

Figure 6.22 Amplitudes of the Fourier transforms for sampling reference functions in the symmetrical five $(4 + 1)$ step algorithm.

and

$$G_2(f) = 2\sin^2\left(\frac{\pi}{2}\frac{f}{f_r}\right)\exp\left(-i\left(\pi\frac{f}{f_r} - \pi\right)\right) \qquad (6.82)$$

These functions are orthogonal at all frequencies, and their amplitudes are equal only at the reference frequency f_r and at its odd harmonics.

We see that the amplitudes of these two functions become equal at values of the frequency signal equal to $f_r, 5f_r, 9f_r$, etc. At these points the curves for the two Fourier transforms touch each other tangentially, making the algorithm insensitive to small frequency detuning.

Using Eq. 5.75, the value of $r(f)$ is given by

$$r(f) = -\frac{\tan\left(\phi + \pi\frac{f}{f_r}\right)}{\sin\left(\pi\frac{f}{f_r}\right)} \qquad (6.83)$$

From Fig. 6.22 we may see that this algorithm has the following properties:

1. It is insensitive to small detuning error, since the two plots for the Fourier transform magnitudes touch each other tangentially at the reference frequency.

2. Signals with frequencies $f_r, 3f_r, 5f_r, 7f_r, 9f_r$, etc. can be detected, since the amplitudes of the Fourier transforms are the same (even if of different sign) at these frequencies.
3. Phase errors may be introduced by the presence in the signal of odd harmonics. It is insensitive to even harmonics.

Hariharan *et al.* (1987) derived this algorithm by assuming that the phase separation between the five sampling points was not known and algebraically representing it by α in Eqs. 6.70. Then the value of α is found by equating to zero the derivative of $\tan \phi_0$ with respect to the angle α. Thus, an angle α equal to 90° is found. In this algorithm a symmetrical sampling point distribution from $-\pi$ to π was used.

6.5.3 Symmetrical Six (5 + 1) Step Algorithm

In this algorithm the irradiance measurements for the six sampling points, as illustrated in Fig. 6.23, are

$$s_1 = a + b\cos(\phi)$$
$$s_2 = a + b\cos(\phi + 72°)$$
$$s_3 = a + b\cos(\phi + 144°)$$
$$s_4 = a + b\cos(\phi + 216°) \tag{6.84}$$
$$s_5 = a + b\cos(\phi + 288°)$$
$$s_6 = a + b\cos(\phi + 360°)$$

From these equations, the phase can be shown to be

$$\tan \phi = -\frac{\displaystyle\sum_{n=1}^{6} \sin\left(\frac{2\pi(n-1)}{5}\right) s_n}{\frac{1}{2}s_1 + \displaystyle\sum_{n=2}^{5} \cos\left(\frac{2\pi(n-1)}{5}\right) s_n + \frac{1}{2}s_6} \tag{6.85}$$

The reference sampling functions are

$$g_1(x) = \sum_{n=1}^{6} \sin\left(\frac{2\pi(n-1)}{5}\right) \delta\left(x - \frac{n-1}{5} X_r\right) \tag{6.86}$$

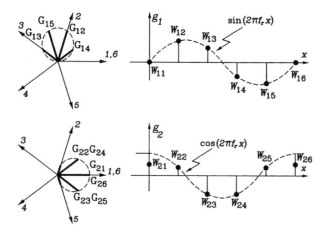

Figure 6.23　Symmetrical six $(5 + 1)$ step algorithm.

and

$$g_2(x) = \frac{1}{2}\delta(x) + \sum_{n=2}^{5} \cos\left(\frac{2\pi(n-1)}{5}\right)\delta\left(x - \frac{n}{5}X_r\right) + \frac{1}{2}\delta(x - X_r) \quad (6.87)$$

The Fourier transforms of the sampling functions, shown in Fig. 6.24, are

$$G_1(f) = 2\left(\sin\left(\frac{2\pi}{5}\right)\sin\left(\frac{3\pi}{5}\frac{f}{f_r}\right) + \sin\left(\frac{\pi}{5}\right)\sin\left(\frac{\pi}{5}\frac{f}{f_r}\right)\right)$$
$$\times \exp\left(-i\left(\pi\frac{f}{f_r} - \frac{\pi}{2}\right)\right) \quad (6.88)$$

and

$$G_2(f) =$$
$$2\left(\cos\left(\frac{\pi}{5}\right)\cos\left(\frac{\pi}{5}\frac{f}{f_r}\right) - \cos\left(\frac{2\pi}{5}\right)\cos\left(\frac{3\pi}{5}\frac{f}{f_r}\right) - \frac{1}{2}\cos\left(\pi\frac{f}{f_r}\right)\right)$$
$$\times \exp\left(-i\left(\pi\frac{f}{f_r} - \pi\right)\right) \quad (6.89)$$

These functions are orthogonal at all frequencies, as expected. The amplitudes of these two functions become equal at values of the frequency signal equal to $f_r, 6f_r$, etc.

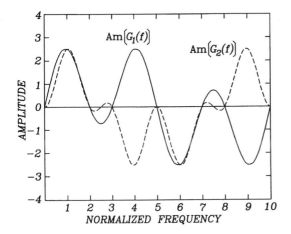

Figure 6.24 Amplitudes of the Fourier transforms for sampling reference functions in the symmetrical six $(5+1)$ step algorithm.

Using Eq. 5.75, the value of $r(f)$ is given by

$$r(f) = \frac{\left(\sin\left(\frac{2\pi}{5}\right)\sin\left(\frac{3\pi}{5}\frac{f}{f_r}\right) + \sin\left(\frac{\pi}{5}\right)\sin\left(\frac{\pi}{5}\frac{f}{f_r}\right)\right)\tan\left(\phi - \pi\frac{f}{f_r}\right)}{\frac{1}{2}\cos\left(\pi\frac{f}{f_r}\right) + \cos\left(\frac{2\pi}{5}\right)\cos\left(\frac{3\pi}{5}\frac{f}{f_r}\right) - \cos\left(\frac{\pi}{5}\right)\cos\left(\frac{\pi}{5}\frac{f}{f_r}\right)}$$

(6.90)

From Fig. 6.24 we may see that this algorithm has the following properties:

1. It is not insensitive to small detuning error, since the two plots for the Fourier transform magnitudes do not touch each other tangentially at the reference frequency, as desired.
2. Signals with frequencies $f_r, 4f_r, 6f_r, 9f_r$, etc. can be detected, since the amplitudes of the Fourier transforms are the same (even if of different sign) at these frequencies.
3. Phase errors may be introduced by the presence in the signal of fourth, sixth, and ninth harmonics. It is insensitive to second, third, fifth, seventh, eighth, and tenth harmonics.

6.5.4 Symmetrical Seven (6 + 1) Step Algorithm

This algorithm was first described by Larkin (1992). The irradiance measurements for the seven sampling points, as illustrated in Fig. 6.25, are

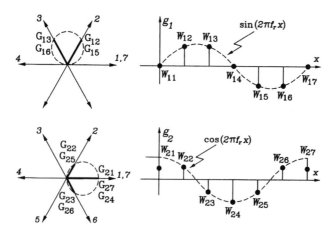

Figure 6.25 Symmetrical seven $(6 + 1)$ step algorithm.

$$s_1 = a + b\cos(\phi)$$
$$s_2 = a + b\cos(\phi + 60°)$$
$$s_3 = a + b\cos(\phi + 120°)$$
$$s_4 = a + b\cos(\phi + 180°)$$
$$s_5 = a + b\cos(\phi + 240°)$$
$$s_6 = a + b\cos(\phi + 300°)$$
$$s_7 = a + b\cos(\phi + 360°)$$

(6.91)

From these equations, the desired solution for the phase is

$$\tan\phi = -\sqrt{3}\,\frac{s_2 + s_3 - s_5 - s_6}{s_1 + s_2 - s_3 - 2s_4 - s_5 + s_6 + s_7}$$

(6.92)

The sampling weights have the values $W_{11} = 0$, $W_{12} = \sqrt{3/2}$, $W_{13} = \sqrt{3/2}$, $W_{14} = 0$, $W_{15} = -\sqrt{3/2}$, $W_{16} = -\sqrt{3/2}$, $W_{17} = 0$, $W_{21} = 1/2$, $W_{22} = 1/2$, $W_{23} = -1/2$, $W_{24} = -1$, $W_{25} = -1/2$, $W_{26} = 1/2$, and $W_{27} = 1/2$. Thus, the reference sampling functions are

$$g_1(x) = \sqrt{3}\left(\delta\left(x - \frac{X_r}{6}\right) + \delta\left(x - \frac{2X_r}{6}\right) - \delta\left(x - \frac{4X_r}{6}\right) - \delta\left(x - \frac{5X_r}{6}\right)\right)$$

(6.93)

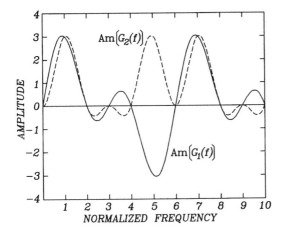

Figure 6.26 Amplitudes of the Fourier transforms for sampling reference functions in the symmetrical seven $(6 + 1)$ step algorithm.

and

$$g_2(x) = \delta(x) + \delta\left(x - \frac{X_r}{6}\right) - \delta\left(x - \frac{2X_r}{6}\right) - 2\delta\left(x - \frac{3X_r}{6}\right)$$

$$- \delta\left(x - \frac{4X_r}{6}\right) + \delta\left(x - \frac{5X_r}{6}\right) + \delta\left(x - \frac{6X_r}{6}\right) \qquad (6.94)$$

The Fourier transforms of the sampling functions, shown in Fig. 6.26, are

$$G_1(f) = \sqrt{3}\left(\sin\left(\frac{2\pi}{3}\frac{f}{f_r}\right) + \sin\left(\frac{\pi}{3}\frac{f}{f_r}\right)\right)\exp\left(-i\left(\pi\frac{f}{f_r} - \frac{\pi}{2}\right)\right) \quad (6.95)$$

and

$$G_2(f) = \left(1 - \cos\left(\pi\frac{f}{f_r}\right) - \cos\left(\frac{2\pi}{3}\frac{f}{f_r}\right) + \cos\left(\frac{\pi}{3}\frac{f}{f_r}\right)\right)\exp\left(-i\pi\frac{f}{f_r}\right)$$

$$(6.96)$$

These functions are orthogonal at all frequencies, as expected. The amplitudes of these two functions become equal at values of the frequency signal equal to f_r, $7f_r$, etc.

Using Eq. 5.75, the value of $r(f)$ is given by

$$r(f) = \frac{\left(\sin\left(\frac{2\pi}{3}\frac{f}{f_r}\right) + \sin\left(\frac{\pi}{3}\frac{f}{f_r}\right)\right)\tan\left(\phi + \pi\frac{f}{f_r}\right)}{\cos\left(\pi\frac{f}{f_r}\right) + \cos\left(\frac{2\pi}{3}\frac{f}{f_r}\right) - \cos\left(\frac{\pi}{3}\frac{f}{f_r}\right) - 1} \qquad (6.97)$$

From Fig. 6.26 we may see that this algorithm has the following properties:

1. It is not insensitive to small detuning error, since the two plots for the Fourier transform amplitudes do not touch each other tangentially at the reference frequency, as desired.
2. Signals with frequencies $f_r, 5f_r, 7f_r$, etc. can be detected, since the amplitudes of the Fourier transforms are the same (even if of different sign) at these frequencies.
3. Phase errors may be introduced by the presence in the signal of fifth and seventh harmonics. It is insensitive to second, third, fourth, sixth, eighth, and ninth harmonics.

6.6 COMBINED ALGORITHMS IN QUADRATURE

We saw at the beginning of this chapter that if the reference function $g_1(x)$ is symmetrical and $g_2(x)$ is antisymmetrical, or vice versa, the two functions are orthogonal at all frequencies. Then, as shown in Chap. 5, in this case the phase error due to detuning oscillates sinusoidally with the value of the phase $(\phi + \psi(f_r))$, as expressed by Eq. 5.110. Thus, if we use two different sampling algorithms of this kind, but with two different values of this phase $(\phi + \psi(f_r))$, the phase errors upon detuning will have the same magnitudes but opposite sign. If the two phase results are averaged, as follows, the phase error due to detuning will cancel out.

$$\phi' = \frac{\tan^{-1}\phi_a + \tan^{-1}\phi_b}{2} \qquad (6.98)$$

Another possibility is to superimpose the two algorithms, as proposed by Schwider *et al.* (1983, 1993). Let us assume that the basic sampling reference functions are $g_1(x)$ and $g_2(x)$. The only requirement is that the phase separation between the sampling points is a submultiple of $\pi/2$. Thus, the shifted algorithm will have the same sampling points, with only a few points being added to the final algorithm. For the initial algorithm the phase equation is

$$\tan \phi_a = -\frac{\displaystyle\sum_{n=1}^{N} g_1(x_n)s(x_n)}{\displaystyle\sum_{n=1}^{N} g_2(x_n)s(x_n)} \tag{6.99}$$

and for the shifted algorithm, from Eqs. 5.121 and 5.122, it is

$$\tan \phi_b = -\frac{\displaystyle\sum_{n=M}^{N+M} g_2\left(x_n - \frac{X_R}{4}\right)s(x_n)}{\displaystyle\sum_{n=M}^{N+M} -g_1\left(x_n - \frac{X_r}{4}\right)s(x_n)} \tag{6.100}$$

Then, the phase equation for the combined algorithm is

$$\tan \phi' = -\frac{\displaystyle\sum_{n=1}^{M} g_1'(x_n)s(x_n)}{\displaystyle\sum_{n=1}^{M} g_2'(x_n)s(x_n)} \tag{6.101}$$

where $x_n = f_r/4$. The reference sampling functions for this combined algorithm are

$$g_1'(x) = g_1(x) + g_2\left(x - \frac{X_r}{4}\right) \tag{6.102}$$

and

$$g_2'(x) = g_2(x) - g_1\left(x - \frac{X_r}{4}\right) \tag{6.103}$$

Then, the Fourier transforms of these functions are

$$G_1'(f) = G_1(f) + G_2(f)\exp\left(-i\frac{\pi}{2}\frac{f}{f_r}\right) \tag{6.104}$$

and

$$G_2'(f) = G_2(f) - G_1(f) \exp\left(-i\frac{\pi}{2}\frac{f}{f_r}\right) \tag{6.105}$$

but this last expression may be transformed into

$$G_2'(f) = \left(G_1(f) + G_2(f) \exp\left(i\left(\frac{\pi}{2}\frac{f}{f_r} - \pi\right)\right)\right) \exp\left(-i\left(\frac{\pi}{2}\frac{f}{f_r} - \pi\right)\right) \tag{6.106}$$

Then, writing the Fourier transforms in terms of their magnitudes and phases, we can find

$$G_1'(f) = \left(|G_1(f)| + |G_2(f)| \exp\left(-i\left(\frac{\pi}{2}\frac{f}{f_r} - \gamma_2 + \gamma_1\right)\right)\right) \exp(i\gamma_1) \tag{6.107}$$

and

$$G_2'(f) = \left(|G_1(f)| + |G_2(f)| \exp\left(i\left(\frac{\pi}{2}\frac{f}{f_r} + \gamma_2 - \gamma_1 - \pi\right)\right)\right)$$
$$\times \exp\left(i\left(\gamma_1 - \frac{\pi}{2}\frac{f}{f_r} + \pi\right)\right) \tag{6.108}$$

where γ_1 and γ_2 are the phases of the complex functions $G_1(f)$ and $G_2(f)$, respectively.

This is a general expression for the combined algorithm, formed by the base algorithm and its 90° shifted version. Here, we have two possible particular cases. The first case is when in the base algorithm the magnitudes of the Fourier transforms $G_1(f)$ and $G_2(f)$ are equal at all frequencies but orthogonal only at the reference frequency. In this case we may show that

$$G_1'(f) = 2|G_1(f)| \cos\left(\frac{\pi}{4}\frac{f}{f_r} - \frac{\gamma_2 - \gamma_1}{2}\right) \exp\left(i\left(-\frac{\pi}{4}\frac{f}{f_r} + \frac{\gamma_2 + \gamma_1}{2}\right)\right) \tag{6.109}$$

and

$$G_2'(f) = 2|G_2(f)| \sin\left(\frac{\pi}{4}\frac{f}{f_r} + \frac{\gamma_2 - \gamma_1}{2}\right)$$
$$\times \exp\left(i\left(-\frac{\pi}{4}\frac{f}{f_r} + \frac{\gamma_2 + \gamma_1}{2} + \frac{\pi}{2}\right)\right) \tag{6.110}$$

We may see that these Fourier transforms are orthogonal at all frequencies, but their magnitudes are equal only at the reference frequency.

A second particular case is when the orthogonality condition in the original algorithm is satisfied at all frequencies ($\gamma_2 = \gamma_1 + \pi/2$), but the magnitudes of $G_1(f)$ and $G_2(f)$ are equal only at the reference frequency. Then we have

$$G_1'(f) = \left(|G_1(f)| + |G_2(f)| \exp\left(-i \left(\frac{\pi}{2} \frac{f}{f_r} - \frac{\pi}{2} \right) \right) \right) \exp(i\gamma_1) \quad (6.111)$$

and

$$G_2'(f) = \left(|G_1(f)| + |G_2(f)| \exp\left(i \left(\frac{\pi}{2} \frac{f}{f_r} - \frac{\pi}{2} \right) \right) \right)$$

$$\times \exp\left(i \left(\gamma_1 - \frac{\pi}{2} \frac{f}{f_r} + \pi \right) \right) \quad (6.112)$$

We may see that the two reference sampling functions of the combined algorithm have equal magnitudes at all frequencies, but they are orthogonal only at the signal frequency. The square magnitude is equal to

$$|G_2'(f)|^2 = |G_1(f)|^2 + |G_2(f)|^2 + 2|G_1(f)||G_2(f)|\cos\left(\frac{\pi}{2} \frac{f}{f_r} \right) \quad (6.113)$$

In both cases, as expected, the combined algorithm is insensitive to a small detuning. The formal mathematical proof is left to the reader as an exercise.

Schmit and Creath (1995) extended this averaging concept to multiple steps. By combining two detuning uncompensated algorithms, they produced an algorithm that is insensitive to small detuning, that is, detuning in a relatively small frequency range. By repeating the same process in sequence, combining an already compensated algorithm and its 90° shifted version, a better compensated algorithm is obtained. These algorithms, called class B algorithms, are detuning-insensitive over a wider frequency range.

Instead of the multiple sequential application of an algorithm and its shifted version, in a process called a *multiple sequential technique*, Schmit and Creath (1996) proposed a method in which several shifted algorithms are combined at the same time, in a process they call a *multiple averaging technique*. Then Eqs. 6.102 and 6.103 become

$$g_1'(x) = g_1(x) + g_2\left(x - \frac{X_r}{4} \right) - g_1\left(x - \frac{X_r}{2} \right) + \cdots \quad (6.114)$$

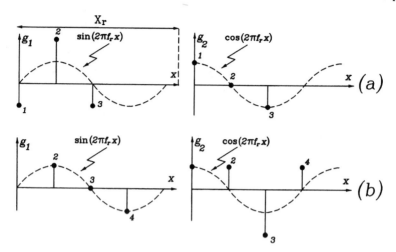

Figure 6.27 Sampling with two combined algorithms in quadrature.

and

$$g_2'(x) = g_2(x) - g_1\left(x - \frac{X_r}{4}\right) - g_2\left(x - \frac{X_r}{2}\right) - \cdots \qquad (6.115)$$

6.6.1 Schwider Algorithm

Schwider *et al.* (1983, 1993) described an algorithm with four sampling points, separated by 90°, which may be considered the sum of two three-point algorithms, separated by 90°. The first algorithm, shown in Fig. 6.27a is the three steps in inverted T algorithm described before, whose phase equation is

$$\tan \phi_a = -\frac{-s_1 + 2s_2 - s_3}{s_1 - s_3} \qquad (6.116)$$

The second algorithm is identical, but shifted by $\epsilon = \pi/2$, as described in Sec. 5.7.2 and illustrated in Fig. 6.27b. Then the reference functions for the second algorithm are as described by Eqs. 5.164 and 5.165, as follows:

$$\tan \phi_b = -\frac{s_2 - s_4}{s_2 - 2s_3 + s_4} \qquad (6.117)$$

Let us now superimpose the two algorithms, to obtain the combined reference functions shown in Fig. 6.28, as follows:

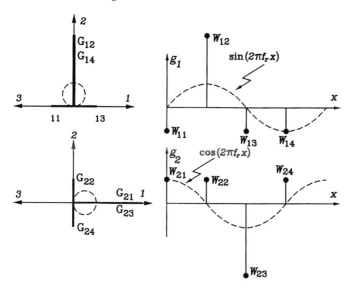

Figure 6.28 Sampling functions for two combined algorithms in quadrature.

$$g_1(x) = -\delta(x) + 3\delta\left(x - \frac{X_r}{4}\right) - \delta\left(x - \frac{2X_r}{4}\right) - \delta\left(x - \frac{3X_r}{4}\right) \quad (6.118)$$

and

$$g_2(x) = \delta(x) + \delta\left(x - \frac{X_r}{4}\right) - 3\delta\left(x - \frac{2X_r}{4}\right) + \delta\left(x - \frac{3X_r}{4}\right) \quad (6.119)$$

The phase is now given by

$$\tan\phi = -\frac{-s_1 + 3s_2 - s_3 - s_4}{s_1 + s_2 - 3s_3 + s_4} \quad (6.120)$$

and the Fourier transforms of the sampling functions become

$$G_1(f) = 4\sin\left(\frac{\pi}{4}\frac{f}{f_r}\right)\left(\sin\left(\frac{\pi}{2}\frac{f}{f_r}\right) + i\right)\exp\left(-i\frac{3\pi}{2}\frac{f}{f_r}\right) \quad (6.121)$$

and

$$G_2(f) = 4\sin\left(\frac{\pi}{4}\frac{f}{f_r}\right)\left(\sin\left(\frac{\pi}{2}\frac{f}{f_r}\right) - i\right)\exp\left(-i\left(\frac{3\pi}{2}\frac{f}{f_r} - \pi\right)\right) \quad (6.122)$$

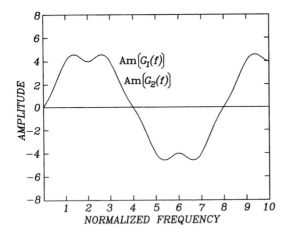

Figure 6.29 Fourier transform amplitudes of sampling functions for two combined algorithms in quadrature.

We may see that the amplitudes of these functions are equal at all frequencies, since the orthogonality condition in the original three-point algorithm was preserved at all frequencies, as shown in Fig. 6.29.

These Fourier transforms are orthogonal only at the reference frequency f_r and all odd harmonics, as shown in Fig. 6.30, where the difference between the phases for the two reference functions is plotted.

We may also notice in this figure that at the signal frequency f_r and all its odd harmonics the slope of this phase difference is zero. Thus, from Eq. 5.112

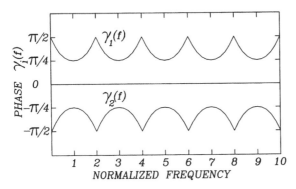

Figure 6.30 Phase difference between the phases for the two Fourier transforms of the reference functions.

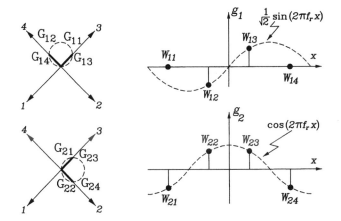

Figure 6.31 Sampling reference functions for shifted Schwider algorithm.

we see that this algorithm has a low detuning sensitivity. There is no sensitivity to the fourth and eighth harmonics.

Another equivalent algorithm with low sensitivity to detuning may be obtained from this one by shifting the sampling points $\pi/4$ to the left, as shown in Sec. 5.7.2. Then applying Eqs. 5.122 and 5.123 we obtain

$$\tan \phi = -2 \frac{-s_2 + s_3}{-s_1 + s_2 + s_3 - s_4} \tag{6.123}$$

There is a singularity and indetermination when $\phi = 0°$ ($s_1 = -s_4$ and $s_2 = -s_3$). The sampling weights have the values $W_{11} = 0, W_{12} = -2, W_{13} = 2, W_{14} = 0, W_{21} = -1, W_{22} = 1, W_{23} = 1, W_{24} = -1$. The sampling reference functions for this algorithm, shown in Fig. 6.31, are

$$g_1(x) = 2[\delta(x + X_r) - \delta(x - X_r)] \tag{6.124}$$

and

$$g_2(x) = -\delta(x + 3X_r) + \delta(x + X_r) + \delta(x - X_r) - \delta(x - 3X_r) \tag{6.125}$$

These Fourier transforms, shown in Fig. 6.32, are thus given by

$$G_1(f) = 4 \sin\left(\frac{\pi}{4} \frac{f}{f_r}\right) \exp\left(i\frac{\pi}{2}\right) \tag{6.126}$$

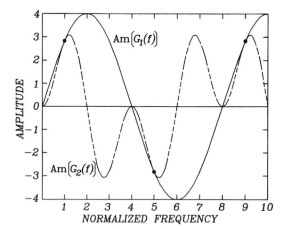

Figure 6.32 Amplitudes of the Fourier transforms of sampling reference functions for shifted Schwider algorithm.

and

$$G_2(f) = 8\cos\left(\frac{\pi}{4}\frac{f}{f_r}\right)\sin^2\left(\frac{\pi}{4}\frac{f}{f_r}\right)\exp(i\pi) \qquad (6.127)$$

As we expected, these two functions are orthogonal at all frequencies, since the original algorithm had the same amplitudes of the Fourier transforms at all frequencies.

Since the two Fourier transform plots touch each other tangentially at the reference frequency, the algorithm has detuning insensitivity. Like the original algorithm, there is no sensitivity to the fourth and eighth harmonics.

The value of $r(f)$ using Eq. 5.75, is given by

$$r(f) = \frac{2\tan\phi}{\sin\left(\dfrac{\pi}{2}\dfrac{f}{f_r}\right)} \qquad (6.128)$$

With this procedure more complex algorithms can be generated by linearly combining several inverted T algorithms instead of only two, each one shifted with respect to the preceding algorithm by 90°. It must be noted, however, that the insensitivity to detuning is obtained only when they are added in such a manner that the sum of all odd coefficients of the linear combination is equal to the sum of all even coefficients.

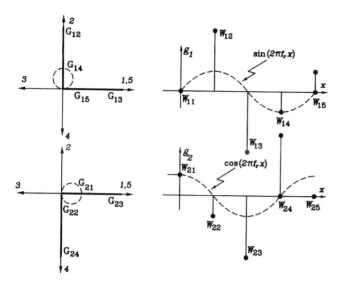

Figure 6.33 Sampling reference functions for Schmit-Creath algorithm.

6.6.2 Schmit-Creath Algorithm

The following class B algorithm with five sampling points has been described by Schmit and Creath (1995). The base algorithm is the Schwider algorithm in Eq. 6.123:

$$\tan \phi_a = -\frac{2s_2 - 2s_3}{s_1 - s_2 - s_3 + s_4} \tag{6.129}$$

and the 90° shifted algorithm is

$$\tan \phi_b = -\frac{s_2 - s_3 - s_4 + s_5}{-2s_3 + 2s_4} \tag{6.130}$$

Hence, the combined algorithm is

$$\tan \phi = -\frac{3s_2 - 3s_3 - s_4 + s_5}{s_1 - s_2 - 3s_3 + 3s_4} \tag{6.131}$$

with the sampling reference functions in Fig. 6.33.

The Fourier transforms of these sampling reference functions, illustrated in Fig. 6.34, are

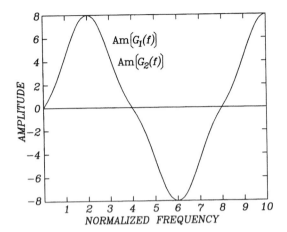

Figure 6.34 Fourier transforms of sampling reference functions for Schmit-Creath algorithm.

$$G_1(f) = 4\sin\left(\frac{\pi}{4}\frac{f}{f_r}\right)\left(\cos\left(\frac{\pi}{4}\frac{f}{f_r}\right) + 2i\sin\left(\frac{\pi}{4}\frac{f}{f_r}\right)\right)$$
$$\times \exp\left(-i\left(\frac{5\pi}{4}\frac{f}{f_r} + \frac{\pi}{2}\right)\right) \tag{6.132}$$

and

$$G_2(f) = 4i\sin\left(\frac{\pi}{4}\frac{f}{f_r}\right)\left(\cos\left(\frac{\pi}{4}\frac{f}{f_r}\right) - 2i\sin\left(\frac{\pi}{4}\frac{f}{f_r}\right)\right)$$
$$\times \exp\left(-i\left(\frac{3\pi}{4}\frac{f}{f_r}\right)\right) \tag{6.133}$$

The amplitudes of these Fourier transforms are equal at all frequencies. The orthogonality condition is valid in a small region about the reference frequency, making the algorithm insensitive to small detuning. As shown in this figure, there is insensitivity only to the fourth and eighth harmonics.

If we shift the sampling points of this algorithm by $\pi/4$ to the left, applying Eqs. 5.122 and 5.123, we obtain

$$\tan\phi = -\frac{-s_1 + 4s_2 - 4s_4 + s_5}{s_1 + 2s_2 - 6s_3 + 2s_4 + s_5} \tag{6.134}$$

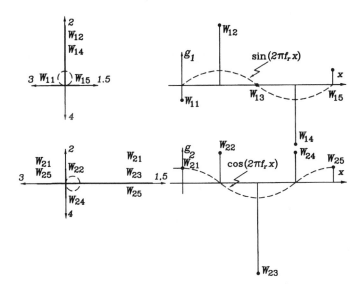

Figure 6.35 Sampling reference functions for shifted Schmit-Creath algorithm.

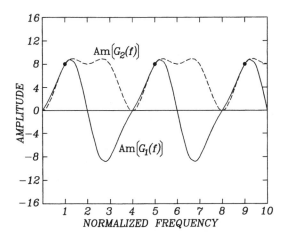

Figure 6.36 Fourier transforms of sampling reference functions for shifted Schmit-Creath algorithm.

with the sampling reference functions as illustrated in Fig. 6.35. The Fourier transforms of these sampling reference functions, illustrated in Fig. 6.36, are

$$G_1(f) = 2\left(4\sin\left(\frac{\pi}{2}\frac{f}{f_r}\right) - \sin\left(\pi\frac{f}{f_r}\right)\right)\exp\left(-i\left(\pi\frac{f}{f_r} + \frac{\pi}{2}\right)\right) \quad (6.135)$$

and

$$G_2(f) = \left(6 - 4\cos\left(\frac{\pi}{2}\frac{f}{f_r}\right) - 2\cos\left(\pi\frac{f}{f_r}\right)\right)\exp\left(-i\left(\pi\frac{f}{f_r} + \pi\right)\right) \tag{6.136}$$

These Fourier transforms are orthogonal at all signal frequencies. The slope of these functions is the same at the reference frequency, where we also have the same amplitudes, making the algorithm insensitive to small detuning. As in the original algorithm, there is insensitivity to the fourth and eighth signal harmonics.

6.7 DETUNING-INSENSITIVE ALGORITHMS FOR DISTORTED SIGNALS

When a signal is distorted and, as a consequence, harmonics are present, a detuning-insensitive algorithm must also be detuning-insensitive to the signal harmonics. The reason is that when detuning is present, not only the fundamental frequency but also its harmonics are detuned. This problem, first studied by Hibino et al. (1995) and a little later by Surrel (1996) and Zhao and Surrel (1995), was described earlier in this book in Sec. 5.6.2.

In order to have an algorithm with detuning sensitivity up the mth harmonic we need enough sampling points to determine the signal bias, the amplitudes of all harmonic components i.e., $S_0, S_1, S_2, \ldots, S_m$, their phases $\phi_1, \phi_2, \ldots, \phi_m$, in Eq. 5.57, plus the linear phase error magnitude. This makes a total of $2m + 2$ unknowns. Thus, a minimum of $2m + 2$ sampling points are needed. It should be pointed out here that Hibino et al. (1995) found a minimum of $2m + 3$ points, but this value was later corrected by Surrel (1996).

For an algorithm with detuning insensitivity up to the mth harmonic, as pointed out before,

1. The phase interval between sampling points must be smaller than $2\pi/(m+2)$.
2. When the maximum phase interval is used, the minimum number of sampling points is $2m + 2$. With a smaller phase interval the number of required sampling points would be larger.

Table 6.1 Minimum Number of Sampling Points for
Detuning-Insensitive Algorithms, with Harmonically
Distorted Signals

Minimum number of samples $N = 2m + 2$	Maximum harmonic m with detuning insensitivity	Maximum phase interval $2\pi/(m + 2)$
4	1	120°
6	2	90°
8	3	72°
10	4	60°
12	5	51.14°
14	6	45°

For example, as described in Table 6.1, taken from the publication by Hibino *et al.* (1995) and with the correction by Surrel (1996), for an algorithm to be detuning-insensitive only up to the second harmonic using the maximum phase interval of 90° there must be at least six sampling points. If this phase interval is reduced, more than six points are needed.

As an example, let us consider the six-sample algorithm (Zhao and Surrel 1995; Surrel 1996), which takes the signal measurements at constant phase intervals equal to 90°, as follows:

$$s_1 = a + b\cos(\phi)$$
$$s_2 = a + b\cos(\phi + 90°)$$
$$s_3 = a + b\cos(\phi + 180°)$$
$$s_4 = a + b\cos(\phi + 240°) \tag{6.137}$$
$$s_5 = a + b\cos(\phi + 360°)$$
$$s_6 = a + b\cos(\phi + 450°)$$

From these equations, the desired solution for the phase satisfying the conditions described earlier is

$$\tan\phi = -\frac{s_1 + 3s_2 - 4s_4 - s_5 + s_6}{s_1 - s_2 - 4s_3 + 3s_5 + s_6} \tag{6.138}$$

Thus, the sampling reference functions, shown in Fig. 6.37, are

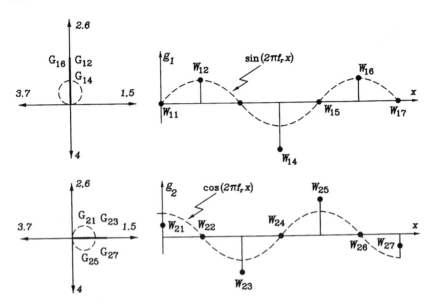

Figure 6.37 Sampling reference functions for seven-sample detuning-insensitive algorithm.

$$g_1(x) = \delta\left(x - \frac{X_r}{4}\right) + 3\delta\left(x - \frac{X_r}{2}\right) - 4\delta(x - X_r) - \delta\left(x - \frac{5X_r}{4}\right)$$

$$+ \delta\left(x - \frac{3X_r}{2}\right) \tag{6.139}$$

and

$$g_2(x) = \delta\left(x - \frac{X_r}{4}\right) - \delta\left(x - \frac{X_r}{2}\right) - 4\delta\left(x - \frac{3X_r}{4}\right) + 3\delta\left(x - \frac{5X_r}{4}\right)$$

$$+ \delta\left(x - \frac{3X_r}{2}\right) \tag{6.140}$$

The Fourier transforms for these sampling reference functions are

$$G_1(f) =$$
$$\left[\cos\left(\frac{5\pi}{4}\frac{f}{f_r}\right) - \cos\left(\frac{3\pi}{4}\frac{f}{f_r}\right) + 2\exp\left(i\frac{3\pi}{4}\frac{f}{f_r}\right) - 2\exp\left(-i\frac{\pi}{4}\frac{f}{f_r}\right)\right]$$
$$\exp\left(-i\frac{7\pi}{4}\frac{f}{f_r}\right) \tag{6.141}$$

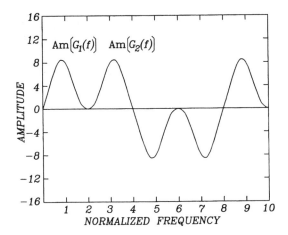

Figure 6.38 Fourier transforms for six-sample detuning-insensitive algorithm.

and

$$G_2(f) =$$

$$\left[\cos\left(\frac{5\pi}{4}\frac{f}{f_r}\right) - \cos\left(\frac{3\pi}{4}\frac{f}{f_r}\right) + 2\exp\left(-i\frac{3\pi}{4}\frac{f}{f_r}\right) - 2\exp\left(i\frac{\pi}{4}\frac{f}{f_r}\right)\right]$$

$$\exp\left(-i\frac{7\pi}{4}\frac{f}{f_r}\right) \quad (6.142)$$

These Fourier transforms have the same amplitudes at all frequencies, but they are orthogonal in the vicinity of the reference frequency and in the vicinity of the second harmonic. This algorithm is shifted $\pi/4$ with respect to the one described in the article by Zhao and Surrel (1995) and and by Surrel (1996), which is orthogonal to all frequencies, but their magnitudes are equal in the vicinity of the reference frequency and its second harmonic. This algorithm is detuning-insensitive up to the second harmonic, but it is not insensitive to the third harmonic.

Another algorithm with small sensitivity to the second harmonic even when detuning is present, using seven sampling points, was described by Hibino *et al.* (1995).

6.8 CONTINUOUS SAMPLING IN A FINITE INTERVAL

When sampling a sinusoidal signal with a finite aperture or a finite sampling interval, this aperture or finite interval acts as a filtering window. This problem

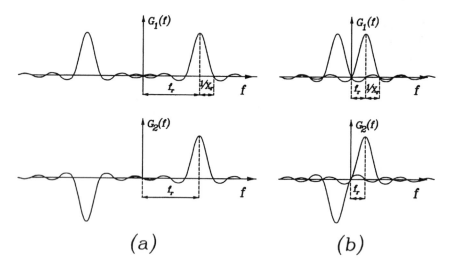

Figure 6.39 Fourier transforms of functions $g_1(x)$ and $g_2(x)$ with continuous sampling in a finite interval.

has been theoretically and experimentally studied by Nakadate (1988a, 1988b), but with a different approach than that presented here. Here, we will use a similar but slightly simpler approach, with the Fourier theory just developed.

The sampling functions using a finite interval of size X may be written as

$$g_1(x) = \sin(2\pi f_r x) \qquad \text{for} \quad -\frac{X}{2} \leq x \leq \frac{X}{2}$$

$$= 0 \qquad\qquad \text{for} \quad |x| > \frac{X}{2} \tag{6.143}$$

and

$$g_2(x) = \cos(2\pi f_r x) \qquad \text{for} \quad -\frac{X}{2} \leq x \leq \frac{X}{2}$$

$$= 0 \qquad\qquad \text{for} \quad |x| > \frac{X}{2} \tag{6.144}$$

Then the Fourier transforms of these functions, shown in Fig. 6.39, are

$$G_1(f) = i[\text{sinc}(\pi(f + f_r)X) - \text{sinc}(\pi(f - f_r)X)] \tag{6.145}$$

and

$$G_2(f) = \text{sinc}(\pi(f + f_r)X) + \text{sinc}(\pi(f - f_r)X) \qquad (6.146)$$

We may see, as shown in Fig. 6.39, that the separation between these two sinc functions is equal to twice the reference frequency f_r. When the reference frequency is large compared with $1/X$, the two sinc functions are quite separated from each other. Then the side lobes of one will not overlap those of the other, as in Fig. 6.39a. On the other hand, if the reference frequency is low compared with $1/X$, the side lobes of one sinc function will overlap the other sinc function, as in Fig. 6.39b, where $X = X_r = 1/f_r$.

Since the functions $G_i(f)$ are the sum of the two sinc functions, we will see that the $G_i(f_r)$ will not change and will remain equal to each other when

$$f_r X = \frac{n}{2} \qquad (6.147)$$

where n is any positive integer. Then there will not be any error in the phase detection. This result means that the sampling interval (or aperture) should be an integral number of half the spatial period of the fringes, as we saw in Sec. 5.1.2. This property was used by Morimoto and Fujisawa (1994).

A peak in the error will occur, however, at intermediate positions, given by

$$f_r X = \frac{n}{2} + \frac{1}{4} \qquad (6.148)$$

If a phase-detecting algorithm uses the sampling interval X_r, the phase ϕ is given by

$$\tan \phi = -\frac{\displaystyle\int_{x=0}^{X_r} s(x) \sin(2\pi f x)\, dx}{\displaystyle\int_{x=0}^{X_r} s(x) \cos(2\pi f x)\, dx} \qquad (6.149)$$

with the sampling reference functions, as illustrated in Fig. 6.40.

The Fourier transforms of the sampling reference functions are

$$G_1(f) = i\left[\text{sinc}\left(\pi\left(\frac{f}{f_r} + 1\right)\right) - \text{sinc}\left(\pi\left(\frac{f}{f_r} - 1\right)\right)\right] \qquad (6.150)$$

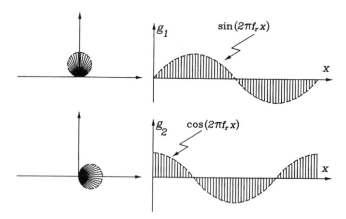

Figure 6.40 Sampling reference functions $g_1(x)$ and $g_2(x)$ for a continuous sampling interval $X_r = 1/f_r$.

and

$$G_2(f) = \text{sinc}\left(\pi\left(\frac{f}{f_r}+1\right)\right) + \text{sinc}\left(\pi\left(\frac{f}{f_r}-1\right)\right) \qquad (6.151)$$

which are illustrated in Fig. 6.41.

As shown in Fig. 6.41, the Fourier transforms are orthogonal at all signal frequencies, but they have the same amplitude only at the reference frequency.

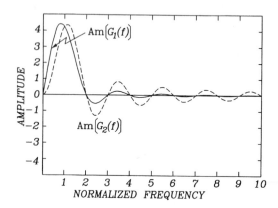

Figure 6.41 Fourier transforms of functions $g_1(x)$ and $g_2(x)$ for a continuous sampling interval $X_r = 1/f_r$.

Thus, this algorithm is sensitive to detuning. It is quite interesting to see that there is no sensitivity to any harmonics, provided there is no detuning. Insensitivity to small detuning may be obtained if the additional sampling points at the ends of the sampling interval, as described in Sec. 6.5, are used.

This is a limiting case for discrete sampling algorithms when the number of sampling steps tends to infinity.

6.9 ASYNCHRONOUS PHASE DETECTION ALGORITHMS

In synchronous detection we have assumed that the frequency of the detected signal and the phase steps taken during the measurements are known. It may be, however, that the phase steps or the frequency of the measured signal are unknown. In that case, before calculating the phase, the signal frequency must be determined. To do so, we need a minimum of four sampling points.

If we examine the expression for $r(f)$ in Eq. 5.61, we see that if we require that the two Fourier transforms $G_1(f)$ and $G_2(f)$ have the same phase ϕ, instead of being orthogonal to each other and we also remove the condition that their magnitudes must be equal, using Eq. 5.75 we have

$$r(f) = \frac{|G_1(f_r)|}{|G_2(f_r)|} = \frac{\displaystyle\int_{-\infty}^{\infty} s(x)g_1(x)\,dx}{\displaystyle\int_{-\infty}^{\infty} s(x)g_2(x)\,dx} \qquad (6.152)$$

This is possible if the two reference functions are both antisymmetrical and different.

Then we may see that the value of $r(f)$ is not a function of the signal phase ϕ as before. Instead, it is a function of the signal frequency. The value of $r(f)$ may be calculated for a given sampling algorithm satisfying this condition, allowing a signal frequency determination. A simple way to obtain the Fourier transforms with the same phase is by requiring the sampling reference functions $g_1(x)$ and $g_2(x)$ to be both antisymmetrical or both symmetrical. Thus, they must have different frequencies, normally equal to f_r and $2f_r$, respectively.

We may see that if the reference functions $g_1(x)$ and $g_2(x)$ are antisymmetrical and the signal is symmetrical, or vice versa, both integrals in this expression become equal to zero. Then, with symmetrical reference functions the value of $r(f)$ becomes undetermined when the signal is symmetrical, that is, when the phase has a value equal to $n\pi$, n being an integer. On the other hand, with antisymmetrical reference functions the value of $r(f)$ becomes undetermined when the signal is antisymmetrical, that is, when the phase has a value equal to $n\pi/2$, n being an odd integer.

6.9.1 Carré Algorithm

The classic asynchronous algorithm was developed by Carré (1966). Four measurements of the signal are taken at equally spaced phase increments. The sampling points are symmetrically placed with respect to the origin, as expressed by

$$s_1 = a + b\cos(\phi - 3\beta)$$
$$s_2 = a + b\cos(\phi - \beta)$$
$$s_3 = a + b\cos(\phi + \beta) \tag{6.153}$$
$$s_4 = a + b\cos(\phi + 3\beta)$$

where the phase increment is 2β. If the reference frequency f_r and the signal frequency f are different, the phase increment would have a different value when referred to the reference function or to the signal phase scales. When measured with respect to the signal phase scale, its value is β, but if measured with respect to the reference function phase scale, its value is α. In synchronous phase detection we have $\alpha = \beta$, but in general we have

$$\beta = \alpha \frac{f}{f_r} \tag{6.154}$$

The value of β is unknown, either because the value of α or the frequency f of the signal is unknown. The most common phase step used in this algorithm is $\alpha = \pi/4$. The value of β may be calculated with the following expression obtained from Eqs. 6.153:

$$\tan^2 \beta = \frac{3(s_2 - s_3) - (s_1 - s_4)}{(s_1 - s_4) + (s_2 - s_3)} \tag{6.155}$$

or, alternatively, defining a value of $r_\beta(f)$ given by

$$r_\beta(f) = -\frac{\sin 2\beta \cos 2\beta \sin \phi}{\cos 2\beta \sin \beta \sin \phi} = -\frac{\tan 2\beta}{\tan \beta}$$
$$= \frac{-s_1 - s_2 + s_3 + s_4}{s_1 - s_2 + s_3 - s_4} \tag{6.156}$$

with the reference functions whose sampling weights have the values $W_{11} = -1$, $W_{12} = -1$, $W_{13} = 1$, $W_{14} = 1$, $W_{21} = 1$, $W_{22} = -1$, $W_{23} = 1$, and $W_{24} = -1$. We see that there is a singularity and indetermination when $\sin \phi = 0$, since then $s_2 = s_3$ and $s_1 = s_4$. Also, another singularity and indetermination occurs when $\beta = \pi/2$. The sampling reference functions, illustrated in Fig. 6.42 for

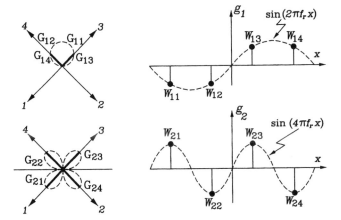

Figure 6.42 Sampling in the Carré algorithm, with $\alpha = \pi/4$, to obtain the signal frequency.

$\alpha = \pi/4$, are

$$g_1(x) = -\delta\left(x + \frac{3X_r}{8}\right) - \delta\left(x + \frac{X_r}{8}\right) + \delta\left(x - \frac{X_r}{8}\right) + \delta\left(x - \frac{3X_r}{8}\right)$$

(6.157)

and

$$g_2(x) = \delta\left(x + \frac{3X_r}{8}\right) - \delta\left(x + \frac{X_r}{8}\right) + \delta\left(x - \frac{X_r}{8}\right) - \delta\left(x - \frac{3X_r}{8}\right)$$

(6.158)

The Fourier transforms of the sampling functions, illustrated in Fig. 6.43 for $\alpha = \pi/4$, are

$$G_1(f) = 4\cos\left(\frac{\pi}{4}\frac{f}{f_r}\right)\sin\left(\frac{\pi}{2}\frac{f}{f_r}\right)\exp\left(-i\frac{\pi}{2}\right)$$

(6.159)

and

$$G_2(f) = 4\sin\left(\frac{\pi}{4}\frac{f}{f_r}\right)\cos\left(\frac{\pi}{2}\frac{f}{f_r}\right)\exp\left(-i\frac{\pi}{2}\right)$$

(6.160)

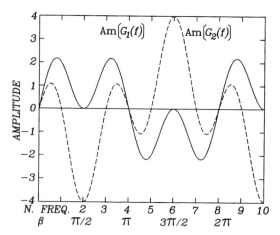

Figure 6.43 Amplitudes of the Fourier transforms of the reference functions for the Carré algorithm for $\alpha = \pi/4$ to find the signal frequency.

We may observe in Fig. 6.43 that these functions are symmetrical about the value of the normalized frequency equal to 2, which corresponds to $\beta = \pi/2$. Hence, the measurement of β can be performed without uncertainty only if it is in the range $0 < \beta < \pi/2$. Hence, the value of the reference frequency f_r should in principle be chosen so that the values of α and β are as close as possible to each other. In other words, the reference frequency should be higher than half the signal frequency but as close as possible to this value. This condition may also be expressed by saying that the four sampling points must be separated by at least one-fourth of the period of the signal. Nevertheless, if we take into account the presence of additive noise in the measurements, it can be shown that the noise influence is minimized when $\beta = 110°$, as pointed out by Carré (1966) and Freischlad and Koliopoulos (1990).

In Fig. 6.43 we may observe the singularity and indetermination that occurs when $\beta = \pi$, since both Fourier transform amplitudes are zero. This algorithm is quite sensitive to signal harmonics.

Once the value of β has been calculated, the signal phase ϕ may be found with another algorithm, with the same sampling points, hence the same measured values, as follows:

$$\tan \phi = \frac{(s_1 - s_4) + (s_2 - s_3)}{(s_2 + s_3) - (s_1 + s_4)} \tan \beta \qquad (6.161)$$

As in the previous algorithm, this expression has an indetermination when $\phi = 0$, since then $s_1 = s_3$ and $s_1 = s_4$. Hence, when ϕ is small, large errors may occur.

Having calculated the value of β with a set of four sampling points, the same value of β may be used to calculate the phase for several signal points with different locations if the signal frequency is the same everywhere. This is the case of temporal phase shifting, where the signal frequency is frequently the same for all points in the interferogram. Alternatively, if the frequency is not constant, as in space phase shifting, when the wavefront is not aberration-free, the value of β has to be calculated for every point where the phase is to be determined.

Let us consider the first case in which the value of β is a constant. Then, we may write Eq. 6.161 as

$$\tan \phi = -\tan \beta \frac{s_1 + s_2 - s_3 - s_4}{s_1 - s_2 - s_3 + s_4} \tag{6.162}$$

with the sampling weights values $W_{11} = \tan \beta$, $W_{12} = \tan \beta$, $W_{13} = -\tan \beta$, $W_{14} = -\tan \beta$, $W_{21} = 1$, $W_{22} = -1$, $W_{23} = -1$, $W_{24} = 1$. The sampling reference functions, illustrated in Fig. 6.44, are

$$g_1(x) = \delta \left(x + \frac{3X_r}{8} \right) + \delta \left(x + \frac{X_r}{8} \right) - \delta \left(x - \frac{X_r}{8} \right) - \delta \left(x - \frac{3X_r}{8} \right) \tag{6.163}$$

and

$$g_2(x) = \delta \left(x + \frac{3X_r}{8} \right) - \delta \left(x + \frac{X_r}{8} \right) - \delta \left(x - \frac{X_r}{8} \right) + \delta \left(x - \frac{3X_r}{8} \right) \tag{6.164}$$

The Fourier transforms of the sampling functions with $\alpha = \pi/4$ are thus given by

$$G_1(f) = 4 \sin \left(\frac{\pi}{2} \frac{f}{f_r} \right) \cos \left(\frac{\pi}{4} \frac{f}{f_r} \right) \tan \beta \exp \left(i \frac{\pi}{2} \right) \tag{6.165}$$

and

$$G_2(f) = 4 \sin \left(\frac{\pi}{2} \frac{f}{f_r} \right) \sin \left(\frac{\pi}{4} \frac{f}{f_r} \right) \tag{6.166}$$

which are illustrated in Fig. 6.45.

We may see that this algorithm is insensitive to all even harmonics only if $\beta/\alpha = 1$, which is not frequent, and it is always quite sensitive to all odd

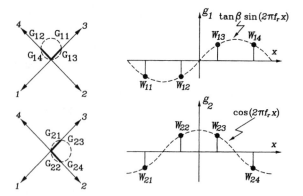

Figure 6.44 Sampling in the reference function for the Carré algorithm with $\alpha = \pi/4$ and a constant value of β to find the phase.

harmonics. It must be pointed out here that this is for the second part, after β has been calculated, but errors due to the presence of harmonics may also appear in the calculation of β, as we pointed out before. We also can see that it is quite sensitive to detuning, but that is not a serious problem, since the frequency has been previously calculated in the first step. Notice that this algorithm is identical to the four points in X algorithm, described before, when $\beta/\alpha = 1$.

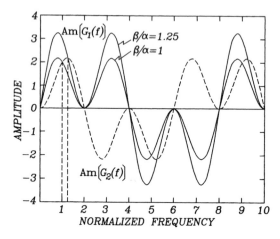

Figure 6.45 Amplitudes of the Fourier transforms of the reference functions in the Carré algorithm using $\alpha = \pi/4$ and two different constant values of β.

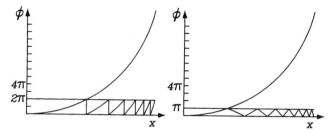

Figure 6.46 Wrapping of the phase in the Carré algorithm (right side), compared with other phase-detecting algorithms (left side).

There is a problem here, however, if the value of β is not a constant for all locations where it is measured. Then the frequency is not a constant, and it is better to recalculate β every time the phase is to be obtained. Then we may combine Eqs. 6.156 and 6.61, with the result

$$\tan \phi = \frac{[3(s_2 - s_3)^2 - (s_1 - s_4)^2 + 2(s_1 - s_4)(s_2 - s_3)]^{1/2}}{(s_2 + s_3) - (s_1 + s_4)} \qquad (6.167)$$

thus removing the indetermination.

We may see that in this case substituting the value of β in Eq. 6.156 into expression 6.165 for $G_1(f)$, the two Fourier transforms $G_1(f)$ and $G_2(f)$ become equal at all frequencies. This is to be expected, because now there is no detuning error, since the algorithm is self-calibrating.

One problem with this algorithm is that the numerator in Eq. 6.167 is the square root of a number. Thus, the sign of $\sin \phi$ is lost. As a consequence, the phase is wrapped modulo π instead of modulo 2π as in most phase-detecting algorithms. Figure 6.46 shows the phase wrapping in the Carré algorithm compared with the phase wrapping in other algorithms.

The Carré algorithm was adapted by Rastogi (1993) to the study of four-wave holographic interferometry.

6.9.2 Schwider Asynchronous Algorithm

The Schwider asynchronous algorithm is a four-sampling-point asynchronous algorithm (Schwider *et al.* 1983; Cheng and Wyant 1985). The four points are at phases -2β, $-\beta$, β, and 2β, with β as defined in Eq. 6.154, and a value of $\alpha = \pi/4$. The cosine of the phase increment becomes

$$r_\beta(f) = \cos \beta = \frac{-s_1 + s_4}{-2s_2 + 2s_3} \qquad (6.168)$$

and the sampling reference functions, illustrated in Fig. 6.47, are

$$g_1(x) = -\delta\left(x - \frac{X_r}{2}\right) + \delta\left(x + \frac{X_r}{2}\right) \tag{6.169}$$

and

$$g_2(x) = -2\delta\left(x - \frac{X_r}{4}\right) + 2\delta\left(x + \frac{X_r}{4}\right) \tag{6.170}$$

The Fourier transforms of these sampling reference functions, illustrated in Fig. 6.48, are

$$G_1(f) = 2\sin\left(\frac{\pi}{2}\frac{f}{f_r}\right)\exp\left(i\frac{\pi}{2}\right) \tag{6.171}$$

and

$$G_2(f) = 2\sin\left(\frac{\pi}{4}\frac{f}{f_r}\right)\exp\left(i\frac{\pi}{2}\right) \tag{6.172}$$

In this algorithm the reference frequency may be as low as one-eighth of the signal frequency. However, there are singularities and indeterminations at f/f_r equal to 4 and 8. Ideally, the reference frequency should be as close as possible to the signal frequency. There is a large sensitivity to the presence of signal harmonics.

6.9.3 Two Algorithms in Quadrature

We have seen in Sec. 6.6 that two algorithms in quadrature produce phases with opposite errors in the phase. Hence, by averaging their phases as in Eq. 6.85 the error-free phase may be calculated. The error in the phase may be obtained if instead of averaging the two phases, their difference is taken as

$$\delta\phi = \frac{\tan^{-1}\phi_a - \tan^{-1}\phi_b}{2} \tag{6.173}$$

Now, from Eq. 5.111, if the base (non-shifted) algorithm is orthogonal at all frequencies, we have

$$\frac{|G_1(f)|}{|G_2(f)|} = \rho = 1 + \frac{2\sin\phi}{\sin 2\phi} \tag{6.174}$$

where the phase ϕ is calculated with Eq. 6.85.

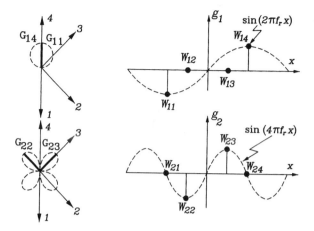

Figure 6.47 Sampling reference functions for the Schwider asynchronous algorithm.

Once the value of $\rho(f)$ has been obtained, the normalized frequency f/f_r may be calculated. For example, if the inverted T algorithm has been used, we have

$$\frac{f}{f_r} = \frac{4}{\pi} \tan^{-1} \rho \tag{6.175}$$

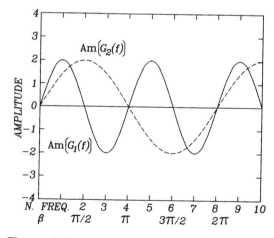

Figure 6.48 Amplitudes of the Fourier transforms of the sampling reference functions for the Schwider asynchronous algorithm.

6.9.4 A Zero Bias and Three Sampling Points Algorithm

We have seen that four measurements are necessary to determine the four parameters of a sinusoidal signal, i.e., a, b, ϕ_0, and ω. Ransom and Kokal (1986) and later, independently, Servín and Cuevas (1995) described a method in which the DC or bias term is first eliminated from the signal by means of a convolution with a high pass filter, as described in Sec. 2.6.1. Then the only problem is that the whole signal interval must be sampled and processed before the phase measuring points are sampled.

Thus, after eliminating the bias (coefficient a), the signal may be expressed by

$$s(x) = b\cos(\omega x + \phi) \tag{6.176}$$

If three sampling points at x positions $x_0, 0, -x_0$ are used, we have

$$s_1 = b\cos(-\omega x_0 + \phi) \tag{6.177}$$

$$s_2 = b\cos(\phi) \tag{6.178}$$

and

$$s_3 = b\cos(\omega x_0 + \phi) \tag{6.179}$$

But these three expressions may also be written as

$$s_1 = b\cos(\omega x_0)\cos\phi + b\sin(\omega x_0)\sin\phi \tag{6.180}$$

$$s_2 = b\cos\phi \tag{6.181}$$

and

$$s_3 = b\cos(\omega x_0)\cos\phi - b\sin(\omega x_0)\sin\phi \tag{6.182}$$

Then it is easy to see that

$$\frac{s_1 + s_3}{2s_2} = \cos(\omega x_0) \tag{6.183}$$

and

$$\frac{s_1 - s_3}{2s_2} = \sin(\omega x_0)\tan\phi \tag{6.184}$$

Now, from Eq. 6.183,

$$\sin(\omega x_0) = \left[1 - \left(\frac{s_1 + s_3}{2s_2}\right)^2\right]^{1/2} \tag{6.185}$$

Thus, it is easy to show from Eqs. 6.184 and 6.185 that

$$\tan\phi = \frac{s_1 - s_3}{(\text{sinc } s_2)[4s_2^2 - (s_1 + s_3)^2]^{1/2}} \tag{6.186}$$

We can see that this phase expression is insensitive to the signal frequency and therefore the result is not affected by detuning. The unknown signal frequency may then be found with

$$\omega = \frac{1}{x_0}\cos^{-1}\left(\frac{s_1 + s_3}{2s_2}\right) \tag{6.187}$$

6.9.5 Correlation with Two Sinusoidal Signals in Quadrature

In Chap. 5 we studied the synchronous detection method by multiplication of the signal by two orthogonal sinusoidal reference functions with the same frequency as the signal. Let us now assume that the two reference orthogonal functions have a different frequency ω_r than the signal. The parameters S and C are not constants. Instead, we now have

$$S(x) = s(x)\sin\omega_r x = a\sin\omega_r x + b\cos(\phi + \omega x)\sin\omega_r x$$

$$= a\sin\omega_r x + \frac{b}{2}\sin(\phi + (\omega + \omega_r)x) - \frac{b}{2}\sin(\phi + (\omega - \omega_r)x) \tag{6.188}$$

and

$$C(x) = s(x)\cos\omega_r x = a\cos\omega_r x + b\cos(\phi + \omega x)\cos\omega_r x$$

$$= a\cos\omega_r x + \frac{b}{2}\cos(\phi + (\omega + \omega_r)x)$$

$$+ \frac{b}{2}\cos(\phi + (\omega - \omega_r)x) \tag{6.189}$$

These two functions contain three spatial frequencies: the reference frequency, the sum of the reference and signal frequencies, and their difference. If we apply a low pass filter, so that only the term with the frequency difference remains, we obtain the filtered versions of $S(x)$ and $C(x)$ as

$$\bar{S}(x) = S(x) \sin \omega_{rx} \tag{6.190}$$

and

$$\bar{C}(x) = C(x) \cos \omega_{rx} \tag{6.191}$$

and thus we may obtain

$$\tan(\phi + (\omega_S - \omega_r)x) = -\frac{\bar{S}(x)}{\bar{C}(x)} \tag{6.192}$$

which is possible only if the reference frequency f_r is higher than half the signal frequency,

$$\omega_r > \frac{\omega}{2} \tag{6.193}$$

However, ideally the two frequencies should be equal.

The low pass filtering process is performed by means of a convolution with a filtering function $h(x)$. Then, the values of $S(x)$ and $C(x)$ may be expressed by

$$\bar{S}(x) = \int_{-\infty}^{\infty} s(\alpha) \sin(\omega_r\alpha)h(x - \alpha) \, d\alpha \tag{6.194}$$

and

$$\bar{C}(x) = \int_{-\infty}^{\infty} s(\alpha) \cos(\omega_r\alpha)h(x - \alpha) \, d\alpha \tag{6.195}$$

The filtering function must be selected so that the term with the lowest frequency (the difference term) remains. Hence, we may also write

$$\phi + (\omega - \omega_r)x = -\tan^{-1}\left(\frac{S(x)}{C(x)}\right) \tag{6.196}$$

6.10 OPTIMIZATION OF PHASE DETECTION ALGORITHMS

Given a number of sampling points and their phase positions, there are an infinite number of sampling weight sets that can define the algorithm. In this chapter we have developed some methods to find the algorithms with the desired properties, but mainly methods to evaluate them. Another approach is to use optimization techniques to find the optimum sampling weights for some desired algorithm properties (Servín *et al.* 1997).

To simplify the analysis we assume that the sampling reference functions $g_1(x)$ and $g_2(x)$ are antisymmetrical and symmetrical, respectively. There is no loss in generality, since, as described in Sec. 5.7, any algorithm can be shifted, without losing its properties, until these symmetry conditions are satisfied. Then it is possible to show that the Fourier transforms of the reference functions are given by

$$G_1(f) = -2i \sum_{n=1}^{N/2} W_{1n} \sin \left(\alpha_n \frac{f}{f_r} \right) \tag{6.197}$$

and

$$G_2(f) = 2 \sum_{n=1}^{N/2} W_{2n} \sin \left(\alpha_n \frac{f}{f_r} \right) + \sigma_1 W_{2\left(\frac{N+1}{2}\right)} \tag{6.198}$$

with

$$\alpha_n = \frac{2\pi}{N} \left(n - \sigma_2 \frac{1}{2} \right) \tag{6.199}$$

where

$$
\begin{aligned}
\sigma_1 &= 0 \quad \text{and} \quad \sigma_2 = 1 \quad \text{for } N \text{ even} \\
\sigma_1 &= 1 \quad \text{and} \quad \sigma_2 = 0 \quad \text{for } N \text{ odd}
\end{aligned}
\tag{6.200}
$$

These symmetries ensure that the two sampling functions are orthogonal at all signal frequencies. The sampling weights values can now be found by minimizing the merit function $U(W_1, W_2, \ldots, W_N)$ defined by

$$U(W_1, W_2, \ldots, W_N) = \rho_0 G_2(0)^2$$

$$+ \rho_1 \int_{f=f_r-\Delta_1}^{f_r+\Delta_1} [G_1(f) - G_2(f)]^2 \, df$$

$$+ \rho_2 \int_{f=2f_r-\Delta_2}^{2f_r+\Delta_2} [G_1(f)^2 + G_2(f)^2] \, df + \cdots$$

$$(6.201)$$

The first term minimizes the bias (DC) component of the second sampling function. The bias of the second reference function is zero due to its antisymmetry. The second term minimizes the difference between the magnitudes of the sampling reference functions at the reference frequency. The third term minimizes the sensitivity of the algorithm to the second signal harmonic. More terms may be added if insensitivity to other signal harmonics is desired. The constants ρ_m are the weights assigned to each term. The constants Δ_m are the widths of the frequency intervals on which the optimizations for each signal harmonic are desired.

The optimum values of the sampling weights W_n may now be obtained by minimizing the merit function $U(W_1, W, \ldots, W_N)$ for the parameters W_n by solving the linear system of equations

$$\frac{\partial U(W_1, W_2, \ldots, W_N)}{\partial W_n} = 0 \qquad (6.202)$$

When solving the linear system, analytical or numerical integration may be used in the expression for the merit function. For practical convenience, numerical integration was preferred.

To optimize the algorithm, a minimum of four sampling points are required. Servín et al. (1997) obtained optimized algorithms with four, five, and seven sampling points. As an example of the method let us consider seven equally spaced sampling points with a phase interval $\pi/2$ and optimize for detuning, using the following weights:

$$\rho_0 = \rho_1 = 1$$

$$\rho_3 = 0.01$$

$$\rho_2 = \rho_4 = \rho_5 = \rho_6 = \cdots = 0 \qquad (6.203)$$

$$\Delta_1 = 0.8$$

$$\Delta_2 = 0.1$$

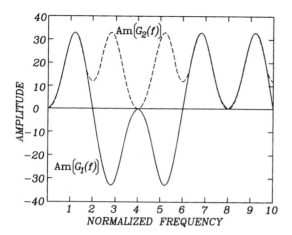

Figure 6.49 Fourier transforms of sampling reference functions for optimized seven-sample algorithm.

With these parameters we define an algorithm with attenuation in the third harmonic. The solution of the linear system with five phase steps α_i at $-3\pi/2$, $-\pi$, $-\pi/2$, 0, $\pi/2$, π, and $3\pi/2$ produce the phase equation

$$\tan \phi = -\frac{1s_1 + 4.3s_2 - 14s_3 + 14s_5 - 4.3s_6 - 1s_7}{1.5s_1 - 6s_2 - 4.5s_3 + 18s_4 - 4.5s_5 - 6s_6 + 1.5s_7} \tag{6.204}$$

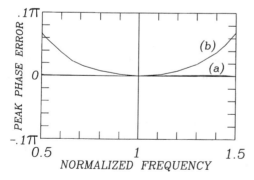

Figure 6.50 Detuning sensitivity of (a) optimized seven-sample algorithm compared with the (b) detuning sensitivity of the Schwider-Hariharan algorithm.

Figure 6.49 shows the Fourier transforms of the sampling reference functions, illustrating the frequency response and detuning insensitivity of this algorithm.

For comparison purposes, Fig. 6.50 shows the detuning insensitivity of the Schwider-Hariharan algorithm compared with this algorithm.

It should be pointed out that the detuning insensitivity obtained in the algorithms just presented has been obtained at the expense of some harmonic leaks (if any).

6.11 INFLUENCE OF WINDOW FUNCTION OF SAMPLING ALGORITHMS

If the signal has harmonics that the signal algorithms cannot eliminate, they can be reduced by a suitable additional filtering function, sometimes called a window function, as described by de Groot (1995) and Schmit and Creath (1996). Any algorithm with reference sampling reference functions $g_1(x)$ and $g_2(x)$ may be modified by means of a window function $h(x)$. Then, the new sampling reference functions $g_1'(x)$ and $g_2'(x)$ would be given by

$$g_1'(x) = h(x)g_1(x) \tag{6.205}$$

and

$$g_2'(x) = h(x)g_2(x) \tag{6.206}$$

With the convolution theorem, the Fourier transforms of these functions are

$$G_1'(f) = H(f) * G_1(f) \tag{6.207}$$

and

$$G_2'(f) = H(f) * G_2(f) \tag{6.208}$$

These new reference sampling functions must satisfy at the reference frequency the orthogonality and equal-magnitudes conditions in Eq. 5.92. Hence, we require

$$G_1'(f_r) \pm i G_2'(f_r) = (H(f) * [G_1(f) \pm i G_2(f)])_{f=f_r} = 0 \tag{6.209}$$

The zero bias condition in Eq. 5.74 must also be satisfied. Thus, from Eqs. 5.106 and 5.107 we may write

$$\sum_{n=1}^{N} W_{1n}' = \sum_{n=1}^{N} h(x_n) W_{1n} = 0 \tag{6.210}$$

and

$$\sum_{n=1}^{N} W'_{2n} = \sum_{n=1}^{N} h(x_n) W_{2n} = 0 \tag{6.211}$$

Any window function satisfying these conditions transforms one algorithm into another with different properties. A formal mathematical derivation of the general conditions required by the window function is possible using these relations. Nevertheless, we will restrict ourselves to the simple particular case of an algorithm with sampling points in two periods of the reference function, with an identical distribution on each of the two periods, so that if the sampling function for the basic one-period algorithm is $g_{bi}(x)$, the sampling function $g_i(x)$ for the two periods is

$$g_i(x) = g_{bi}(x) + g_{bi}(x + 2\pi) \tag{6.212}$$

A particular case of this kind of algorithm is when the points are equally spaced in the two periods and the number of points is even. Thus, its Fourier transform is

$$G_i(f) = G_{bi}(f)\left(1 + \exp\left(i2\pi\frac{f}{f_r}\right)\right) \tag{6.213}$$

It is relatively simple to prove either mathematically or graphically that any window function that satisfies the condition

$$h(x) = 1 - h\left(x + \frac{1}{f_r}\right) \tag{6.214}$$

preserves the magnitude and phase of the Fourier transforms of the sampling reference functions at the reference frequency as well as the zero bias. Figure 6.51 illustrates a particular case of these functions.

Thus, this window function can be expressed by a Fourier series as

$$h(x) = 1 + \sum_{m=1}^{\infty} A_m \cos(m\pi f_r x) = 1 + \sum_{m=1}^{\infty} A_m \cos\left(\frac{\alpha_m}{2}\right) \tag{6.215}$$

where m is an odd integer. The Fourier transform of this filter function thus becomes

$$H(f) = 2\delta(f) + \frac{1}{2}\sum_{m=-\infty}^{\infty} A_m \delta\left(f - \frac{mf_r}{2}\right) \tag{6.216}$$

where $m \neq 0$ and $A_m = A_{-m}$.

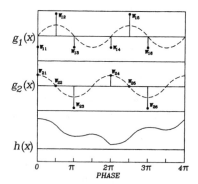

Figure 6.51 Reference sampling functions $g_1(x)$ and $g_2(x)$ and window function $h(x)$ when two periods of the signal are sampled.

Using the merit function defined in the preceding section, the best value for these A_m coefficients can be calculated.

Schmit and Creath (1996) described with some detail a triangular function and a bell function, which can be considered particular cases of the one described here. If these window functions are applied to the eight sampling points diagonal least squares algorithms, with an even number of points, improved algorithms are obtained. These window functions, shown in Fig. 6.52, improve the algorithms' characteristics.

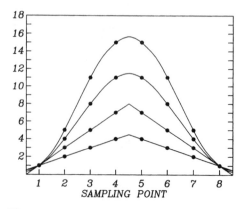

Figure 6.52 Triangular and bell window functions, described by Schmit and Creath (1996), for an eight sampling points diagonal least squares algorithm.

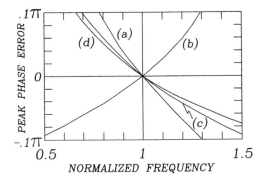

Figure 6.53 Detuning sensitivity for four algorithms. (a) Three points at three 120°; (b) three steps in inverted T; (c) four steps in X; (d) five steps.

Schmit and Creath (1996) proved that the triangular window produces the same effect as the multiple sequential technique, while the bell window produces the same effect as the multiple averaging technique. De Groot (1995) also studied the effect of a window function, with a concept closer to the filtering function studied in Chap. 5.

6.12 ALGORITHM SUMMARY

In this section some of the main properties of phase detecting algorithms are described.

6.12.1 Detuning Sensitivity

We saw in Chap. 4 that by shifting the sampling point locations we may obtain an algorithm in which the Fourier transforms of the reference sampling functions are either orthogonal or have the same magnitudes at all frequencies. We have also seen that the sensitivity to detuning is not affected by this shifting of the sampling points.

The detuning sensitivity for some of the main algorithms described in this chapter will now be described. In the following figures the peak phase error is represented by the quantity in front of the sine function in Eq. 5.114. In Fig. 6.53 the detuning error for four algorithms is illustrated. The first plot (a) is for the three sampling points algorithm at 120°. This is the algorithm with the largest error. The second plot (b) is for the algorithm with three steps in an inverted T. In this case the sign of the error is opposite that of the former. The third plot (c) is for the four steps in X algorithm. The fourth plot (d) is for the

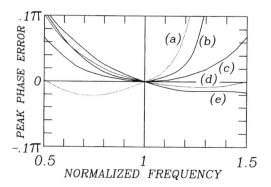

Figure 6.54 Detuning sensitivity for five symmetrical $(N + 1)$ algorithms. (a) Uncompensated four steps $(3 + 1)$; (b) compensated four steps; (c) Schwider-Hariharan five steps $(4 + 1)$; (d) uncompensated six steps $(5 + 1)$; (e) uncompensated seven steps $(6 + 1)$.

five-step algorithm. This phase error is the smallest of these four algorithms, but not by much.

Figure 6.54 shows the detuning phase error for some symmetrical $(N + 1)$ algorithms. The first plot (a) is for the four $(3 + 1)$ step algorithm. We can see that there is sensitivity to detuning. If this algorithm is compensated with the extra sampling weights described earlier, as in (b), the sensitivity to detuning is reduced, since the slope of the curve is zero at the origin. The next plot (c) is for the popular Schwider-Hariharan five $(4 + 1)$ step algorithm, where the insensitivity to detuning is clearly seen to be better than in the $(3 + 1)$

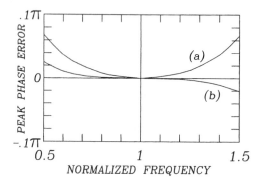

Figure 6.55 Detuning sensitivity for (a) Schwider and (b) Schmit-Creath algorithms.

Table 6.2 Sensitivity to Signal Harmonics of Some Algorithms

Algorithm	Harmonics being suppressed								
	2	3	4	5	6	7	8	9	10
Three points at 120° or in T		×			×			×	
Wyant's three points			×				×		
Four points in X or cross	×		×		×	×			
Five points	×	×		×		×	×	×	
Symmetrical 3 + 1		×			×		×		
Symmetrical 4 + 1	×	×	×		×		×	×	
Symmetrical 5 + 1	×	×	×		×		×	×	
Symmetrical 6 + 1	×	×	×		×		×	×	×
Schwider			×				×		
Schmit and Creath			×				×		

algorithm. The six $(5 + 1)$ step algorithm is not compensated with the extra sampling weights. Thus, some detuning sensitivity is present.

Finally, the seven $(6 + 1)$ step algorithm also has some detuning sensitivity because it is also uncompensated. If compensated, this is the algorithm with the lowest detuning sensitivity.

Figure 6.55 shows the detuning sensitivity for the Schwider (a) and the Schmit-Creath (b) algorithms.

6.12.2 Harmonic Sensitivity

The harmonic sensitivity for some algorithms described in this chapter is summarized in Table 6.2.

REFERENCES

Angel J. R. P and P. L. Wizinowich, "A Method of Phase Shifting in the Presence of Vibration," *Eur. Southern Obs. Conf. Proc.*, **30**, 561 (1988).

Bhushan B., J. C. Wyant, and C. L. Koliopoulos, "Measurement of Surface Topography of Magnetic Tapes by Mirau Interferometry," *Appl. Opt.*, **24**, 1489–1497 (1985).

Carré P., "Installation et Utilisation du Comparateur Photoelectrique et Interferentiel du Bureau International des Poids et Measures," *Metrologia*, **2**, 13–23 (1966).

Cheng Y.-Y. and J. C. Wyant, "Phase Shifter Calibration in Phase-Shifting Interferometry," *Appl. Opt.*, **24**, 30–49 (1985).

Creath K., "Comparison of Phase Measuring Algorithms," *Proc. SPIE*, **680**, 19–28 (1986).

Creath K., "Phase Measuring Interferometry: Beware These Errors," *Proc. SPIE*, **1553**, 213–220 (1991).

de Groot P., "Derivation of Algorithms for Phase Shifting Interferometry Using the Concept of a Data-Sampling Window," *Appl. Opt.*, **34**, 4723–4730 (1995).

Freischlad K. and C. L. Koliopoulos, "Fourier Description of Digital Phase Measuring Interferometry," *J. Opt. Soc. Am. (A)*, **7**, 542–551 (1990).

Greivenkamp J. E. and J. H. Bruning, "Phase Shifting Interferometers," in *Optical Shop Testing*, D. Malacara, Ed., John Wiley and Sons, New York, 1992.

Hariharan P., B. F. Areb, and T. Eyui, "Digital Phase-Shifting Interferometry: A Simple Error-Compensating Phase Calculation Algorithm," *Appl. Opt.*, **26**, 3899 (1987).

Hibino K., B. F. Oreb, and D. I. Farrant, "Phase Shifting for Non-Sinusoidal Waveforms with Phase Shift Errors," *J. Opt. Soc. Am. A*, **12**, 761–768 (1995).

Joenathan C., "Phase Measuring Interferometry: New Methods and Error Analysis," *Appl. Opt.*, **33**, 4147–4155 (1994).

Larkin K. G., "New Seven Sample Symmetrical Phase-Shifting Algorithm," *Proc. SPIE*, **1755**, 2–11 (1992).

Larkin K. G. and B. F. Oreb, "Design and Assessment of Symmetrical Phase-Shifting Algorithm," *J. Opt. Soc. Am.*, **9**, 1740–1748 (1992).

Mendoza-Santoyo F., D. Kerr, and J. R. Tyrer, "Interferometric Fringe Analysis Using a Single Phase Step Technique," *Appl. Opt.*, **27**, 4362–4364 (1988).

Morimoto Y. and M. Fujisawa, "Fringe Pattern Analysis by a Phase-Shifting Method Using Fourier Transform," *Opt. Eng.*, **33**, 3709–3714 (1994).

Nakadate S., "Phase Detection of Equidistant Fringes for Highly Sensitive Optical Sensing, I. Principle and Error Analysis," *J. Opt. Soc. Am. A*, **5**, 1258–1264 (1988a).

Nakadate S., "Phase Detection of Equidistant Fringes for Highly Sensitive Optical Sensing, II. Experiments," *J. Opt. Soc. Am. A*, **5**, 1265–1269 (1988b).

Ransom P. L. and J. B. Kokal, "Interferogram Analysis by a Modified Sinusoid Fitting Technique," *Appl. Opt.*, **25**, 4199–4204 (1986).

Rastogi, P. K., "Modification of the Carré Phase-Stepping Method to Suit Four-Wave Hologram Interferometry," *Opt. Eng.*, **32**, 190–191 (1993).

Schmit J. and K. Creath, "Extended Averaging Technique for Derivation of Error-Compensating Algorithms in Phase Shifting Interferometry," *Appl. Opt.*, **34**, 3610–3619 (1995).

Schmit J. and K. Creath, "Window Function Influence on Phase Error in Phase-Shifting Algorithms," *Appl. Opt.*, **35**, 5642–5649 (1996).

Schwider J., R. Burow, K.-E. Elssner, J. Grzanna, R. Spolaczyk, and K. Merkel, "Digital Wave-Front Measuring Interferometry: Some Systematic Error Sources," *Appl. Opt.*, **22**, 3421–3432 (1983).

Schwider J., O. Falkenstörfer, H. Schreiber, A. Zöller, and N. Streibl, "New Compensating Four-Phase Algorithm for Phase-Shift Interferometry," *Opt. Eng.*, **32**, 1883–1885 (1993).

Servín M. and F. J. Cuevas, "A Novel Technique for Spatial Phase-Shifting Interferometry," *J. Mod. Opt.*, **42**, 1853–1862 (1995).

Servín M., D. Malacara, J. L. Marroquín, and F. J. Cuevas, "Complex Linear Filters for Phase Shifting with Very Low Detuning Sensitivity," *J. Mod. Opt.*, **44**, 1269–1278 (1997).

Surrel Y., "Phase Stepping: A New Self-Calibrating Algorithm," *Appl. Opt.*, **32**, 3598–3600 (1993).

Surrel Y., "Design of Algorithms for Phase Measurements by the Use of Phase Stepping," *Appl. Opt.*, **35**, 51–60 (1996).

Wizinowich P. L., "Phase Shifting Interferometry in the Presence of Vibration: A New Algorithm and System," *Appl. Opt.*, **29**, 3271–3279 (1990).

Wyant J. C., C. L. Koliopoulos, B. Bushan and O. E. George, "An Optical Profilometer for Surface Characterization of Magnetic Media," *ASLE Trans.*, **27**, 101 (1984).

Zhao B. and Y. Surrel, "Phase Shifting: Six-Sample Self-Calibrating Algorithm Insensitive to the Second Harmonic in the Fringe Signal," *Opt. Eng.*, **34**, 2821–2822 (1995).

7

Phase Shifting Interferometry

7.1 PHASE SHIFTING BASIC PRINCIPLES

Phase shifting interferometric techniques have their first indirect antecedent in the works by Carré (1966), but they really began with Crane (1969), Moore (1973), Bruning *et al.* (1974), and some others. These techniques have also been applied to speckle pattern interferometry (Nakadate and Saito 1985; Creath 1985; Robinson and Williams 1986) and to holographic interferometry (Nakadate *et al.* 1986; Stetson and Brohinski 1988). Many reviews on this field have been written, for example by Greivenkamp and Bruning (1992). In phase shifting interferometers the reference wavefront is moved along the direction of propagation, with respect to the wavefront under test, changing their phase difference.

By measuring the irradiance changes for different phase shifts, it is possible to determine the phase for the wavefront under test, relative to the reference wavefront, for the measured point on the wavefront. The irradiance signal $s(x, y)$ at the point (x, y) in the detector changes with the phase, as

$$s(x, y, \alpha) = a(x, y) + b(x, y) \cos{(\alpha + \phi(x, y))} \tag{7.1}$$

where $\phi(x, y)$ is the phase at the origin and α is a known phase shift with respect to the origin. By measuring the phase for many points over the wavefront, the complete wavefront shape is thus determined.

If we consider any fixed point in the interferogram, the phase difference between the two wavefronts has to be changed. We might wonder how this is possible, because relativity does not permit any of the two wavefronts to move faster than the other, because the phase velocity is c for both waves. However, it has been shown (Malacara *et al.* 1969) that what really takes place is the Doppler effect, with a shift both in the frequency and in the wavelength. Then, both beams with different wavelengths interfere, producing beats. These beats can also be interpreted as the changes in the irradiance due to the continuously changing phase difference. These two conceptually different models are physically equivalent.

Thus, the change in the phase may be accomplished if the frequency of one of the beams is modified during the process. This is possible in a continuous fashion using certain devices, but only for a relatively short period of time with some other devices. This fact has led to a semantic problem: If the frequency can be modified in a permanent way, some people refer to these instruments as AC, heterodyne, or frequency shift interferometers. Otherwise, it is a phase shifting interferometer. Here, we will refer to all of these instruments as phase shifting interferometers.

7.2 SOME METHODS TO INTRODUCE PHASE SHIFTING

The procedure just described can be implemented in almost any kind of two-beam interferometer, for example, in the Twyman- Green or Fizeau interferometers. The phase may be shifted in several different ways, as reviewed by Creath (1988).

7.2.1 Moving Mirror with a Linear Transducer

A method that can be used to shift this phase is by moving the mirror for the reference beam along the light trajectory, as shown in Fig. 7.1 for a Twyman-Green interferometer, by means of an electromagnetic or piezoelectric transducer.

The transducer moves the mirror and thus the phase is changed to a new value, as shown in Fig. 7.2a. An alternative way to interpret the phase shift, as pointed out before, is to think of the reflected light as Doppler shifted light.

A piezoelectric transducer (PZT) typically has a linear displacement over a 1 μm (2λ). A voltage from zero to a few hundred volts is used to produce the displacement.

7.2.2 Rotating Glass Plate

Another method to shift the phase is to insert a plane parallel glass plate in the light beam (Wyant and Shagam 1978), as shown in Fig. 7.2b. The phase shift α

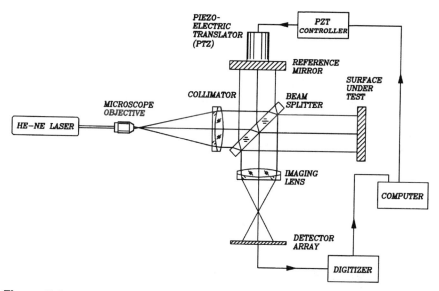

Figure 7.1 A Twyman-Green interferometer with a phase shifting transducer.

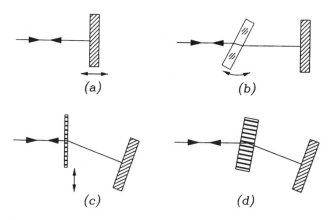

Figure 7.2 Some methods to shift the phase in an interferometer. (a) By means of a mirror moving along the light path; (b) by means of a rotating glass plate; (c) by means of a moving diffraction grating; (d) by means of a Bragg cell.

introduced by this glass plate when it is tilted by an angle θ with respect to the optical axis is given by

$$\alpha - \frac{t}{k}(n\cos(\theta') - \cos(\theta)) \tag{7.2}$$

where t is the plate thickness, n is its refractive index, and $k = 2\pi/\lambda$. The angles θ and θ' are the angles between the normal to the glass plate and the light rays outside and inside the plate, respectively. A rotation of the plate increasing the angle θ increases the optical path difference. Thus, if the plate is rotated a small angle $\Delta\theta$, the phase shift α is given by

$$\alpha = \frac{t}{k}\left(1 - \frac{1}{n}\frac{\cos(\theta)}{\cos(\theta')}\right)\sin(\theta)\Delta\theta \tag{7.3}$$

An important requirement in this method is that the plate has to be inserted in a collimated light beam to avoid introducing aberrations.

7.2.3 Moving Diffraction Grating

Another way to shift the phase is to use a diffraction grating or ruling moving perpendicularly to the light beam (Suzuki and Hioki 1967; Stevenson 1970; Bryngdahl 1976; Srinivasan *et al.* 1985) as shown in Fig. 7.2c. It is easy to notice that the phase of the diffracted light beam is shifted n times 2π the number of slits that pass through a fixed point, where the letter n represents the order of diffraction. Thus, the shift in the frequency is equal to n times the number of slits in the grating that pass through a fixed point in the unit of time. To say this in a different manner, the shift in the frequency is equal to the speed of the grating, divided by the period d of the grating. It is interesting to note that the frequency is increased for the light beams diffracted in the same direction as the movement of the grating. The light beams diffracted in the opposite direction to the movement of the grating decrease in frequency. As is to be expected, the direction of the beam is changed because the first-order beam has to be used, and the zero-order beam must be blocked by means of a properly placed diaphragm.

If the diffraction grating is moved a small distance Δy, the phase changes by an amount α given by

$$\alpha = \frac{2\pi n}{d}\Delta y \tag{7.4}$$

where d is the period of the grating and n is the order of diffraction.

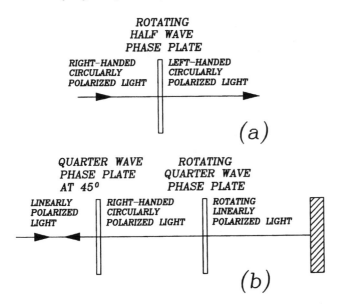

Figure 7.3 A polarized light device to shift the phase. (a) Single pass and (b) double pass.

A Ronchi ruling moving perpendicularly to its lines in the Ronchi test is a particular case of a moving diffraction grating. This method has been used by several researchers, for example, by Indebetow (1978), with the name of running projection fringes.

A similar method uses an acoustic optic Bragg cell to diffract the light (Massie and Nelson 1978; Wyant and Shagam 1978; Shagam 1983) as shown in Fig. 7.2d. In this cell an acoustic transducer produces ultrasonic vibrations in the liquid of the cell. These vibrations produce periodic changes in the refractive index, inducing the cell to act as a thick diffraction grating. The thickness effect enables this diffraction device to have a high efficiency in the desired order of diffraction.

7.2.4 Rotating Phase Plate

The phase may also be shifted by means of a rotating plane parallel glass plate (Crane 1969; Okoomian 1969; Bryngdahl 1972; Sommargren 1975; Shagam and Wyant 1978; Hu 1983; Zhi 1983; Kothiyal and Delisle 1984, 1985; Salbut and Patorski 1990) as shown in Fig. 7.3. If a beam of circularly polarized light goes through a half-wave phase plate, the handedness of the circular polarization is reversed, as shown in Fig. 7.3a. If the half-wave phase plate rotates, the frequency of the light changes. If the plate rotates in a continuous manner, the frequency change $\Delta \nu$ is equal to twice the frequency of rotation of the plate. If the phase plate is rotated a small angle $\Delta \theta$, the phase changes by α as

$$\alpha = 2\Delta \theta \tag{7.5}$$

This arrangement works if the light goes through the phase plate only once. However, in a Twyman-Green interferometer the light passes twice through the system. Then, it is possible to use the configuration in Fig. 7.3b. The first quarter-wave retarding plate is stationary, with its slow axis at $45°$ with respect to the plane of polarization of the incident linearly polarized light. This plate also transforms the returning circularly polarized light back to linearly polarized. The second phase retarder is also a quarter-wave plate. It is rotating, and the light goes through it twice; hence it really acts as a half-wave plate.

7.2.5 Moiré in an Interferogram with a Linear Carrier

Let us consider an interferogram with a large linear carrier, that is, with many fringes produced by means of a reference wavefront tilt. If a Ronchi ruling or a similar linear ruling with about the same number of fringes is placed on top of this interferogram, a moiré fringe appears, as will be described in Chap. 9. This moiré pattern represents the interferogram with the linear carrier removed. The phase of this interferogram can be changed by moving the superimposed linear ruling. The phase changes by an amount equal to 2π if the linear ruling is moved perpendicularly to the fringes a distance equal to its period. The phase shifting scheme has been described by Kujawinska *et al.* (1991) and by Dorrio *et al.* (1995a, 1995b).

The Ronchi ruling is placed on top of the interferogram, thus producing the multiplication of the interferogram irradiance by the ruling transmission. In principle, this ruling can be implemented by software in the computer, but then information about very high spatial frequencies has to be stored in the computer memory, making the system quite inefficient. It is thus advisable to use a real Ronchi ruling and perform the spatial filtering of high frequencies before the light detector. The low pass filtering may be performed by defocusing the lens that forms the interferogram image on the light detector.

7.2.6 Frequency Changes in Laser Light Source

Another method to produce the phase shift is by shifting the frequency of the laser light source. This shift can be done in either of two ways.

One method is to illuminate the interferometer with a Zeeman split laser line. The frequency of the laser is split into two orthogonally polarized output frequencies by means of a DC magnetic field (Burgwald and Kruger 1970). The frequency separation of the two spectral lines is of the order of 2 to 5 MHz in a helium-neon laser. In the interferometer system the state of polarization of the two spectral lines is modified to produce the interference between them.

Another method is to use an unbalanced interferometer, that is, one with a large optical path difference, and a laser diode with its frequency controlled by the injection electric current, as proposed by Ishii *et al.* (1991) and later studied by Onodera and Ishii (1996). This method is based on the fact that the phase difference in an interferometer is proportional to the product of the OPD (optical path difference) and its temporal frequency; variation of one of them will produce a piston phase change.

7.2.7 Simultaneous Phase Shift Interferometry

Phase shifting methods in an environment with vibrations cannot give good results due to the long time needed to take all measurements. This problem has been avoided by the use of interferometer systems in which all the needed interferometer frames are taken at the same time (Kujawinska and Robinson 1988, 1989; Kujawinska 1987, 1993; Kujawinska *et al.* 1990).

One approach is the use of multichannel interferometers (Kwon 1984), in which an interferometer in a Mach-Zehnder configuration produces three frames at the same time by means of a diffraction grating. Kwon and Shough (1985) and Kwon *et al.* (1987) used radial shear interferometers, also in Mach-Zehnder or triangular configurations, with a diffraction grating. Bareket (1985) and Koliopoulos (1991) also designed other simultaneous or multiple channel phase shift interferometers.

The great disadvantage of these arrangements is that more complicated and expensive hardware is needed. Also, exact pixel-to-pixel correlation between the images is required.

7.3 PHASE SHIFTING SCHEMES AND PHASE MEASUREMENT

We saw in Chap. 1 that the signal is a sinusoidal function of the phase, as shown in Fig. 1.2. In phase shifting interferometers the wavelength of the signal to be detected is equal to the wavelength of the illuminating light. The basic problem is that of determining the non-shifted phase difference between the two waves with the highest possible precision. This may be done by means of any of several procedures to be described next.

The best method to determine the phase depends on many factors, mainly on how the phase shift was performed. The phase may be changed in a continuous manner, by introducing a permanent frequency shift in the reference beam. Some

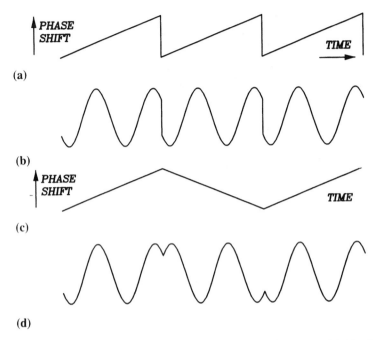

(a)

(b)

(c)

(d)

Figure 7.4 Signals obtained in phase shifting interferometry. (a) Continuous sawtooth phase stepping; (b) signal with continuous sawtooth phase stepping; (c) continuous triangular phase stepping; (d) signal with triangular phase stepping.

authors call this a heterodyne interferometer. As described by Moore (1973), in heterodyne interferometry, there are three basic possibilities: (1) The frequency is permanently shifted and the signal output is continuous; (2) the phase is changed in a sinusoidal manner, as shown in Fig. 7.4a, thus obtaining the signal in Fig. 7.4b; or (3) the phase is changed in a triangular manner, as in Fig. 7.4c, obtaining the symmetrical signal in Fig. 7.4d.

When the synchronous phase detection algorithms in Chap. 5 are used, the phase may also be changed in a discontinuous manner, increasing or decreasing the phase in several steps.

The digital phase stepping method measures the signal values at several known increments of the phase. The measurement of the signal at any given phase takes some time, since there is a time response for the detector. Hence, the phase has to be stationary for a short time to allow the measurement to be taken. Between two consecutive measurements, the phase may change as fast as desired in order to get to the next phase with the smallest delay.

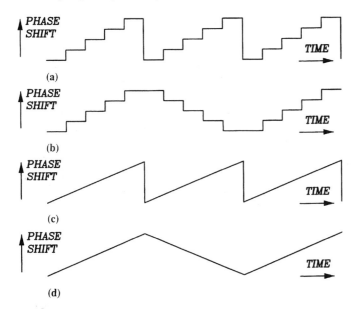

Figure 7.5 Four different methods for periodically shifting the phase. (a) Stepped sawtooth phase stepping; (b) stepped triangular phase stepping; (c) continuous sawtooth phase stepping; (d) continuous triangular phase stepping.

One problem with the phase stepping method is that the sudden changes in the mirror position may introduce some vibrations into the system. In the integrating bucket method, the phase changes continuously and not by discrete steps. Then, the detector continuously measures the irradiance during a fixed time interval without stopping the mirror. Hence, the average value during the measuring time interval is measured, as described in Chap. 3.

This change of phase can thus be made in any of several different schemes, as illustrated in Fig. 7.5.

There are some analog methods to measure the relative irradiance phase at different interferogram points—for example, by detection of the zero crossing point of the phase (Crane 1969) or by means of the phase lock method (Moore *et al.* 1978). In the zero crossing method the phase is detected by locating the phase point where the signal passes through the axis of symmetry of the function, not really zero, which has a signal value equal to *a*. The points crossing the axis of symmetry can be found by amplifying of the signal function to saturation levels. In this manner, the sinusoidal signal becomes a square function.

Digital phase stepping methods are more extensively used than analog methods.

7.4 HETERODYNE INTERFEROMETRY

When the phase shift is continuous we speak of heterodyne or DC interferometry. As we pointed out before, two equivalent models may describe the phase shift, i.e., a change in the optical path difference or a change in the frequency of one of the two interfering light beams. In this case the most common interpretation is that of two different interfering frequencies. Then we think of heterodyning beats. If we measure the relative phase of these beats at different points over the wavefront, we obtain the wavefront deformations.

The phase of the detected beats is measured in real time using electronics hardware instead of sampling the irradiance (Wyant 1975; Massie 1978, 1980, 1987; Massie and Nelson 1978; Massie *et al.* 1979; Sommargren 1981; Hariharan *et al.* 1983; Hariharan 1985; Thalmann and Dändliker 1985). The great advantage of this approach is that a fast measurement is achieved, which is important in many applications, as in many dynamical systems. Beats frequencies of the order of 1 MHz. may be obtained. A high speed detector is thus necessary. A standard television camera cannot be used. A high frame rate image tube, also called an image dissector tube, may be used.

Smythe and Moore (1983, 1984) proposed an alternative heterodyne interferometric system in which the beats are not measured. Instead, by an optical procedure which is not described here, using polarizing optics, two orthogonal bias-free signals are generated. Each of these two signals comes from one of the two arms of the interferometer. Then the phase difference between these two orthogonal signals is the phase difference between the two interferometer optical paths. If we represent these two orthogonal signals in a polar diagram, one along the vertical axis and the other along the horizontal axis, the path described in this diagram when the phase is continually changed is a circle. The angle with respect to the optical axis is the phase. This heterodyning procedure can be easily implemented to measure wavefront deformations in two dimensions.

7.5 PHASE LOCK DETECTION

In the phase lock method to detect the signal, the phase reference wave is phase-modulated with a sinusoidally oscillating mirror (Moore 1973; Moore *et al.* 1978; Moore and Truax 1979; Johnson *et al.* 1979). Two phase components δ_0 and $\delta_1 \sin \omega t$ are added to the signal phase $\phi(x, y)$. One of the additional phase components has a fixed value and the other a sinusoidal time oscillation. Both components are independent and can have any desired value. Omitting the x, y dependence for notational simplicity, the total time-dependent phase is

$$\phi + \delta_0 + \delta_1 \cos(2\pi f t) \tag{7.6}$$

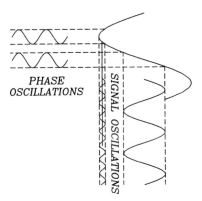

Figure 7.6 Phase lock detection of the signal phase.

thus the signal is

$$s(t) = a + b \cos(\phi + \delta_0 + \delta_1 \cos(2\pi f t)) \tag{7.7}$$

The phase modulation is carried out only in an interval smaller than π, as illustrated in Fig. 7.6.

The output signal may be interpreted as the phase modulating signal after being harmonically distorted by the signal to be detected. This harmonic distortion is a function of the phase ϕ, as shown in Fig. 7.7.

This function is periodic and symmetrical. Thus, to find the harmonic distortion using Eqs. 2.6 and 2.7, this function can now be expanded in series as

$$s(t) = \frac{c_0}{2} + \sum_{n=1}^{\infty} c_n \cos(2\pi n f t) \tag{7.8}$$

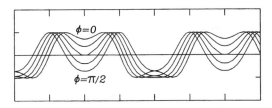

Figure 7.7 Output of a harmonically distorted signal, with $\delta = 0.75\pi$.

where

$$c_n = \frac{1}{t_0} \int_{-t_0}^{t_0} s(t) \cos(2\pi n f t) \, dt \qquad (7.9)$$

Then, making the variable substitution $\theta = 2\pi f t$ we may show that

$$
\begin{aligned}
c_n = {} & \frac{b}{\pi} e^{i(\phi+\delta_0)} \int_0^\pi e^{i(\delta_1 \cos\theta)} \cos(n\theta) \, d\theta \\
& + \frac{b}{\pi} e^{-i(\phi+\delta_0)} \int_0^\pi e^{-i(\delta_1 \cos\theta)} \cos(n\theta) \, d\theta
\end{aligned}
\qquad (7.10)
$$

On the other hand, the Bessel function of the first kind of order n is given by

$$J_n(\delta) = \frac{1}{\pi} e^{-in\pi/2} \int_0^\pi e^{i(\delta_1 \cos\theta)} \cos(n\theta) \, d\theta \qquad (7.11)$$

and using this expression in Eq. 7.10, it is possible to obtain

$$c_n = 2b J_n(\delta) \cos\left(\phi + n\frac{\pi}{2}\right) \qquad (7.12)$$

Hence, the output signal is given by

$$
\begin{aligned}
s(x, y) = {} & a \\
& + b \cos(\phi(x, y) + \delta_0)(J_0(\delta_1) - 2J_2(\delta_1) \cos(2\omega t) + \cdots \\
& + b \sin(\phi(x, y) + \delta_0)(2J_1(\delta_1) \sin(\omega t) - 2J_3(\delta_1) \sin(3\omega t) + \cdots
\end{aligned}
$$

$$\qquad (7.13)$$

where $\omega = 2\pi f$. The first part of this expression represents harmonic components of even order, and the second part represents harmonic components of odd order.

Let us now assume that the amplitude of the phase oscillations component $\delta_1 \sin \omega t$ are much smaller than π. Then, if we adjust the δ_0 component to a value such that $\phi + \delta_0 = n\pi$, then $\sin(\phi + \delta_0)$ is zero and only even harmonics remain. This effect is illustrated in Fig. 7.6, near one of the minima of the signal $s(x, y)$. This is done in practice by slowly changing the value of the phase component δ_0 while keeping the oscillation $\delta_1 \sin(\omega t)$, until the minimum amplitude of the first harmonic (fundamental frequency) is obtained. Then, we have $\phi + \delta_0 = n\pi$ and since the value of δ_0 is known, the value of ϕ has been determined.

This method may also be used at the inflection point for the sinusoidal signal function, as shown in Fig. 7.6, by changing the fixed phase component until the first harmonic has its maximum amplitude.

From Eq. 7.12 we can obtain

$$\tan\phi = \frac{c_1}{c_2}\left(\frac{J_2(\delta)}{J_1(\delta)}\right) \tag{7.14}$$

Thus, since the Bessel function values are known, if the value of δ is also known, the signal phase can be determined if the ratio of the amplitudes of the fundamental component to the second harmonic component is measured. This measurement can be performed analogically by means of electronic hardware. Matthews *et al.* (1986) used this method with a null detecting method instead of a maximum detection procedure.

One disadvantage of this method is that a two-dimensional array of detectors cannot be used. A single detector must move to scan the whole picture.

7.6 SINUSOIDAL PHASE OSCILLATION DETECTION

Sasaki and Okasaki (1986a, 1986b) and Sasaki *et al.* (1987) proposed a sinusoidal phase modulating interferometer in which the reference wave is phase-modulated with a sinusoidally oscillating mirror, as in the phase lock method just described. The main difference is that the phase determination is performed with a signal digital sampling procedure. The modulated phase is

$$\phi + \delta\cos(2\pi f t + \theta) \tag{7.15}$$

where the difference with Eq. 7.6 is that the constant phase value is not present and an extra term θ has been added. The value of θ is the phase of the phase shifter oscillation at $t = 0$. It will be shown later that $\theta = 0$ is not the best value. Sasaki and Okasaki (1986a) added an extra random phase term $n(t)$ to this expression, to consider the presence of multiplicative noise due to disturbing effects such as system vibrations. They derived the optimum values of the amplitude δ and the phase θ of the oscillating driving signal, considering a minimization of the effects of noise. For notational simplicity, we did not add this term here. Thus, the modulated signal to be measured is

$$s(t) = a + b\cos\left(\phi + \delta\cos(2\pi f t + \theta)\right) \tag{7.16}$$

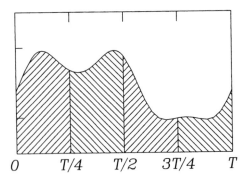

Figure 7.8   Interval integrating sampling of harmonic distorted signal, at four points.

This function is periodic, but it is asymmetrical, $\theta = 0$, and can be written as

$$s(t) = a + b\cos(\phi)\cos{(\delta\cos(2\pi ft + \theta))}$$
$$- b\sin(\phi)\sin{(\delta\cos(2\pi ft + \theta))} \qquad (7.17)$$

This signal contains a large amount of signal harmonics. A phase detecting sampling algorithm different from those studied in Chap. 6 may be used to take into account the presence of harmonics. Four sampling measurements with 90° separation, with interval averaging as described in Chap. 2, are used. The integrating interval has a width of 90°, equal to the sampling points separation. This integration eliminates most harmonic content above the third harmonic. The first zero of the associated filter function has its first zero at the frequency of the fourth harmonic. The second and third harmonics remain. As shown in Fig. 7.8, the averaged signal measurements are

$$\bar{s}_i = a + (b\cos(\phi))\,C_i - (b\sin(\phi))\,S_i \qquad (7.18)$$

with

$$C_i = \frac{1}{4T}\int_{(i-1)T/4}^{iT/4} \cos(\delta\cos(2\pi ft + \theta)) \qquad (7.19)$$

and

$$S_i = \frac{1}{4T}\int_{(i-1)T/4}^{iT/4} \sin(\delta\cos(2\pi ft + \theta)) \qquad (7.20)$$

where T is the signal period.

Sasaki and Okasaki (1986a) found the expressions for C_i to be

$$C_1 = C_3 = J_0(\delta) + \frac{4}{\pi} \sum_{n=1}^{\infty} \frac{J_{2n}(\delta)}{2n} [1 - (-1)^n] \sin(2n\pi) \tag{7.21}$$

and

$$C_2 = C_4 = J_0(\delta) - \frac{4}{\pi} \sum_{n=1}^{\infty} \frac{J_{2n}(\delta)}{2n} [1 - (-1)^n] \sin(2n\pi) \tag{7.22}$$

and the values of S_i

$$S_1 = -S_3$$

$$= -\frac{4}{\pi} \sum_{n=1}^{\infty} \frac{J_{2n-1}(\delta)}{(2n-1)} [(-1)^n \sin(2(n-1)\pi) + \cos(2(n-1)\pi)] \tag{7.23}$$

and

$$S_2 = -S_4$$

$$= -\frac{4}{\pi} \sum_{n=1}^{\infty} \frac{J_{2n-1}(\delta)}{(2n-1)} [(-1)^n \sin(2(n-1)\pi) - \cos(2(n-1)\pi)] \tag{7.24}$$

Then the signal phase may be proved to be

$$\tan \phi = \frac{C_1 - C_2}{S_1 + S_2} \left(\frac{\bar{s}_1 - \bar{s}_2 + \bar{s}_3 - \bar{s}_4}{\bar{s}_1 + \bar{s}_2 - \bar{s}_3 - \bar{s}_4} \right) \tag{7.25}$$

the optimum values of δ and θ are $\delta = 0.78\pi = 2.45$ and $\theta = 56°$.

According to Sasaki et al. (1987), this interferometric phase demodulation system yields a measurement accuracy of the order of 1.0 to 1.5 nm. Sasaki et al. (1990a) used a laser diode as a light source with a reference fringe pattern and electronic feedback to the laser current. In this manner they eliminated the noise due to laser intensity variations and to object vibrations.

Sinusoidal phase modulating schemes may be implemented in Twyman-Green as well as in Fizeau interferometers (Sasaki et al. 1990b).

7.7 PRECISION AND SOURCES OF PHASE ERROR

There are many sources of error in phase shifting interferometry. These errors have been studied by several researchers, for example by Creath (1986, 1991), Schwider et al. (1983), Cheng and Wyant (1985), Ohyama et al. (1988), Brophy (1990), and many others.

Wingerden *et al.* (1991) made a general study of many phase errors in phase detecting algorithms. They classify these errors as follows:

1. *Systematic errors.* The values of systematic errors vary sinusoidally with respect to the signal phase with a frequency of twice the signal frequency. These errors have a constant amplitude and phase. Averaging the measurements made with two algorithms whose sampling points in one algorithm are displaced 90° with respect to those on the other algorithm to reduce these errors.
2. *Random errors with sinusoidal phase dependence.* Random additive noise affects the signal measurements in such a manner that there is statistical independence of the noise errors corresponding to any two different signal measurements. Also, the noise is independent of the signal frequency. Thus, we can think that the noise amplitude and phase are random, not constant. Like the systematic errors, these have a sinusoidal phase dependence. The effect of the presence of additive noise on sampling algorithms has been studied in detail by Surrel (1997a). Mechanical vibrations introduce this kind of noise if its frequency is not too high, as will be seen later.

    Hibino (1997) proved that when a phase detection algorithm is designed to compensate for systematic phase errors, it becomes more susceptible to random noise and gives larger random errors in the phase.
3. *Random errors without phase dependence.* The value of these errors is independent of the phase of the measured signal. The case of additive random errors with a Gaussian distribution has been studied in detail by Rathjen (1995).

We have seen in Sec. 5.6 that the phase error when any of the four conditions are not fulfilled can be calculated by means of the general expression 5.107. In this section several particular cases were considered. Expressions for the analysis of phase errors were given, which may be applied to the calculation of errors in phase shifting interferometry, as will be described in the next few sections.

7.8 PHASE SHIFTER MISCALIBRATION AND NONLINEARITIES

The phase shifter device may not be well calibrated, or its response may not be linear, so that the target phase shift α is not the real phase shift α'. This effect may be represented by the expression

$$\alpha' = \alpha + \sigma_1\alpha + \sigma_2\alpha^2$$
$$= \alpha + \Delta\alpha \tag{7.26}$$

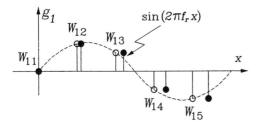

Figure 7.9 Displaced sampling points due to linear phase error.

where α is the target or reference value of the phase shift and α' is the real obtained value. The linear and quadratic error coefficients are σ_1 and σ_2, respectively.

If we require that the quadratic error become zero at the beginning ($\alpha = 0$) and at the end ($\alpha = 2\pi$) of the reference period, we write Eq. 7.26 as

$$\alpha' = \alpha + \epsilon_1\alpha + \epsilon_2(\alpha - 2\pi)\alpha_r$$
$$= \alpha + \epsilon_1\alpha + \epsilon_2((\alpha - \pi)^2 - \pi^2) \tag{7.27}$$

where $\epsilon_2 = \sigma_2$ and $\epsilon_1 = \sigma_1 - \pi\sigma_2$.

Substituting the value of α' in Eqs. 7.26 and 7.27 into the expression for the signal in Eq. 7.1, we see that it produces an error on the measured signal. However, this effect may be interpreted in two alternative ways:

1. *The error is in the sampling reference functions.* There is an error in the actual phase shift, or equivalently, in the interferometer optical path difference, so that the sampling points are displaced from their correct positions, as shown in Fig. 7.9, but the signal to be detected remains unmodified. The phase shift α with the error being introduced is related to the optical path difference OPD and the signal frequency f. In this case we consider the error to be produced by a deviation in OPD; thus we may write

$$2\pi f \text{OPD}_r = \alpha \tag{7.28}$$

Then, when the phase error is present, the sampling reference functions in Eqs. 5.85 and 5.86, as pointed out by Freischlad and Koliopoulos (1990), can be written

$$g_1(x) = \sum_{n=1}^{N} W_{1n}\delta(x - x_n - \Delta x_n) \tag{7.29}$$

and

$$g_2(x) = \sum_{n=1}^{N} W_{2n}\delta(x - x_n - \Delta x_n) \tag{7.30}$$

where $\Delta x_n = \Delta\alpha_n/2\pi f_r$. Thus, from Eqs. 5.90 and 5.91, the Fourier transform of these sampling reference functions is

$$G_1(f) = \sum_{n=1}^{N} W_{1n} \exp\left(-i\left(\alpha_n + \Delta\alpha_n \frac{f}{f_r}\right)\right) \tag{7.31}$$

and

$$G_2(f) = \sum_{n=1}^{N} W_{2n} \exp\left(-i\left(\alpha_n + \Delta\alpha_n \frac{f}{f_r}\right)\right) \tag{7.32}$$

The error-free Fourier transforms are orthogonal to each other and have the same magnitude at the reference frequency. Nevertheless, with the phase error added, either of the two conditions or both fail. Then these modified Fourier transforms allow us to compute the phase error, as explained in Sec. 5.6.

2. *The error is in the measured signal.* In this model we consider that the signal is phase-modulated by the error and that the sampling points positions are correct. Then we may write

$$2\pi f_r \text{OPD} = \alpha \tag{7.33}$$

where f_r is the reference frequency.

If we think of a phase-modulated signal, we may see that the phase modulation is a nonperiodic function of α. Thus, the signal is not periodic and the Fourier transform of the signal is no longer discrete but continuous. Figure 7.10a shows the error-free signal and the signal phase modulated with the error. The difference between these two signals is shown in Fig. 7.10b.

Since the Fourier transform is not discrete, in order to find the correct phase the correlations between the reference sampling functions and the signal have to be found with the integrals in Eq. 5.62.

The phase errors would have no importance at all if their values were independent of the signal phase. In that case the error would be just a constant

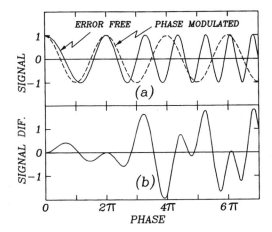

Figure 7.10 (a) Plots of the error-free signal (dashed curve) and the signal with error (continuous curve); (b) difference between these two signals. A value $\epsilon_2 = 0.05$ was used.

piston term on the measured wavefront. Unfortunately, this is not the case. We have seen in Chap. 5 that the phase errors have a value that varies sinusoidally with the signal frequency.

The presence of phase error may be detected by measuring a flat wavefront with a large linear carrier introduced with tilt fringes. If there is phase error, a sinusoidally corrugated wavefront will be detected, with twice the spatial frequency of the tilt fringes being introduced, as shown in Fig. 7.11.

The presence of phase shifter error may also be detected with a procedure suggested by Cheng and Wyant (1985). Tilt fringes are introduced, and measurements of the signal are taken across the interferogram in a direction perpendicular to the fringes. These measurements are then plotted to obtain a sinusoidal curve. This plot is repeated $N + 1$ times, with shift increments of $2\pi/N$. The first and $(N + 1)$th measurements should overlap each other, unless there is phase error, as shown in Fig. 7.12.

Another interesting method to detect phase errors was proposed by Kinnstaetter *et al.* (1988). In the fringe pattern two points in quadrature (phase difference equal to 90°) are selected. Then the signal values at these two points are plotted in a diagram for several values of the phase shift. These are Lissajous displays, in which the following characteristics appear, as illustrated in Fig. 7.13:

1. If there are not phase errors and the points being selected have the same signal amplitude and are exactly in quadrature, the diagram is a circle.

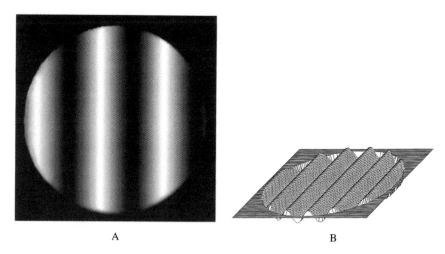

A   B

Figure 7.11 Detection of phase error by the presence of a corrugated wavefront.

2. If there is no phase error, but the interferogram points being selected do not have the same signal amplitude or are not in perfect quadrature, the diagram is an ellipse.
3. If there is linear error present, the ellipse or circle either does not close or it leaves a gap open. In other words, the first dot and the last are not at the same diagram place.
4. If there is nonlinear error, the distance between the dots is not constant.
5. If there is nonlinear response or saturation in the light detector, the ellipse is deformed, with some parts having different local curvature.
6. If there is vibrational noise, the curve is smaller and irregular.

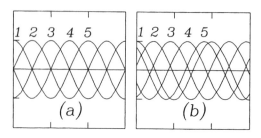

Figure 7.12 Plots to detect phase error.

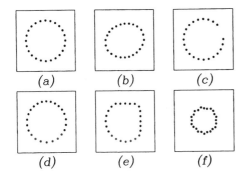

Figure 7.13 Lissajous curves with different types of phase errors.

Sometimes the measurement of the phase difference between any two inter-ferograms with different phases is difficult because of a large amount of noise. In this case the phase difference between two fixed interferograms can be measured directly, if many tilt fringes are present, using a method described by Wang *et al.* (1996).

Another method to eliminate phase shift errors is to directly measure the phase shift every time the phase is shifted. Lai and Yatagay (1991) propose an interferometer in which the phase is measured in an extra calibration fringe interference pattern with many tilt fringes. This auxiliary interferogram is projected on one side of the interferogram to be measured, using a high precision tilted mirror

We explained in Chap. 6 that these phase errors are sinusoidally dependent on the measured phase with twice the signal frequency. This fact was used to design special detuning-insensitive algorithms. Special algorithms can be devised to detect or reduce the phase errors due to phase shifter miscalibration and nonlinearity, as described by Joenathan (1994). Schwider (1989) also used this sinusoidal variation of the phase error by calculating an error function which is subtracted from the calculated phase values to substantially reduce the error.

7.8.1 Linear Phase Error

When there is no quadratic (nonlinear) error and only the linear error is present, we have $\epsilon_2 = 0$. The net effect in this case is a change in the frequency of the signal, i.e., a detuning. The error may be easily calculated for each algorithm, as described in Chap. 6.

To eliminate the linear error it is necessary to calibrate the phase shifter using an asynchronous algorithm, as described by Cheng and Wyant (1985) or any other of the methods mentioned before in this chapter.

7.8.2 Nonlinear Phase Error

If the linear error ϵ_1 has been properly eliminated by calibration of the phase shifter, the quadratic error may still be present.

With the phase error expression we can use any of the two previously described models. We may modify the sampling points position and calculate the Fourier transforms of the reference sampling functions. Alternatively, we may think of the measured signal, as being phase modulated by the phase error.

To use the first model it is convenient to express the phase error in such a way that the quadratic error becomes zero at the first sampling point ($n = 1$) and at the last sampling point ($n = N$). Thus, we can write

$$\Delta\alpha_n = \epsilon_1\alpha_n + \epsilon_2(\alpha_n - \alpha_N)\alpha_r$$

$$= -\epsilon_2\left(\frac{\alpha_N}{2}\right)^2 + \epsilon_1\alpha_n + \epsilon_2\left(\alpha_n - \frac{\alpha_N - \alpha_1}{2}\right)^2 \tag{7.34}$$

where $\epsilon_2 = \sigma_2$ and $\epsilon_1 = \sigma_1 - \pi\sigma_2$. The first term is a piston or phase offset term without any practical importance. We see that in this expression the quadratic error is symmetrical about the central point between the first and last sampling points.

Thus, the significant term for the quadratic error may be written as

$$\Delta\alpha_n = \epsilon_2\left(\alpha_n - \frac{\alpha_N - \alpha_1}{2}\right)^2 \tag{7.35}$$

which leads us to

$$\Delta\alpha_n = \Delta\alpha_{N-n+1} \tag{7.36}$$

An example of application of these concepts is illustrated in Fig. 7.14 for the algorithm with four sampling points in X. We can see that this algorithm is insensitive to nonlinear phase error. Other algorithms may be analyzed in a similar manner.

To use the second model, the signal may be represented by

$$s(z) = a + b\cos(2\pi fz + 4\pi^2\epsilon_2(fz - 1)fz + \phi) \tag{7.37}$$

where for notational simplicity the x-y dependence has been omitted and OPD has been replaced by z. Also, since there is no change in the signal period introduced by the compensated nonlinear error in Eq. 7.37, there is no detuning, and the reference frequency f_r becomes equal to the signal frequency f.

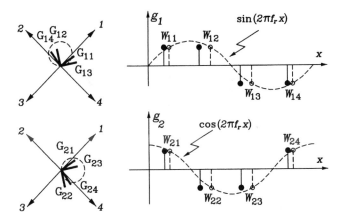

Figure 7.14 Effect of quadratic phase error in an algorithm.

In the results from the Fourier theory developed in Chap. 5 we assumed the signal to be periodic, so that its Fourier transform is discrete. If we assume that the phase value α is applied to each period of the signal, taking the beginning of each period as the new origin, we obtain a periodicity of the signal as shown in Fig. 7.15 and its Fourier transform is discrete. This approach is valid only when, as in most phase detecting algorithms, the sampling points are within one signal period.

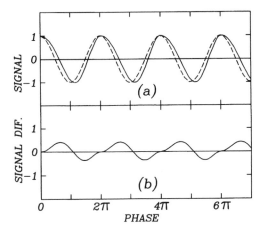

Figure 7.15 Periodic distorted signal due to non linear phase error.

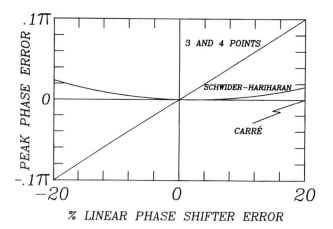

Figure 7.16 Non linear phase error for three common phase detecting algorithms. (From Creath 1988).

The Fourier coefficients in Eq. 2.6 may then be found with Eqs. 2.7 and 2.8. Unfortunately the evaluation of these integrals is not simple and leads to Fresnel integrals, as shown by Ai and Wyant (1987). Creath (1988) made numerical simulations to get insight into the nature of this phase error, as shown in Fig. 7.16.

7.9   OTHER SOURCES OF PHASE ERRORS

There are many other sources of phase errors in phase shifting interferometry. A few of these possible errors are described in this section.

7.9.1   Light Detector Nonlinearities

The light detector being used may have an electric output with a nonlinear relationship with the signal, even though these detectors are normally adjusted to work in the most linear region. If s' is the detector signal output and s is the input signal, we can write

$$s' = s + \epsilon s^2 \tag{7.38}$$

where ϵ is the nonlinearity error coefficient. Thus, the output from the detector is

$$s' = a(1 + \epsilon a) + (1 + 2\epsilon a)b\cos(\alpha_n + \phi)$$
$$+ \frac{1}{2}\epsilon b^2(1 + \cos^2 2(\alpha_n + \phi)) \tag{7.39}$$

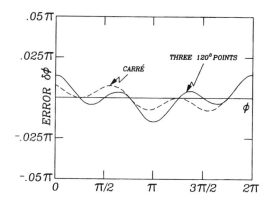

Figure 7.17 Phase error as a function of the phase, due to detector second order nonlinearities, for two common phase detecting algorithms. (From Creath 1991.)

We see that a second harmonic component appears in the signal. If the value of the coefficient ϵ for this nonlinearity is known, the compensation can be made. Otherwise, a phase error appears. As pointed out by Creath (1991), there is no error of this nature present for the algorithms with four and five samples. However, the three samples algorithm and the Carré's algorithm have noticeable errors with four times the fringe frequency.

Some corrections can be made on the video camera after the image has been digitized. Care must be taken to avoid saturating the detector, because that increases the harmonic content. Creath has made numerical calculations of this phase error. Figure 7.17 shows the peak phase error as a function of the phase, due to detector second order nonlinearities, in some common phase detecting algorithms.

The peak phase error for different amounts of nonlinear error, due to detector second-order nonlinearities, for some common phase detecting algorithms is illustrated in Fig. 7.18.

Third-order detector nonlinearities may also appear. Figure 7.19 shows the peak phase error as a function of the phase, due to detector third-order nonlinearities, in some common phase detecting algorithms.

The peak phase error for different amounts of nonlinear error, due to detector third-order nonlinearities, for some common phase detecting algorithms is in Fig. 7.20.

7.9.2 Vibration and Air Tubulence

Two of the most important sources of errors in phase shifting interferometry are vibrations and air turbulence. Their nature and consequences have been studied

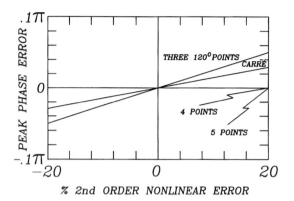

Figure 7.18 Peak phase error, as a function of amount of nonlinear error, due to detector second-order nonlinearities, for some common phase detecting algorithms. (From Creath 1988.)

by many researchers, for example, Kinnstaeter *et al.* (1988), Crescentini (1989), Wingerden *et al.* (1991), de Groot (1995), de Groot and Deck (1996), Deck (1996), and many others.

It is desirable to take as many preventive measures as possible to minimize these two disturbing factors.

If the vibration frequency is high enough, with an average period higher than the integration time of the detector, which is of the order of 1/60 s, the interference fringes are washed out, reducing their contrast.

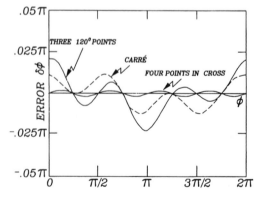

Figure 7.19 Phase error, as a function of the phase, due to detector third-order nonlinearities, for some common phase detecting algorithms. (From Creath 1988.)

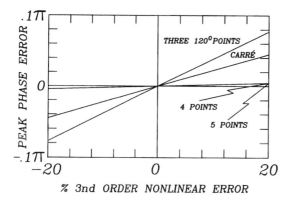

Figure 7.20 Peak phase error as a function of amount of nonlinear error, due to detector third-order nonlinearities, for some common phase detecting algorithms. (From Creath 1988.)

In close analogy with the mathematical treatment for phase lock and sinusoidal phase oscillation detection, de Groot and Deck (1996) studied the effect of noise by considering the signal to be phase-modulated with the noise as follows:

$$s(t) = a + b\cos(\alpha + \phi + n(t)) \tag{7.40}$$

This expression is not restricted to any particular case of vibrational noise. However, some insight may be gained by assuming that the noise is of a sinusoidal nature, with amplitude δ and phase offset θ, as follows:

$$s(t) = a + b\cos(\alpha + \phi + \delta\cos(2\pi f t + \theta)) \tag{7.41}$$

In a linear approximation, if the noise is not sinusoidal but its amplitude is small, we can sum, as pointed out by de Groot (1996), the contributions from each of the Fourier components of the vibration. When the noise amplitudes are not small, there may be nonlinear couplings between these components. In general the phase of the noise vibration is not coherent, but its phase varies at random. Thus, it is more logical to express the phase error as the root mean square (rms) value of the disturbed phase.

This rms error varies sinusoidally with the phase of the signal, with twice the frequency of the signal.

DeGroot (1996) made numerical simulations to calculate the effect of vibrational noise for several phase detecting algorithms. Figure 7.21 show the rms error for two of these algorithms. Here we may note the following general interesting facts, valid for most algorithms:

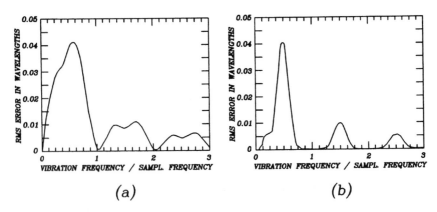

Figure 7.21 Vibrational root mean square error for two different algorithms. (a) For three sampling points algorithm; (b) for seven sampling points algorithm. (From de Groot and Deck 1996.)

1. The maximum vibrational sensitivity occurs when the vibration has a frequency of one-half the sampling frequency.
2. Zeros of the sensitivity occur at vibration frequencies which are multiples of the sampling frequency.
3. The sensitivity decreases exponentially for high vibrational frequencies. If the frequency is extremely high, only the contrast is reduced, but its dependence on the signal phase is lost.

Brophy (1990) studied the effect of additive noise and, as particular cases, mechanical vibrations whose frequencies are extremely high or of the order of the sampling rate.

An immediate practical consequence of these results is that to reduce the effect of the vibrations the sampling rate has to be as high as possible with respect to the vibration frequency. Unfortunately, high sampling rates require light detectors with a low integration time, which are quite expensive. Because of this, Deck (1996) proposed an interferometer with two light detectors, one with a fast integration time and the other with a low integration time, to reduce the interferometer sensitivity to vibrations.

Another approach to eliminate the effect of vibrations is to take the necessary irradiance samples at the same time, not in sequence (Kwon 1984; Kwon and Shough 1985; Kujawinska 1987; Kujawinska *et al.* 1990; Kujawinska and Robinson 1988, 1989).

7.9.3 Multiple Beam Interference and Frequency Mixing

Signal harmonics may also occur in the interference process if there are more than two beams interfering. In many cases this effect is due to the nature of the interferometer; in some other cases it is by accident. Typical examples of multiple beam interferometers are the Ronchi test and the Newton or Fizeau interferometers with high reflection beam splitters. However, even if the beam splitter in the Fizeau interferometer has a very low reflectance it is impossible to reduce multiple reflections to absolute zero.

Multiple reflections may also occur by accident, due to spurious unwanted reflections. The influence of these spurious reflections has been considered by several authors, for example, by Bruning *et al.* (1974) Schwider *et al.* (1983), Hariharan *et al.* (1987), Ai and Wyant (1988), and Dorrío *et al.* (1996).

We have seen in Chap. 1 the signal (irradiance) due to two beams with amplitudes A_1 and A_2. If, following Schwider *et al.* (1983) and Chaiu and Wyant (1988), we add a third coherent beam with amplitude B due to the coherent noise we obtain

$$E = A_1 \exp(i\phi) + A_2 \exp(i\alpha) + B \exp(i\beta) \tag{7.42}$$

where ϕ is the signal phase, α is the sampling reference function phase, and β is the extraneous coherent wave phase. The phases of these beams are referred to the same origin as the sampling reference functions. We also assume that there is no detuning, so that the reference wavefront may be considered to have the same phase as the sampling reference function. Thus the signal (irradiance) in the presence of coherent noise is given by

$$\begin{aligned} s' = E \cdot E^* = A_1^2 + A_2^2 + B^2 + 2A_1 A_2 \cos(\phi - \alpha) \\ + 2A_1 B \cos(\phi - \beta) + 2A_2 B \cos(\beta - \alpha) \end{aligned} \tag{7.43}$$

or

$$\begin{aligned} s' &= s + B^2 + 2A_1 B \cos(\phi - \beta) + 2A_2 B \cos(\beta - \alpha) \\ &= s + B^2 + 2A_1 B \cos(\phi - \beta) + 2A_2 B \cos(\beta) \cos(\alpha) \\ &\quad + 2A_2 B \sin(\beta) \sin(\alpha) \end{aligned} \tag{7.44}$$

Now we will study the particular case of algorithms with equally and uniformly spaced sampling points. In this case, the phase of the signal without coherent noise, from Eq. 5.19 or 5.102, is

$$\tan \phi = \frac{\displaystyle\sum_{n=1}^{N} s \sin(\alpha_n)}{\displaystyle\sum_{n=1}^{N} s \cos(\alpha_n)} \tag{7.45}$$

where α_n is the value of the phase α for the sampling point n. With the presence of coherent noise, the signal phase is

$$\tan \phi' = \frac{\displaystyle\sum_{n=1}^{N} s' \sin(\alpha_n)}{\displaystyle\sum_{n=1}^{N} s' \cos(\alpha_n)} \tag{7.46}$$

thus, using Eqs. 5.11, 5.13, and 5.14 we may find

$$\tan \phi' = \frac{\sin \phi + \dfrac{B}{A_1} \sin(\beta)}{\cos \phi + \dfrac{B}{A_1} \cos(\beta)} \tag{7.47}$$

Thus, the phase error is given by

$$\tan(\phi' - \phi) = -\frac{\dfrac{B}{A_1} \sin(\phi - \beta)}{1 + \dfrac{B}{A_1} \cos(\phi - \beta)} \tag{7.48}$$

As we see, this phase error is a periodic, although not exactly sinusoidal, function of the signal phase. Its period is equal to that of the signal frequency. This phase error is illustrated in Fig. 7.22.

This phase error can thus be substantially reduced by averaging two sets of measurements with a phase difference $\phi - \beta$ of π between them. This is possible only if another phase shifter is placed in the object beam. A phase shift in the reference beam does not change the phase difference $\phi - \beta$. Ai and Wyant (1988) point out that if the spurious light comes from the reference arm in the interferometer or by reflection from the test surface, this method does not work, and they propose an alternative way to eliminate the error.

In a Fizeau interferometer, as explained by Hariharan (1987), the spurious light appears due to multiple reflections between the object under test and the

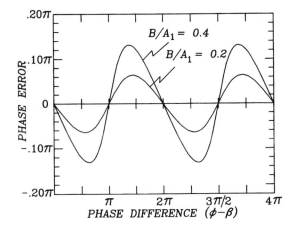

Figure 7.22   Phase error due to the presence of spurious coherent light beams.

reference surface (beam splitter). In this case the error can be minimized by proper selection of the sampling algorithm to eliminate the signal harmonics being generated.

Speckle noise is another kind of coherent noise that may become important in some applications, for example, speckle interferometry. This kind of noise may also be reduced in some cases (Creath 1985; Slettemoen and Wyant 1986).

7.9.4   Spherical Reference Wavefronts

When the reference wavefront in phase shifting interferometry is not plane but spherical, as in the spherical Fizeau interferometer, the spherical surface under test is shifted to introduce the phase shift. Then, if the phase shift at the center of the fringe pattern is 90°, the phase shift at the edge of the pupil would be slightly smaller. This introduces a phase error, as pointed out by Moore and Slaymaker (1980) and by Schwider *et al.* (1983). Nevertheless, this error is not large. For spherical test surfaces with numerical apertures smaller than 0.8, the phase error introduced can be smaller than one-hundredth of a wavelength. If this error becomes important, it can be minimized using Carré's algorithm.

7.9.5   Quantization Noise

As we studied in Sec. 2.4, when digitizing images the number of bits used to digitize the image defines the number of gray levels. A simple method to evaluate the quantization error has been given by Brophy (1990), showing that there is some correlation between signal samples taken 90° apart. Then he showed that for algorithms whose samples are taken at $90°$ intervals, the rms error σ due to quantization into Q gray levels is given by

$$\sigma = \frac{a}{\sqrt{3}bQ} \qquad (7.49)$$

where the a and b are the bias and amplitude, respectively, of the signal. As an example, if 8 bits are used, Q is equal to 256 gray levels. Then, if a/b is equal to 1, the rms quantization error σ is equal to 0.00036 wavelength or about $\lambda/2777$. This value is so small that it is difficult to reach this limit. Zhao and Surrel (1977) have made a detailed study of quantization noise for several algorithms.

Of course, the fringe contrast is not always perfect, and the ratio a/b may be much larger than 1. To avoid this error the signal must cover as much as possible of the detector dynamic range.

7.9.6 Photon Noise Phase Errors

There are other random phase errors, for example photon noise. This source of errors has been studied by Koliopoulos (1981), Brophy (1990), and Freischlad and Koliopoulos (1990). This error occurs due to fluctuations in the arrival of photons to the light detector, when the number of photon is not large. In other words, this noise appears where the signal is relatively small.

7.9.7 Laser Diode Intensity Modulation

When the phase shift is produced by phase current modulating a laser diode in an unbalanced interferometer, an amplitude modulation also occurs simultaneously with the phase modulation as described in Sec. 7.2.6. The phase error introduced by this undesired intensity modulation has been studied by Onodera and Ishii (1996) and by Surrel (1997b), assuming that the irradiance variation is linear with the phase shift.

REFERENCES

Ai C. and J. C. Wyant, "Effect of Piezoelectric Transducer Nonlinearity on Phase Shift Interferometry," *Appl. Opt.*, **26**, 1112–1116 (1987).

Ai C. and J. C. Wyant, "Effect of Spurious Reflection on Phase Shift Interferometry," *Appl. Opt.*, **27**, 3039–3045 (1988).

Bareket N., "Three Channel Phase Detector for Pulsed Wavefront Sensing," *Proc. SPIE*, **551**, 12 (1985).

Brophy C. P., "Effect of Intensity Error Correlation on the Computed Phase of Phase Shifting," *J. Opt. Soc. Am. A*, **7**, 537–541 (1990).

Bruning J. H., D. R. Herriott, J. E. Gallagher, D. P. Rosenfeld, A. D. White, and D. J. Brangaccio, "Digital Wavefront Measuring Interferometer for Testing Surfaces and Lenses," *Appl. Opt.*, **13**, 2693–2703 (1974).

Bryngdahl O., "Polarization Type Interference Fringe Shifter," *J. Opt. Soc. Am.*, **62**, 462–464 (1972).

Bryngdahl O., "Heterodyne Shearing Interferometers Using Diffractive Filters with Rotational Symmetry," *Opt. Commun.*, **17**, 43 (1976).

Burgwald G. M. and W. P. Kruger, "An Instant-On Laser for Distant Measurement," *Hewlett-Packard J.*, **21**, 14 (1970).

Carré, P., "Installation et Utilisation du Comparateur Photoelectrique et Interferentiel du Bureau International des Poids et Measures," *Metrologia*, **2**, 13–23 (1966).

Cheng Y.-Y. and J. C. Wyant, "Phase Shifter Calibration in Phase- Shifting Interferometry," *Appl. Opt.*, **24**, 30–49 (1985).

Crane R., "Interference Phase Measurement," *Appl. Opt.*, **8**, 538– 542 (1969).

Creath K., "Phase-Shifting Speckle Interferometry," *Appl. Opt.*, **24**, 3053–3058 (1985).

Creath K., "Comparison of Phase Measuring Algorithms," *Proc. SPIE*, **680**, 19–28 (1986).

Creath K., "Phase-Measurement Interferometry Techniques," in *Progress in Optics*, Vol. XXVI, E.Wolf, Ed., Elsevier Science Publishers, Amsterdam, 1988.

Creath K., "Phase Measurement Interferometry: Beware These Errors," *Proc. SPIE*, **1553**, 213–220 (1991).

Crescentini L., "Fringe Pattern Analysis in Low-Quality Interferograms," *Appl. Opt.*, **28**, 1231–1234 (1989)

Deck L., "Vibration-Resistant Phase-Shifting Interferometry," *Appl. Opt.*, **34**, 6555–6662 (1996).

de Groot P., "Vibration in Phase-Shifting Interferometry," *J. Opt. Soc. Am. A*, **12**, 354–365 (1995). "Errata," **12**, 2212 (1995).

de Groot P. and L. L. Deck, "Numerical Simulations of Vibration in Phase-Shifting Interferometry," *Appl. Opt.*, **35**, 2173–2181 (1996).

Dorrío B. V., A. F. Doval, C. López, R. Soto, J. Blanco-García, J. L. Fernández, and M. Pérez Amor, "Fizeau Phase-Measuring Interferometry Using the Moiré, Effect," *Appl. Opt.*, **34**, 3639–3643 (1995a).

Dorrío B. V., J. Blanco-García, A. F. Doval, C. López, R. Soto, J. Bugarín, J. L. Fernández, and M. Pérez Amor, "Surface Evaluation Combining the Moiré, Effect and Phase-Stepping Techniques in Fizeau Interferometry," *Proc. SPIE*, **2730**, 346–349 (1995b).

Dorrío B. V., J. Blanco-García, C. López, A. F. Doval, R. Soto, J. L. Fernández, and M. Pérez Amor, "Phase Error Calculation in a Fizeau Interferometer by Fourier Expansion of the Intensity Profile," *Appl. Opt.*, **35**, 61–64 (1996).

Freischlad K. and C. L. Koliopoulos, "Fourier Description of Digital Phase Measuring Interferometry," *J. Opt. Soc. Am. A*, **7**, 542–551 (1990).

Greivenkamp J. E. and J. H. Bruning, "Phase Shifting Interferometry," in *Optical Shop Testing*, 2nd Edition, D. Malacara, Ed., John Wiley and Sons, New York, 1992.

Hariharan P., "Quasi-Heterodyne Hologram Interferometry," *Opt. Eng.*, **24**, 632–638 (1985).

Hariharan P., "Digital Phase-Stepping Interferometry: Effects of Multiply Reflected Beams," *Appl. Opt.*, **26**, 2506–2507 (1987).

Hariharan P., B. F. Oreb, and N. Brown, "Real Time Holographic Interferometry: A Microcomputer System for the Measurement of Vector Displacements," *Appl. Opt.*, **22**, 876–880 (1983).

Hariharan P., B. F. Oreb, and T. Eiju, "Digital Phase Shifting Interferometry: A Simple Error Compensating Phase Calculator Algorithm," *Appl. Opt.*, **26**, 2504–2506 (1987).

Hibino K., "Suceptibility of Systematic Error-Compensating Algorithms to Random Noise in Phase-Shifting Interferometry," *Appl. Opt.*, **36**, 2084–2093 (1997).

Hu H. Z., "Polarization Heterodyne Interferometry Using Simple Rotating Analyzer, 1: Theory and Error Analysis," *Appl. Opt.*, **22**, 2052–2056 (1983).

Indebetow G., "Profile Measurement Using Projection of Running Fringes," *Appl. Opt.*, **17**, 2930–2933 (1978).

Ishii Y., J. Chen, and K. Murata, "Digital Phase Measuring Interferometry with a Tunable Laser Diode," *Opt. and Lasers in Eng.*, **14**, 293–309 (1991).

Joenathan C., "Phase-Measurement Interferometry: New Methods and Error Analysis," *Appl. Opt.*, **33**, 4147–4155 (1994).

Johnson G. W., D. C. Leiner, and D. T. Moore, "Phase Locked Interferometry," *Opt. Eng.*, **18**, 46–52 (1979).

Kinnstaetter K., A. Lohmann, W. Schwider, and J. N. Streibl, "Accuracy of Phase Shifting Interferometry," *Appl. Opt.*, **27**, 5082–5089 (1988).

Koliopoulos C. L., Interferometric Optical Phase Measurement Techniques, Ph.D. Dissertation, University of Arizona, Tucson, AZ, 1981.

Koliopoulos C. L., "Simultaneous Phase Shift Interferometer," *Proc. SPIE*, **1531**, 119, 133 (1991).

Kothiyal M. P. and C. Delisle, "Optical Frequency Shifter for Heterodyne Interferometry Using Counterrotating Wave Plates," *Opt. Lett.*, **9**, 319–321 (1984).

Kothiyal M. P. and C. Delisle, "Rotating Analyzer Heterodyne Interferometer: Error Analysis," *Appl. Opt.*, **24**, 2288–2290 (1985).

Kujawinska M., "Multichannel Grating Phase-Stepped Interferometers," *Opt. Appl.*, **17**, 313–332 (1987).

Kujawinska M., "Spatial Phase Measurement Methods," in *Interferogram Analysis*, D. W. Robinson and G. T. Reid, Eds., Institute of Physics Publ., Bristol and Philadelphia, 1993.

Kujawinska M. and D. W. Robinson, "Multichannel Phase-Stepped Holographic Interferometry," *Appl. Opt.*, **27**, 312–320 (1988).

Kujawinska M. and D. W. Robinson, "Comments on the Error Analysis and Adjustment of the Multichannel Phase-Stepped Holographic Interferometers," *Appl. Opt.*, **28**, 828–829 (1989).

Kujawinska M., L. Salbut, and K. Patorski, "Three Channel Phase Stepped System for Moiré Interferometry," *Appl. Opt.*, **29**, 1633–1636 (1990).

Kujawinska M., L. Salbut, and R. Jozwicki, "Moiré and Spatial Carrier Approaches to Phase Shifting Interferometry," *Proc. SPIE*, **1553**, 44–54 (1991).

Kwon O. Y., "Multichannel Phase Shifted Interferometer," *Opt. Lett.*, **9**, 59–61 (1984).

Kwon O. Y. and D. M. Shough, "Multichannel Grating Phase Shift Interferometer," *Proc. SPIE*, **599**, 273–279 (1985).

Kwon O. Y., D. M. Shough, and R. A. Williams, "Stroboscopic Phase-Shifting Interferometry," *Opt. Lett.*, **12**, 855–857 (1987).

Lai G. and T. Yatagai, "Generalized Phase Shifting Interferometry," *J. Opt. Soc. Am. A*, **8**, 822–827 (1991).

Malacara D., I. Rizo and A. Morales, "Interferometry and the Doppler Effect," *Appl. Opt.*, **8**, 1746–1946 (1969).

Massie N. A., "Heterodyne Interferometry," in *Optical Interferograms–Reduction and Interpretation*, ASTM Symp., Tech. Publ. 666, A. H. Guenther and D. H. Liedbergh, Eds., American Society for Testing and Materials, West Consohocken, PA, 1978.

Massie N. A., "Real Time Digital Heterodyne Interferometry: A System," *Appl. Opt.*, **19**, 154–160 (1980).

Massie N. A., "Digital Heterodyne Interferometry," *Proc. SPIE*, **816**, 40–48 (1987).

Massie N. A. and R. D. Nelson, "Beam Quality of Acousto Optic Phase Shifters," *Opt. Lett.*, **3**, 46–47 (1978).

Massie N. A., R. D. Nelson, and S. Holly, "High Performance Real-Time Heterodyne Interferometry," *Appl. Opt.*, **18**, 1797–803 (1979).

Matthews H. J., D. K. Hamilton, and C. J. R. Sheppard, "Surface Profiling by Phase Locked Interferometry," *Appl. Opt.*, **25**, 2372–2374 (1986).

Moore D. T., Gradient Index Optics and Tolerancing, Ph.D. Thesis, University of Rochester, 1973.

Moore D. T. and B. E. Truax, "Phase Locked Moiré Fringe Analysis for Automated Contouring of Diffuse Surfaces," *Appl. Opt.*, **18**, 91–96 (1979).

Moore D. T., R. Murray, and F. B. Neves, "Large Aperture AC Interferometer for Optical Testing," *Appl. Opt.*, **17**, 3959–3963 (1978).

Moore R. C. and F. H. Slaymaker, "Direct Measurement of Phase in a Spherical Fizeau Interferometer," *Appl. Opt.*, **19**, 2196–2200 (1980).

Nakadate S. and H. Saito, "Fringe Scanning Speckle Pattern Interferometry," *Appl. Opt.*, **24**, 2172–2180 (1985).

Nakadate S., H. Saito, and T. Nakajima, "Vibration Measurement Using Phase-Shifting Stroboscopic Holographic Interferometry," *Opt. Acta*, **33**, 1295–1309 (1986).

Ohyama N., S. Kinoshita, A. Cornejo-Rodríguez,. T. Honda, and J. Tsujiuchi, "Accuracy of Determination with Unequal Reference Phase Shift," *J. Opt. Soc. Am. A*, **5**, 2019–2025 (1988).

Okoomian H. J., "A Two Beam Polarization Technique to Measure Optical Phase," *Appl. Opt.*, **8**, 2363–2365 (1969).

Onodera R. and Y. Ishii, "Phase-Extraction Analysis of Laser- Diode Phase-Shifting Interferometry That Is Insensitive to Changes in Laser Power," *J. Opt. Soc. Am. A*, **13**, 139–146 (1996).

Rathjen C., "Statistical Properties of Phase-Shift Algorithms," *J. Opt. Soc. Am.*, **12**, 1997 (1995).

Robinson D. and D. Williams, "Digital Phase Stepping Speckle Interferometry," *Opt. Commun.*, **57**, 26 (1986).

Salbut L. and K. Patorski, "Polarization Phase Shifting Method for Moiré Interferometry and Flatness Testing," *Appl. Opt.*, **29**, 1471–1476 (1990).

Sasaki O. and H. Okasaki, "Sinusoidal Phase Modulating Interferometry for Surface Profile Measurement," *Appl. Opt.*, **25**, 3137–3140 (1986a).

Sasaki O. and H. Okasaki, "Analysis of Measurement Accuracy in Sinusoidal Phase Modulating Interferometry," *Appl. Opt.*, **25**, 3152–3158 (1986b).

Sasaki O., H. Okasaki, and M. Sakai, "Sinusoidal Phase Modulating Interferometer Using the Integrating-Bucket Method," *Appl. Opt.*, **26**, 1089–1093 (1987).

Sasaki O., T. Okamura, and T. Nakamura, "Sinusoidal Phase Modulating Fizeau Interferometer," *Appl. Opt.*, **29**, 512–515 (1990a).

Sasaki O., K. Takahashi, and T. Susuki, "Sinusoidal Phase Modulating Laser Diode Interferometer with a Feedback Control System to Eliminate External Disturbance," *Opt. Eng.*, **29**, 1511–1515 (1990b).

Schwider J., "Phase Shifting Interferometry: Reference Phase Error Reduction," *Appl. Opt.*, **28**, 3889–3892 (1989).

Schwider J., R. Burow, K.-E. Elssner, J. Grzanna, R. Spolaczyk, and K. Merkel, "Digital Wave-Front Measuring Interferometry: Some Systematic Error Sources," *Appl. Opt.*, **22**, 3421–3432 (1983).

Shagam R. N., "A.C. Measurement Technique for Moiré Interferograms," *Proc. SPIE*, **429**, 35 (1983).

Shagam R. N. and J. C. Wyant, "Optical Frequency Shifter for Heterodyne Interferometers Using Multiple Rotating Polarization Retarders," *Appl. Opt.*, **17**, 3034–3035 (1978).

Slettemoen G. Å. and J. C. Wyant, "Maximal Fraction of Acceptable Measurements in Phase Shifting Interferometry: A Theoretical Study," *J. Opt. Soc. Am. A*, **3**, 210–214 (1986).

Smythe E. R. and R. Moore, "Instantaneous Phase Measuring Interferometry," *Proc. SPIE*, **429**, 16–21 (1983).

Smythe E. R. and R. Moore, "Instantaneous Phase Measuring Interferometry," *Opt. Eng.*, **23**, 361–364 (1984).

Sommargren G. E., "Up-Down Frequency Shifter for Optical Heterodyne Interferometry," *J. Opt. Soc. Am.*, **65**, 960–661 (1975).

Sommargren G. E., "Optical Heterodyne Profilometry," *Appl. Opt.*, **20**, 610–618 (1981).

Srinivasan V., H. C. Liu, and M. Halioua, "Automatic Phase- Measuring Profilometry: A Phase Measuring Approach," *Appl. Opt.*, **24**, 185–188 (1985).

Stetson K. A. and W. R. Brohinsky, "Phase Shifting Technique for Numerical Analysis of Time Average Holograms of Vibrating Objects," *J. Opt. Soc. Am. A*, **5**, 1472–1476 (1988).

Stevenson W. H., "Optical Frequency Shifting by Means of a Rotating Diffraction Grating," *Appl. Opt.*, **9**, 649–652 (1970).

Surrel Y., "Additive Noise Effect in Digital Phase Detection," *Appl. Opt.*, **36**, 271–276 (1997a).

Surrel Y., "Design of Phase Detection Algorithms Insensitive to Bias Modulation," *Appl. Opt.*, **36**, 1–3 (1997b).

Susuki T. and R. Hioki, "Translation of Light Frequency by a Moving Grating," *J. Opt. Soc. Am.*, **57**, 1551–1551 (1967).

Thalmann R. and R. Dändliker, "Holographic Contouring Using Electronic Phase Measurement," *Opt. Eng.*, **24**, 930–935 (1985).

Wang, Z., M. S. Graça, P. J. Bryanston-Cross, and D. J. Whitehouse, "Phase Shifted Image Matching Algorithm for Displacement Measurement," *Opt. Eng.*, **35**, 2327–2332 (1996).

Wingerden J. van, H. J. Frankena, and C. Smorenburg, "Linear Approximation for Measurement Errors in Phase Shifting Interferometry," *Appl. Opt.*, **30**, 2718–2729 (1991).

Wyant J. C., "Use of an A.C. Heterodyne Lateral Shear Interferometer with Real Time Wavefront Correction Systems," *Appl. Opt.*, **14**, 2622–2626 (1975).

Wyant J. C. and R. N. Shagam, "Use of Electronic Phase Measurement Techniques in Optical Testing," *Proc ICO-11, Madrid*, 659–662 (1978).

Zhao B. and Y. Surrel, "Effect of Quantization Error on the Computed Phase of Phase-Shifting Measurements," *Appl. Opt.*, **36**, 2070–2075 (1997).

Zhi H., "Polarization Heterodyne Interferometry Using a Simple Rotating Analyzer 1: Theory and Error Analysis," *Appl. Opt.*, **22**, 2052–2056 (1983).

8

Spatial Linear and Circular
Carrier Analysis

8.1 SPATIAL LINEAR CARRIER ANALYSIS

In phase shifting techniques several frames have to be measured. This requires the shifting of the phase by means of piezoelectric crystals or any other equivalent device. In the spatial carrier methods described in this chapter, only a single frame is necessary to obtain the wavefront. Of course, if desired, several wavefronts may be averaged to improve the result. There are several important practical differences between these two basic methods:

1. In phase shifting methods a minimum of three interferogram frames are needed. In spatial carrier methods only one is necessary.
2. In phase shifting interferometry, to avoid the effect of vibrations, three or more frames have to be taken simultaneously. In spatial carrier analysis there is no problem, since only one frame is taken.
3. In phase shifting methods the sign of the wavefront deformations is determined. In spatial carrier methods the sign cannot be determined, since only one frame is taken. To determine the sign it is necessary to know the sign of at least one of the aberration wavefront components, for example, the sign of the tilt introducing the carrier.
4. In phase shifting methods hardware requirements are greater, since an accurately calibrated phase shifter is needed. In spatial carrier methods a more sophisticated mathematical computer processing is necessary.

5. If a stable environment free of vibrations and turbulence is available (sometimes this is impossible), more accuracy and precision are possible with phase shifting methods than with spatial carrier methods.

8.1.1 Introduction of a Linear Carrier

A large tilt about the y axis in an interferogram may be considered a linear carrier in the x direction. Interferograms with a spatial linear carrier may be analyzed to obtain the wavefront shape by processing the information in the interferogram plane (space domain) or in the Fourier plane (frequency domain). We will study both methods in this chapter. Reviews on the analysis of interferograms using a spatial carrier have been published by Takeda (1987), Kujawinska (1993), and Vlad and Malacara (1994).

The irradiance in an interferogram with a large tilt, along a line parallel to the x axis, is a perfectly sinusoidal function if the two interfering wavefronts are flat. In other words, if the reference wavefront is flat and the wavefront under test is also flat, the fringes are straight, parallel to the y axis, and equidistant. If the wavefront under test is not perfect, this irradiance function is a nearly sinusoidal function with phase modulation. The phase modulation is due to the wavefront deformation $W(x, y)$. If a tilt θ about the y axis between the two wavefronts is introduced, the signal (irradiance) $s(x, y)$ may be written from Eq. 1.4 as

$$s(x, y) = a + b\cos[2\pi f x - kW(x, y)]$$
$$= a + 0.5b\exp\{i[2\pi f x - kW(x, y)]\}$$
$$+ 0.5b\exp\{-i[2\pi f x - kW(x, y)]\} \tag{8.1}$$

where the coefficients a and b may vary for different points on the interferogram, that is, they are functions of x and y, but for notational simplicity this dependence has been omitted. The carrier spatial frequency introduced by the tilt is $f = (\sin\theta)/\lambda$. An example of an interferogram with a linear carrier is presented in Fig. 8.1. Here, the wavefront deformations $W(x, y)$ are for the non-tilted wavefront, before the linear carrier is introduced. To be more precise, a wavefront is said to have no tilt about the x axis when the maximum positive and negative slopes in the x direction have the same magnitudes. The phase modulating function $W(x, y)$ may be obtained using standard communication techniques that are quite similar to holographic techniques.

To achieve this demodulation it is necessary that for a fixed value of y inside the aperture the phase modulating function $W(x, y)$ increases in a monotonic manner with the value of x. This is possible only if the tilt θ between the two wavefronts is chosen so that the slope of the fringes does not change sign inside the interferogram aperture. An immediate consequence of this is that no

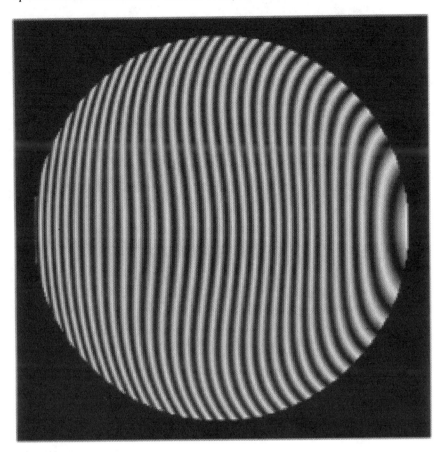

Figure 8.1 Interferogram with a linear carrier.

closed fringes appear in the interferogram and that no fringe in the interferogram aperture crosses more than once any scanning line parallel to the x axis. Thus, if the tilt has a positive value, we have the condition that

$$\frac{\partial(x \sin\theta - W(x, y))}{\partial x} > 0 \tag{8.2}$$

without any change in sign for all points inside the interferogram, or equivalently

$$\sin\theta > \left(\frac{\partial W(x, y)}{\partial x}\right)_{max} \tag{8.3}$$

This result may be interpreted by saying that the slope (tilt) of the reference wavefront has to be greater than the maximum (positive) slope of the wavefront under test in the x direction. If the wavefront under test is almost flat, the tilt can be almost anything between a relatively small value and the Nyquist limit (two pixels per fringe). On the other hand, Macy (1983) and Hatsuzawa (1985) showed that increasing the tilt increases the amount of measured information but reduces the precision. They found that an optimum value for the tilt is about four pixels per fringe.

An interesting point of view is to regard the interferogram with a linear carrier as an off-axis hologram. Then, Eq. 8.3 is equivalent to the condition for the image spot of the first order of diffraction to be separated without any overlapping from the zero-order point at the optical axis.

A problem when setting up the interferogram is the selection of the tilt angle θ that satisfies this condition. This tilt does not have to be very precise, but it is always better to be on the upper side, as long as the Nyquist limit for the detector being used is not exceeded, as will be described in detail later in this chapter. In the case of aspheric surfaces it is easy to approach the Nyquist limit due to the uneven separation between the fringes. In this case we are bounded between the lower limit for the tilt imposed by condition 8.3 and the upper limit imposed by the Nyquist condition.

The lower limit for the tilt in Eq. 8.3 was derived from purely geometrical considerations. However, in any real case the finite size or any uneven illumination of the pupil widens the diameter of the spectrum, because of diffraction. Then, the zero-order image is not a point but an Airy diffraction image (if the pupil is evenly illuminated) and the first-order image is the convolution of this Airy function with the geometrical image. This effect due to the finite size of the pupil introduces some artifacts in the results, mainly near the edge of the interferogram, but they can be minimized by any of several procedures to be described in Sec. 8.1.3.

The approximate minimum required amount of tilt may be experimentally obtained by several different methods, for example:

1. The interferogram tilt is first adjusted to obtain the maximum rotational symmetry. Then the tilt is slowly introduced until the minimum local slope of a fringe in the interferogram has zero value (parallel to the x axis) at the edge of the fringe, as shown in Fig. 8.2. The magnitude of this tilt may be found from the interferometer adjustment.

2. Another procedure is to take the fast Fourier transform of the irradiance and adjust the tilt in an iterative manner until the first-order lobe is clearly separated from the zero-order lobe. Then the distance from the centroid of the first order to that of the zero order is the minimum amount of tilt to introduce, from a geometrical point of view. Later we will see that

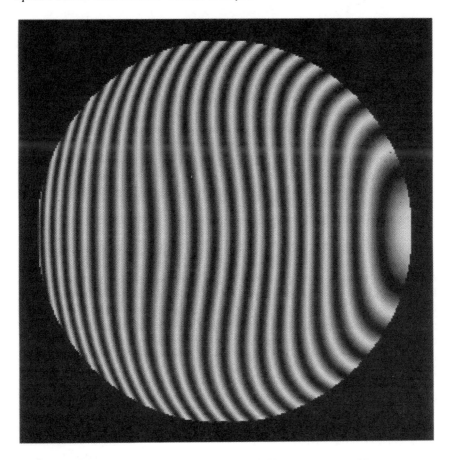

Figure 8.2 Interferogram in which the minimum fringe slope on the interferogram is zero.

a slightly greater tilt might be necessary to avoid phase errors due to diffraction effects.

8.1.2 Holographic Interpretation of the Interferogram

An interferogram with a large linear carrier is formed by the interference of the wavefront to be measured with a flat wavefront forming an angle θ between them, as in Fig. 8.3.

This interferogram can be interpreted as an off-axis hologram of the wavefront $W(x, y)$. The similarity between a hologram and an interferogram has been

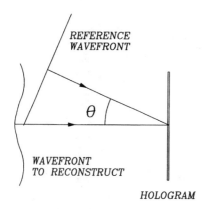

REFERENCE
WAVEFRONT

θ

WAVEFRONT
TO RECONSTRUCT

HOLOGRAM

Figure 8.3   Recording a hologram.

recognized for many years (Horman 1965). The wavefront can be reconstructed by illumination of the hologram with a flat reference wavefront with amplitude $r(x, y)$ and with tilt θ_r. This reference reconstructing wavefront does not necessarily have the same inclination θ as the original flat wavefront used when taking the hologram. It may be almost the same as shown in Fig. 8.4, but it may be different if desired. It will be seen later that the condition in Eq. 8.3 is still valid even if these angles are very different.

The complex amplitude $r(x, y)$ of the reconstructing reference wavefront may be written as

$$r(x, y) = \exp[i(2\pi f_r x)]$$
$$= \cos(2\pi f_r x) + i \sin(2\pi f_r x) \tag{8.4}$$

where $f_r = (\sin\theta_r)/\lambda$. Thus, the amplitude $e(x, y)$ in the hologram plane is given by

$$e(x, y) = r(x, y)s(x, y) = s(x, y)\exp[i(2\pi f_r x)]$$
$$= a\exp[i(2\pi f_r x)] + 0.5b\exp\{i[2\pi(f + f_r)x - kW(x, y)]\}$$
$$+ 0.5b\exp\{-i[2\pi(f - f_r)x - kW(x, y)]\} \tag{8.5}$$

These diffracted wavefronts, as expressed here, are completely general, independently of the relative magnitude of the angles used during the hologram formation and during the reconstruction step.

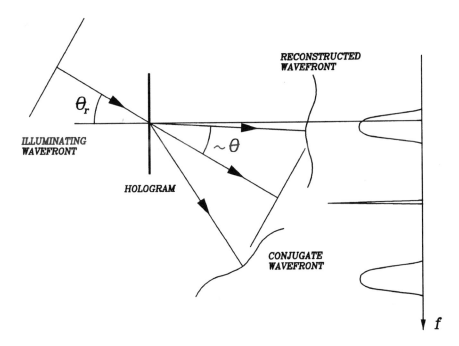

Figure 8.4 Reconstruction of a wavefront with a hologram.

These wavefronts and their frequency distribution in the Fourier plane (spectra) will now be studied. To begin, let us first remember that the phase ϕ of a sinusoidal function $\exp \phi$, its frequency f, and the angular spatial frequency ω are related by

$$\omega = 2\pi f = \frac{\partial \phi}{\partial x} \tag{8.6}$$

where a positive slope for the phase and hence for the wavefront is related to a positive spatial frequency. Thus, according to this sign convention, the axes on the Fourier plane must have opposite directions to those on the interferogram.

Then the linear carrier spatial frequency introduced by the tilt in the flat wavefront used when forming the hologram is

$$f = \frac{\omega}{2\pi} = \frac{\sin \theta}{\lambda} \tag{8.7}$$

The spatial frequency spectrum produced by the wavefront $W(x, y)$ in a direction parallel to the x axis is given by

$$f_W(x, y) = \frac{\omega_W(x, y)}{2\pi} = \frac{1}{\lambda} \frac{\partial W(x, y)}{\partial x} \tag{8.8}$$

Thus, the spatial frequency is directly proportional to the wavefront slope in the x direction at the point (x, y). The first term in Eq. 8.5 represents the flat nondiffracted wavefront with tilt θ_r. The spatial frequency of this term, with zero order, is the reference frequency f_r and has a delta distribution in the Fourier plane. As pointed out before, this frequency is not necessarily equal to that of the carrier, as obtained with relation 8.6 and shown in Fig. 8.4, given by

$$f_r = \frac{\omega_r}{2\pi} = \frac{\sin \theta_r}{\lambda} \tag{8.9}$$

This reference spatial frequency was defined when the multiplying function $r(x, y)$, or in other words the angle for the reference wavefront in Eq. 8.4, was determined.

The second term, of order -1, represents a wave with deformations conjugate to those of the wavefront being reconstructed. The spatial frequency of this function in a direction parallel to the x axis is $f_{-1}(x, y)$, given by

$$f_{-1}(x, y) = \frac{\omega_{-1}(x, y)}{2\pi} = \frac{\sin \theta + \sin \theta_r}{\lambda} - \frac{1}{\lambda} \frac{\partial W(x, y)}{\partial x} \tag{8.10}$$

Its deviation from this average value depends on the wavefront slope in the x direction at the point (x, y) on the interferogram, that is, on the frequency $f_W(x, y)$.

The third term, of order $+1$, represents the wavefront under test, with a frequency $f_{+1}(x, y)$ in the x direction, given by

$$f_{+1}(x, y) = \frac{\omega_{+1}(x, y)}{2\pi} = \frac{\sin \theta - \sin \theta_r}{\lambda} - \frac{1}{\lambda} \frac{\partial W(x, y)}{\partial x} \tag{8.11}$$

8.1.3 Fourier Spectrum of Interferogram and Filtering

The expression for the spatial frequency content in the interferogram derived in the preceding section gives us the basics for an understanding of the Fourier spectrum. As pointed out before, this spectrum is geometrical, that is, this model does not take into account diffraction effects due to the pupil boundaries or any unevenness in the pupil illumination.

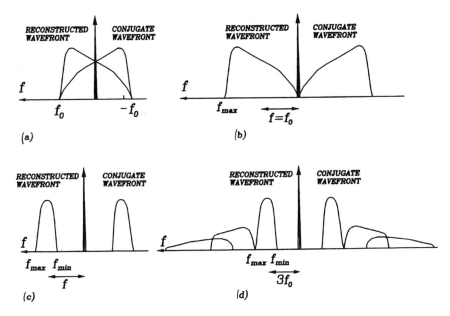

Figure 8.5 Spatial frequency distribution along the x axis in an interferogram with a linear carrier slightly larger than the minimum.

From Eq. 8.8 we can see that the half-bandwidth f_0 along the x axis for the first-order lobe is

$$f_0 = \frac{1}{\lambda}\left(\frac{\partial W}{\partial x}\right)_{\text{max}}$$

(8.12)

as illustrated in Fig. 8.5a. Let us now assume that a spatial linear carrier with frequency f along the x axis is introduced. Then the maximum and minimum frequencies f_{max} and f_{min}, respectively, along the x axis are

$$f_{\text{max}} = f + f_0$$

(8.13)

and

$$f_{\text{min}} = f - f_0$$

(8.14)

When the minimum tilt required by Eq. 8.3 is introduced, we have a spectrum like that in Fig. 8.5b, with a minimum fringe frequency equal to zero (fringe slope zero).

It is desirable to set the linear carrier spatial frequency to its minimum allowed value if a highly aberrated wavefront is under measurement, to avoid having the maximum fringe frequency exceed the Nyquist limit. On the other hand, if the wavefront has small deformations compared with the wavelength, it is convenient, as will be described in the next section, to select a spatial carrier with a spatial frequency much larger than the required minimum, as in Fig. 8.5c.

The minimum allowed linear carrier spatial frequency f has been found with the assumption that we have a sinusoidal phase-modulated signal without any harmonic components (equivalently, we can say that the carrier is not sinusoidal, but distorted). Nevertheless, quite frequently the signal (or carrier) contains harmonics. This is the case when measuring Ronchi patterns or multiple beam interferograms or when using a light detector with a nonlinear response. Then, the maximum allowed linear carrier is three times the former value, as illustrated in Fig. 8.5d.

It is important to remember that the finite size of the detector element acts as a low pass filter, removing some of the harmonic frequencies, before the sampling process is finished. This low pass filtering may be quite important to help avoid having some high frequency components exceed the Nyquist limit, producing aliasing noise.

If the linear carrier in the interferogram is larger than the allowed minimum, the first-order lobe can always be isolated with a suitable bandpass filter without regard to the selected reference frequency. For practical reasons that will become clear later in this chapter, it is desirable for simplicity to use a low pass filter, in other words, a bandpass centered at the origin.

Figure 8.6 shows the minimum widths of the low passbands that should be used when filtering three common Fourier spectrum distributions. Here, a reference frequency equal to the carrier frequency has been assumed.

We see that in order to perform good low pass filtering we must previously determine the values of the two parameters, i.e., the carrier frequency f and the band half-width f_0 of the first-order lobe. Alternatively, we must determine the maximum and minimum fringe frequencies f_{max} and f_{min}, respectively. There are several methods to obtain these values, as described by Kujawinska (1993) and Lai and Yatagai (1994), for example:

1. Directly setting or measuring these parameters when adjusting the interferometer to obtain the desired interferogram.
2. Calculating the fast Fourier transform of the interferogram and isolating the first-order lobe, automatically or with operator intervention.
3. Automatically estimating the fringe frequencies along the x axis, with a zero crossing algorithm, after high pass filtering is used to remove constant or very low frequency terms.

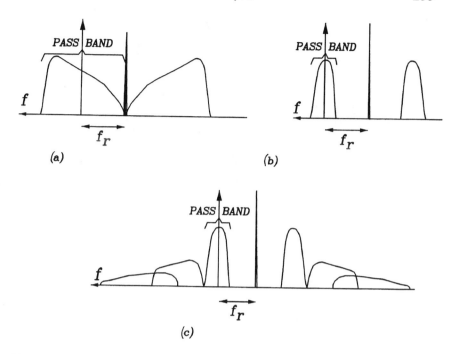

Figure 8.6 Carrier frequencies and pass band widths for three common cases.

4. With a simple rough estimation of the desired parameters, the wavefront may be calculated, even if some errors are introduced. From the calculated wavefront, a better approximation for the desired parameters can be obtained. Then a new iteration would produce better results.

Let us consider that the signal is sinusoidal and phase-modulated, without harmonic components, either because they are not present in the original signal or because they have been filtered out by the sampling procedure with finite-size detectors (pixels). In this case the reference frequency f_r can deviate from the carrier frequency f without introducing any errors, if (1) the reference frequency is within the limits

$$\frac{f + f_0}{2} < f_r \tag{8.15}$$

where f_0 is the band half-width of the first lobe and (2) the filtering band half-width is slightly smaller than the selected reference frequency, which can be larger than f_0.

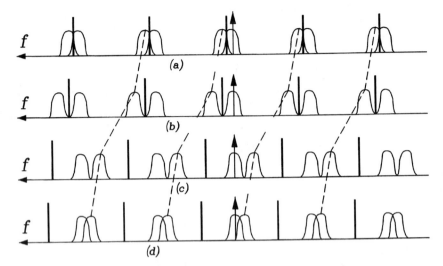

Figure 8.7 Fourier spectrum when the interferogram is sampled. (a) Insufficient carrier frequency, (b) minimum carrier frequency, (c) carrier frequency at Nyquist limit, and (d) Nyquist limit exceeded.

It is interesting to see that if the wavefront deformations are small, so that the carrier frequency f is much larger than the band half-width f_0, this condition is transformed into

$$\frac{f}{2} < f_r \tag{8.16}$$

In conclusion, if the signal is not distorted and the carrier frequency is much larger than the required minimum ($f > f_0$), the reference frequency can have any value larger than half the signal frequency. Even if there are some harmonics, this criterion can help to set a good starting point in an iterative process.

The discrete sampling of the interferogram in the hologram model may be considered as a diffraction grating superimposed on the hologram. Thus, the Fourier spectrum is split into many copies of the hologram spectrum, as shown in Fig. 8.7. We can see in this figure how by increasing the tilt between the two wavefronts the carrier frequency is also increased and approaches the Nyquist limit.

8.1.4 Pupil Diffraction Effects

The pupil of an interferogram is not infinitely extended but finite and most of the time circular. Also, its pupil illumination may be uneven. Thus, the geometrical description of the Fourier spectrum of the interferogram is not complete. The correct Fourier spectrum can be obtained with the convolution of the geometrical spectrum with the Airy function if the pupil illumination is even. This increases the width of all lobes in the spectrum. Then the zero-order lobe is just the Airy function.

The diameter of the first dark ring of the Airy function is equal to $1.22/D$, where D is the diameter of the pupil. With the geometrical model, this spatial frequency $1.22/D$ corresponds to 1.22 tilt fringes. Thus, to obtain more complete separation of the first- and zero-order lobes, an additional tilt of about two to three fringes should be added to the minimum required linear carrier obtained with the geometrical model.

It must be remembered, however, that the rings in the Airy diffraction pattern extend over a large area. Thus, it is frequently convenient to modify the pupil boundaries in some manner, so that the rings are damped down, making possible a good isolation of the first-order lobe. This ring damping may be obtained with either of the following two methods:

1. By extrapolation of the fringes outside the pupil boundaries. This procedure is described in detail in Chap. 3.
2. By softening the edge of the pupil with a two-dimensional Hamming filter, as proposed by Takeda *et al.* (1982). The Hanning or $\cos^4$ filter function may also be used with good results (Malcolm *et al.* 1989; Frankowski *et al.* 1989). The one-dimensional Hamming function was defined in Chap. 3, but a two-dimensional circular Hamming filter may be written as

$$h(x, y) = 0.54 + 0.46 \cos \frac{2\pi \sqrt{(x^2 + y^2)}}{D} \qquad \text{for } (x^2 + y^2) < D^2$$

$$= 0 \qquad\qquad\qquad\qquad\qquad \text{elsewhere}$$

$$(8.17)$$

where D is the pupil diameter.

To better understand this, let us consider Fig. 8.8, where we have some one-dimensional signals on the left and their Fourier transforms on the right. In Fig. 8.8a an infinitely extended sinusoidal signal produces the Fourier transform with only delta functions. In Fig. 8.8b the signal has been limited in extension, as in any finite size interferogram. Then, each of the delta functions is transformed in a sinc function whose width is inversely proportional to the pupil size. In Fig. 8.8c the signal is no longer sinusoidal but has a phase modulation. The diffraction effects had been minimized by artificially extending the pupil in

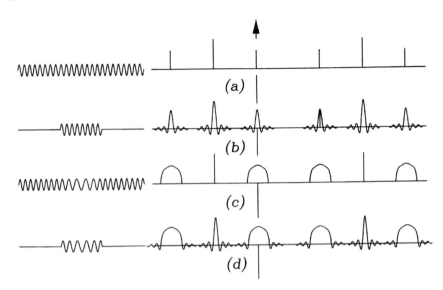

Figure 8.8 Some signals and their Fourier transforms when discretely sampled. (a) Infinitely extended sinusoidal signal; (b) sinusoidal signal with a finite aperture; (c) phase-modulated signal with sinusoidal signal on each side to extend it on both sides; (d) a phase-modulated signal with a finite aperture.

both directions with sinusoidal signals. In this case the Fourier transforms terms corresponding to the orders representing the reconstructed wavefront and its conjugate wavefront are widened, as we have seen before in this chapter. In Fig. 8.8d we have a phase-modulated signal with a finite extension due to the pupil size.

Diffraction effects can introduce some relatively small phase errors at the edge of the pupil when the phase is calculated using space phase demodulation. These errors, however, become more important when using the Fourier transform method. Both of these methods are described later in this chapter.

8.2 SPACE PHASE DEMODULATION WITH A LINEAR CARRIER

The space domain phase demodulation of interferograms with a linear carrier began with the pioneering work of Ichioka and Inuiya (1972). Since then, several other phase demodulation methods have been developed. Some of these procedures are described in the following sections.

8.2.1 Basic Space Phase Demodulation Theory

To describe the space phase demodulation method, let us follow the holographic model, where to separate the three waves the hologram (interferogram) is illuminated (multiplied) with a flat reference wave in Eq. 8.4, obtaining Eq. 8.5, which may be written as

$$z(x, y) = r(x, y)s(x, y) = z_C(x, y) + iz_S(x, y) \tag{8.18}$$

where

$$z_S(x, y) = s(x, y)\sin(2\pi f_r x) \tag{8.19}$$

and

$$z_C(x, y) = s(x, y)\cos(2\pi f_r x) \tag{8.20}$$

or, using Eq. 8.1,

$$
\begin{aligned}
z_S(x, y) &= s(x, y)\sin(2\pi f_r x) \\
&= a\sin(2\pi f_r x) - \frac{b}{2}\sin(2\pi(f - f_r)x - kW(x, y)) \\
&\quad + \frac{b}{2}\sin(2\pi(f + f_r)x - kW(x, y))
\end{aligned}
\tag{8.21}
$$

and

$$
\begin{aligned}
z_C(x, y) &= s(x, y)\cos(2\pi f_r x) \\
&= a\cos(2\pi f_r x) + \frac{b}{2}\cos(2\pi(f - f_r)x - kW(x, y)) \\
&\quad + \frac{b}{2}\cos(2\pi(f + f_r)x - kW(x, y))
\end{aligned}
\tag{8.22}
$$

These expressions are equivalent to Eqs. 5.27 and 5.28 in Chap. 5. An example of the functions $z_S(x, y)$ and $z_C(x, y)$ and their low pass filtered counterparts $\bar{z}_S(x)$ and $\bar{z}_C(x)$ are illustrated in Fig. 8.9. It is interesting to compare these plots with those in Fig. 5.4.

With the holographic model, the terms with frequency f_r and with frequency $2f_r$ can be eliminated with a mask. In practice, however, these two high frequency terms are eliminated by means of a low pass spatial filter. The filter as

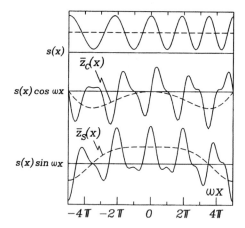

Figure 8.9 Signal along a line in an interferogram with a linear carrier, multiplied by (a) a sine function and (b) a cosine function.

well as the multiplications may be implemented with analog as well as discrete sampling procedures, as will be described in the next few sections.

Once the high frequency terms are filtered out, we may easily find the phase at any point x as

$$[2\pi(f - f_r)x - kW(x, y)] = -\tan^{-1}\left[\frac{\bar{z}_S(x, y)}{\bar{z}_C(x, y)}\right] \tag{8.23}$$

The first term on the left-hand side, $2\pi(f - f_r)x$, is a residual tilt that appears if the carrier and reference frequencies are not exactly equal, but it may easily be removed, if desired, in the final result.

8.2.2 Phase Demodulation with an Aspheric Reference

If the ideal shape of the wavefront being measured is aspheric, this ideal shape is subtracted from the calculated wavefront deformations to obtain the final wavefront error. A slightly different alternative procedure may be employed by using an aspheric wavefront instead of a flat wavefront as a reference. Let us now study this method to assess its relative advantages or disadvantages.

Since the interferogram may be interpreted as a hologram of the wavefront $W(x, y)$, with a reference wavefront with an inclination θ, the flat reference wavefront may be reconstructed if we illuminate this interferogram with the wavefront $W(x, y)$. Hence, a null test may be obtained if we illuminate (reconstruct) with the ideal aspheric wavefront W_r as follows:

$$r(x, y) = \exp[i(2\pi f_r x - kW_r(x, y))] \tag{8.24}$$

Thus we obtain

$$\begin{aligned}
s(x, y)r(x, y) &= s(x, y) \exp[2\pi f_r x - kW_r(x, y)] \\
&= a \exp[i(2\pi f_r x - kW_r(x, y))] \\
&\quad + \frac{b}{2} \exp\{i[2\pi(f + f_r)x - k(W(x, y) + W_r(x, y))]\} \\
&\quad + \frac{b}{2} \exp\{-i[2\pi(f - f_r)x - k(W(x, y) - W_r(x, y))]\}
\end{aligned} \tag{8.25}$$

The first term after the second equals sign represents the tilted ideal aspheric wavefront, with a frequency equal to that of the carrier. The second term represents a wavefront with a large asphericity and a frequency equal to about twice the carrier frequency. The last term represents a wavefront with a shape equal to the difference between the actual measured wavefront and the ideal aspheric wavefront. If all terms in these signals with frequencies equal to or greater than the carrier frequency are removed by means of a low pass filter, only the last term remains, with real and imaginary components given by the signals $z_S(x, y)$ and $z_C(x, y)$ of an ideal aspheric wavefront with tilt (shown in Fig. 8.2), as follows:

$$\bar{z}_S(x, y) = -\frac{b}{2} \sin[2\pi(f - f_r)x - k(W(x, y) - W_r(x, y))] \tag{8.26}$$

and

$$\bar{z}_C(x, y) = \frac{b}{2} \cos[2\pi(f - f_r)x - k(W(x, y) - W_r(x, y))] \tag{8.27}$$

Then the wavefront deformations $W(x, y) - W_r(x, y)$ are given by

$$[2\pi(f - f_r)x - k(W(x, y) - W_r(x, y))] = -\tan^{-1}\left[\frac{\bar{z}_S(x, y)}{\bar{z}_C(x, y)}\right] \tag{8.28}$$

which are the wavefront deviations with respect to the ideal aspheric wavefront.

We can see in Fig. 8.10 that the width of the spectrum of the reconstructed wavefront (under test) is much narrower when an aspheric wavefront is used

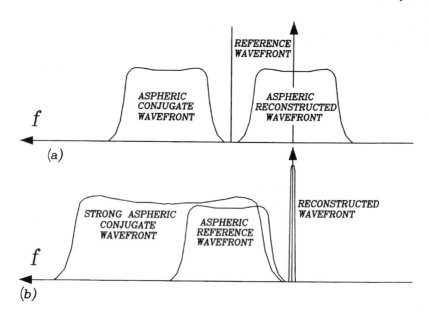

Figure 8.10 Spectra when reconstructing with (a) a flat wavefront and (b) an aspheric wavefront.

as a reference. On the other hand, the width of the spectrum of the conjugate wavefront is duplicated, because its asphericity is duplicated. The Nyquist limit is reached with the same sampling frequency as in the normal case, and thus no improvement is obtained in this respect. However, since the width of the spectrum of the reconstructed wavefront is much narrower, the low pass filter has to be narrower in this case.

8.2.3 Analog and Digital Implementations of Phase Demodulation

As we mentioned before, Ichioka and Inuiya (1972) used analog electronics to implement a simple phase demodulation procedure. Several years later, another phase demodulation method was described by Mertz (1983) in a slightly different manner but again using electronics hardware. He made three measurements in a small interval where the phase could be considered to change linearly with the distance. The measurements were separated 120° in their phase. Macy (1983) used the Mertz method with software calculations instead of hardware.

Commercial interferometers had been constructed that evaluate the two-dimensional wavefront deformations using direct digital phase demodulation (Dörband *et al.* 1990; Küchel 1990; Freischlad *et al.* 1990a, 1990b). The multiplications and spatial filtering are implemented by using dedicated digital electronics hardware. The image was captured with a two-dimensional array of 480 × 480 pixels. Many image frames were obtained at a rate of 30 per second, and then a wavefront averaging technique was used to reduce the effects of atmospheric turbulence. The random wavefront measurement error is inversely proportional to the square root of the number of averaged wavefronts.

Another practical implementation of the digital demodulation of interferograms with a linear carrier was described by Womack (1984).

The interferogram is digitized with a two-dimensional array of light detectors; for example, with a CCD television camera, the irradiance values are sampled at every pixel in the detector. All operations are performed numerically instead of illuminating with a real hologram. The sampled signal values are multiplied by the reference functions $\sin(2\pi f_r x)$ and $\cos(2\pi f_r x)$, obtaining the values of the functions $z_S(x, y)$ and $z_C(x, y)$, respectively. Thus, we may write

$$z_S(x, y) = \sum_{i=1}^{M} s(\alpha_i, y) \sin(2\pi f_r \alpha_i) \delta(x - \alpha_i) \qquad (8.29)$$

and

$$z_C(x, y) = \sum_{i=1}^{M} s(\alpha_i, y) \cos(2\pi f_r \alpha_i) \delta(x - \alpha_i) \qquad (8.30)$$

where M is the number of pixels in a horizontal line to be scanned and sampled.

8.2.4 Spatial Low Pass Filtering

The Fourier theory developed in Chap. 5 is not directly applicable here because we need to calculate the phase for all values of x, not only at the origin, and thus the complete low pass filtering convolution needs to be performed for all values of x.

As we have seen in Sec. 8.1.3, we require the elimination of undesired spatial frequencies at all values of x along the interferogram measured line. Thus, a common filtering function $h(x)$ may be used for $z_S(x)$ and $z_C(x)$. This low pass filter transforms $z_S(x, y)$ and $z_C(x, y)$ into the functions $\bar{z}_S(x)$ and $\bar{z}_C(x)$, respectively, as follows:

$$\bar{z}_S(x, y) = \sum_{i=-N}^{N} s(\alpha_i, y) \sin(2\pi f_r \alpha_i) h(x - \alpha_i) \qquad (8.31)$$

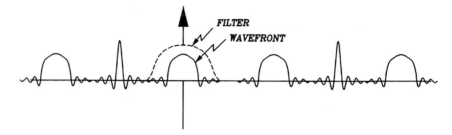

Figure 8.11   Filtering with a low pass filter.

and

$$\bar{z}_C(x, y) = \sum_{i=-N}^{N} s(\alpha_i, y) \cos(2\pi f_r \alpha_i) h(x - \alpha_i) \tag{8.32}$$

where N is the number of pixels taken before and after the point x being considered. We have assumed a finite spatial filter extent of $2N + 1$ pixels for the filtering function, from $i = -N$ to $+N$.

These two functions are evaluated in two steps. First, the interferogram signal values on every pixel are multiplied by the reference functions sine and cosine to obtain $z_S(x, y)$ and $z_C(x, y)$. Then, the spatial low filtering process with the filtering function $h(x)$ is performed. As shown in Fig. 8.11, the purpose of the low pass filter is to filter out all undesired high frequencies in order to isolate the desired first-order lobe in the Fourier spectrum.

The low pass filter may be any symmetric filter, for example, the two-dimensional Hanning, Hamming, $\cos^2$, or any other of the kernel filters described before. In Eqs. 8.31 and 8.32 a kernel with $2N + 1$ elements is assumed.

Since none of the spectral responses of the usual low pass filters has sharp edges, some attenuation of the high spatial frequencies in the wavefront may occur, as illustrated in Fig. 8.12. This attenuation is the same in the real part as in the imaginary part of the Fourier transform of the filtered wavefront, since the same filter is used for both $z_S(x, y)$ and $z_C(x, y)$. Thus, no phase error is introduced for this reason.

Figure 8.13 shows an example of phase demodulation using a linear carrier and discrete sampling of the interferogram.

8.2.5   Sinusoidal Window Filter Demodulation

We will now describe a space domain demodulation method using a sinusoidal filtering window presented by Womack (1984). Let us consider the particular

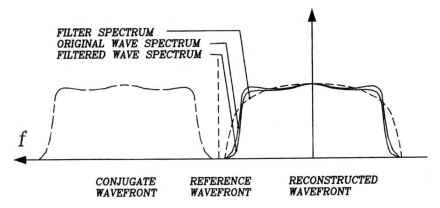

Figure 8.12   Attenuation of high spatial frequencies in the measured wavefront with a low pass filter.

case when the reconstruction frequency is quite different from the carrier frequency and equal to zero. It this case the reconstruction in the hologram is with a flat wavefront perpendicularly impinging on the hologram as shown in Fig. 8.14.

In this case the spectra for the wavefront under reconstruction and the wavefront under test are symmetrically placed with respect to the origin, as shown in Fig. 8.15.

Under these conditions a low pass filter will not allow us to isolate the spectrum of the desired wavefront from the rest. Only the zero-order beam can be isolated with a low pass filter.

A sinusoidal filter $h_S(x)$, permits beam separation. On the other hand, a cosinusoidal filter $h_C(x)$ may be used to eliminate the zero-order beam. That is, we need a set of two filters in quadrature, acting as a bandpass filter, to isolate the first-order beam. The bandpass filtering may then be performed using the relations

$$\bar{z}_S(x, y) = \sum_{i=-N}^{N} s(\alpha_i, y) h_S(x - \alpha_i) \tag{8.33}$$

and

$$\bar{z}_C(x, y) = \sum_{i=-N}^{N} s(\alpha_i, y) h_C(x - \alpha_i) \tag{8.34}$$

as shown in Fig. 8.16.

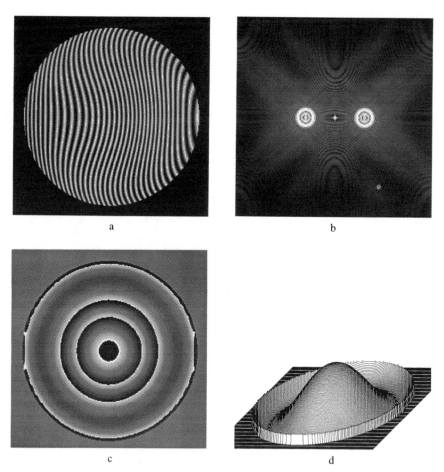

a

b

c

d

Figure 8.13 Phase demodulation with a linear carrier. (a) Interferogram; (b) Fourier transform of interferogram; (c) wrapped phase; (d) unwrapped phase.

An advantage of this method is that the multiplication by the reference functions and the filtering operations are performed in a single step, making use of the appropriate kernel. The frequency width of the filter is given by the space width of the square function, and the frequency position of the filter by the frequency of the sine and cosine functions.

Once the proper convolution kernels for $h_S(x)$ and $h_C(x)$ have been found, the signal phase at the first pixel in the interval is calculated. Then the kernel

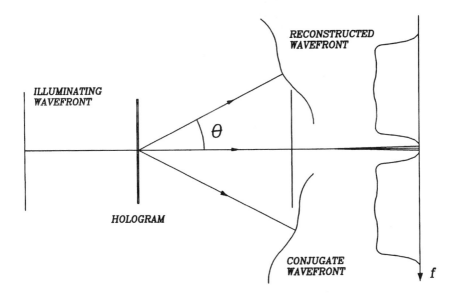

Figure 8.14 Reconstruction with a hologram using a normal reference wavefront.

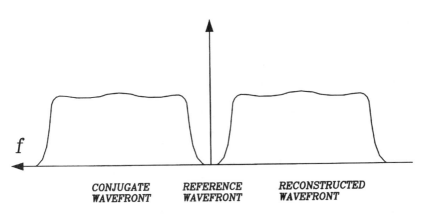

Figure 8.15 Spectra with a hologram using a normal reference wavefront.

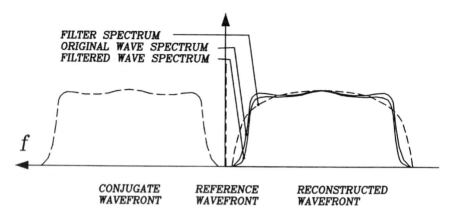

Figure 8.16 Filtering with a sinusoidal window bandpass filter. Note that the origin is not at the same location as in Fig. 8.12.

is moved one pixel to the right and the signal phase is again calculated for this new pixel, until a whole line is scanned.

The wavefront shape may be expressed as

$$W(x, y) = -\frac{1}{k} \tan^{-1} \left[\frac{\bar{z}_S(x, y)}{\bar{z}_C(x, y)} \right] \tag{8.35}$$

8.2.6 Spatial Carrier Phase Shifting Method

The spatial carrier phase shifting method, introduced by Shough *et al.* (1990), is a spatial application of the temporal phase shifting techniques. The basic assumption is that in a relatively small window the wavefront may be considered flat. Then, in a small interval, the phase varies linearly and the phase difference between adjacent pixels is constant.

The interval length is chosen so that the number of pixels it contains is equal to the number of sampling points. The signal phase is calculated at some point in the first interval in a line being scanned, using a phase shifting sampling algorithm. Then the interval is moved one pixel to the right and the signal phase is again calculated. In this manner the procedure continues until a whole line is scanned.

It can be seen that this method is equivalent to the sinusoidal window filter demodulation method described earlier. Here, the chosen phase shifting sampling algorithm defines the filtering functions used in the preceding method. The Fourier theory developed in Chap. 5 is directly applicable, since the phase is to be determined at the local origin of each interval.

Many different phase shifting sampling algorithms can be used. A frequent important requirement is that asynchronous or detuning-insensitive algorithms be used, since the frequency in the interval is not always well known, mainly if the wavefront is aspheric or has strong deformations. A second useful requirement is low sensitivity to harmonics.

The simplest approach when the spatial carrier frequency is well known and the wavefront deviations from sphericity are small is to use the three-step algorithms, for example, the three 120° equally spaced points or Wyant's algorithms, as described by Kujawinska and Wójciak (1991a, 1991b), using a phase step of $\pi/2$ between any two consecutive pixels.

As pointed out before, when the wavefront is defocused or aspheric, the spacing between the fringes is not constant and important detuning errors are likely to appear, because the fringe spacing is quite variable inside the aperture. To solve this problem, Kujawinska and Wójciak (1991a, 1991b) used the Schwider and Hariharan self-calibrating five sampling point method.

Frankowski *et al.* (1989) made a study to experimentally determine the degree of correction obtained with the asynchronous approach originally proposed by Toyooka and Tominaga (1984) and described in Chap. 6.

To test strongly aspheric surfaces it is better to assume that the phase step between adjacent pixels is not constant and has to be determined. Then the phase may be found with an asynchronous algorithm, for example, the Carré algorithm, as proposed by Melozzi *et al.* (1995). Almost any other asynchronous detection algorithm, such as those described in Chap. 6, may be used in this method.

A practical way in which to obtain the signal phase at all points in the pupil is to calculate the two functions $\bar{z}_S(x)$ and $\bar{z}_C(x)$ by means of a convolution of the signal with two one-dimensional kernels $h_S(x)$ and $h_C(x)$ and then use Eq. 8.35. The two kernels are defined by the chosen phase shifting algorithm. Figure 8.17 shows the one-dimensional kernels for three common phase shifting algorithms with the phase equations

$$\tan \phi = \frac{-s_1 + s_3}{s_1 - 2s_2 + s_3} \tag{8.36}$$

with shifts of $-90°$, $0°$, and $+90°$, and

$$\tan \phi = \sqrt{3}\,\frac{-s_1 + s_3}{s_1 - 2s_2 + s_3} \tag{8.37}$$

with shifts of $-120°$, $0°$, and $+120°$.

Küchel (1997) has used in the Zeiss Direct 100 interferometer using a linear carrier with an angular orientation at 45°, with a magnitude such that two consecutive horizontal or vertical pixels have a phase difference of 90°. As pointed

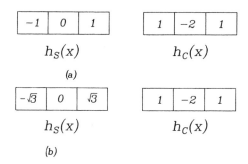

Figure 8.17 Two one-dimensional kernels for phase shifting algorithms with three sampling points.

out earlier by Küchel (1994), the advantages of this linear carrier with this orientation are that (1) a 3×3 convolution kernel measures five steps in a direction perpendicular to the fringes and (2) the distance between pixels in a direction perpendicular to the fringes is $1/\sqrt{2}$ smaller than the distance in a horizontal or vertical direction, thus enhancing spatial resolution.

Figure 8.18a shows a 3×3 kernel suggested by Küchel (1994). This kernel is obtained by the combination of three inverted T algorithms, shifted $90°$, the second with respect to the first and the third with respect to the second. This kernel is symmetrical about its diagonal at $-45°$, due to the inclination of

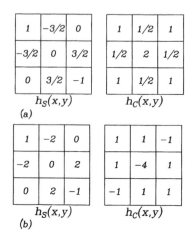

Figure 8.18 Two 3×3 kernel sets for spatial phase shifting phase demodulation.

the carrier fringes at 45°. Unfortunately, complete detuning insensitivity is not obtained as in the Schwider algorithm, since the three algorithms have the same weights when linearly combined. Nevertheless, this kernel has a relatively low sensitivity to detuning. Its phase equation is

$$\tan\phi = \frac{s_1 - 3s_2 + 3s_4 - s_5}{s_1 + s_2 - 4s_3 + s_4 + s_5} \tag{8.38}$$

Better results can be obtained if detuning-insensitive algorithms are used. A similar algorithm, but one that is detuning-insensitive, is obtained if the second algorithm of the combination is given a weight 2 (in the numerator as well as in the denominator of its phase equation), thus obtaining

$$\tan\phi = \frac{s_1 - 4s_2 + 4s_4 - s_5}{s_1 + 2s_2 - 6s_3 + 2s_4 + s_5} \tag{8.39}$$

The kernel for this algorithm is illustrated in Fig. 8.18b.

More flexibility and thus better results can be obtained with a properly designed 5×5 kernel.

It is important to note that the function $\tan^{-1}$ gives the result modulo 2π. This means that in all of these phase demodulation methods the wavefront $W(x, y)$ is calculated modulo λ. This is what is called a wrapped phase. Unwrapping is a general problem in interferogram analysis. Methods to unwrap the phase are studied in detail in Chap. 11.

8.2.7 Phase-Locked Loop Demodulation

Phase-locked loop demodulation is another method for interferogram analysis with a linear carrier, based on the phase-locked loop method used in electrical communications. The phase-locked loop (PLL) technique has been used since 1950 in electronic communications to demodulate electrical signals. However, its use in interferometry has been quite recent (Servín and Rodriguez-Vera 1993; Servín *et al.* 1995).

A PLL may be considered as a narrow bandpass adapting filter whose central frequency tracks the instantaneous fringe pattern frequency along the scanning line. Figure 8.19 shows the building blocks of a typical electronic PLL with its basic components.

The basic principle in this phase tracking loop is the following: The phase changes of a phase-modulated input signal are compared with the output of a voltage-controlled oscillator (VCO) by using a multiplier (see Fig. 8.19). The phase-locked loop works in such a way that the phase difference between the modulated input signal and the VCO's output signal eventually vanishes. This phase following is achieved by means of a closed loop, feeding the VCO's

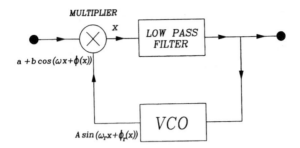

Figure 8.19 Building blocks for an electronic phase-locked loop.

input with the output signal, which is proportional to the modulating signal. In interferogram evaluation this VCO is not a piece of electronics hardware but rather a software-simulated VCO. For convenience the same electronics name will be used here, even though the signals are not voltage signals but numbers. Let us assume that the input phase-modulated signal with amplitude $s(x)$ has a carrier angular frequency ω and a phase modulation $\phi(x)$ given by

$$s(x) = a + b \cos \psi(x) = a + b \cos(\omega x + \phi(x)) \tag{8.40}$$

The VCO is an oscillator tuned to produce a sinusoidal reference signal with angular frequency ω_r in the absence of control voltage. When a control voltage is applied to the VCO, its frequency output changes to a new value. The low pass filter in Fig. 8.19 is a one-pole filter which may be represented by the first-order differential equation

$$\frac{d\phi_r(x)}{dx} = Ag[a + b \cos(\omega x + \phi(x))] \sin(\omega_r x + \phi_r(x)) \tag{8.41}$$

where g is the gain of the PLL's low pass filter. This equation may also be rewritten as

$$\frac{d\phi_r(x)}{dx} = Ag[a + b \cos(\psi(x))] \sin(\psi_r(x)) \tag{8.42}$$

The right-hand term of Eq. 8.42 may be rewritten as

$$\frac{d\phi_r(x)}{dx} = Aag \sin \psi_r(x) + \frac{1}{2} Abg \sin(\psi_r(x) + \psi(x))$$
$$+ \frac{1}{2} Abg \sin(\psi_r(x) - \psi(x)) \tag{8.43}$$

The first-order differential equation filters out all high frequencies. This eliminates the first and second terms, leaving only the last term, with the lowest frequency,

$$\frac{d\phi_r(x)}{dx} = \frac{1}{2} Abg \sin(\psi_r(x) - \psi(x)) \tag{8.44}$$

When the phase-locked loop is operating, the phase difference is small enough to consider a linear approximation valid. Hence we may write

$$\frac{d\phi_r(x)}{dx} = \frac{1}{2} Abg(\psi_r(x) - \psi(x)) \tag{8.45}$$

To understand how this loop works, let us consider the system initially in equilibrium, with $\omega_r = \omega$. Then, due to the phase modulation on the input signal, its frequency changes momentarily, producing a change in its phase. This change produces a change in the input of the low pass filter, which acts on the VCO, increasing its frequency of oscillation. A new equilibrium point is found when the phase of the oscillator matches that of the input. Of course, the change in the phase of the input signal is reflected in an increase in the VCO's input. Thus, the low pass filter output is the demodulated signal.

Equation 8.41 is a differential equation describing the phase-locked loop. If we substitute Eq. 8.45, we find

$$\frac{d\phi_r(x)}{dx} = \frac{1}{2} Abg(\psi_r(x) - \psi(x)) \tag{8.46}$$

which can also be written as

$$\frac{d\phi_r(x)}{dx} = \frac{1}{2} b\tau(\psi_r(x) - \psi(x)) \tag{8.47}$$

where τ is the closed loop gain. This differential equation tells us that the rate of change of the phase of the voltage-controlled oscillator is directly proportional to the demodulated signal. The output phase of the VCO will follow the input phase continuously as long as the input signal does not have any large discontinuities.

If the product of the closed loop gain τ and the signal amplitude b is less than 1 we can compute the modulation signal by using the more precise expression

$$\frac{d\phi_r(x)}{dx} = b\tau \cos(\psi(x)) \sin(\psi_r(x)) \tag{8.48}$$

Since a first-order system with a small closed loop gain τ behaves as a low pass filter, i.e., has a low τ value, no explicit low pass filtering is required.

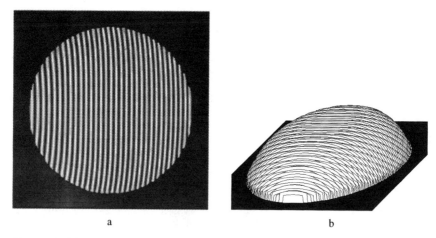

a   b

Figure 8.20 Example of phase demodulation using the phase-locked loop method. (a) Interferogram to be demodulated; (b) two-dimensional demodulated phase.

This theory may be applied to interferogram fringe analysis if the input signal is replaced by signal values along a horizontal scanning line in the interferogram. The variations in the illumination can be filtered out by using a high pass filter. High pass filtering is also convenient because the phase-locked loop low pass filter rejects only an unwanted signal with twice the interferogram's carrier frequency. As pointed out in Chap. 3, a very simple high pass filter is achieved by just substituting the signal function with its derivative with respect to x. Thus, Eq. 8.48 may be written

$$\frac{d\phi_r(x)}{dx} = -b\tau \frac{ds(x)}{dx} \cos(\psi_r(x)) \tag{8.49}$$

One possible way to scan a two-dimensional fringe pattern using a PLL may be found in Servín and Rodriguez-Vera (1993). Figure 8.20 shows an example of phase demodulation using the phase-locked loop method and the two-dimensional scanning strategy proposed in Servín and Rodríguez-Vera (1993).

This demodulation method has been applied to aspheric wavefront measurement and also to the demodulation of Ronchi patterns (Servín *et al.* 1994).

8.3 CIRCULAR SPATIAL CARRIER ANALYSIS

We may find that for some systems of closed fringes the introduction of a linear carrier is not practical for some reason, for example, because the minimum needed carrier is of such a high spatial frequency that the Nyquist limit is exceeded. This situation can arise when the wavefront under measurement is highly aspheric or aberrated. In this case the demodulation has to be performed without the linear carrier. Instead of a linear carrier, sometimes the alternative may be a circular carrier, which introduces a large defocusing, as shown in the interferogram in Fig. 8.21. Then the irradiance function in the interferogram produced by the interference between a reference spheric wavefront and the wavefront under test is

$$s(x, y) = a + b \cos[k(D(x^2 + y^2) - W(x, y))]$$

$$= a + b \cos[k(DS^2 - W(x, y))]$$

$$= a + \frac{b}{2} \exp[ik(DS^2 - W(x, y))]$$

$$+ \frac{b}{2} \exp[-ik(DS^2 - W(x, y))] \tag{8.50}$$

where $S^2 = x^2 + y^2$. The radial carrier spatial frequency is

$$f(x, y) = \frac{2DS}{\lambda} \tag{8.51}$$

Using again the holographic analogy, we can interpret the interferogram as an on-axis or Gabor hologram. This hologram can be demodulated by illuminating it with a reference spherical (or flat) wavefront. This demodulation may be achieved only if the phase in the irradiance function increases or decreases in a monotonic manner from the center toward the edge of the pupil. Thus, if the defocusing term is positive, we require that

$$\frac{\partial(DS^2 - W(x, y))}{\partial S} > 0 \tag{8.52}$$

or

$$D > \frac{1}{2S} \frac{\partial W(x, y)}{\partial S} \tag{8.53}$$

This condition assures us that there are not two fringes in the interferogram aperture with the same order of interference. In other words, no fringe crosses more than once any line traced from the center of the interferogram to its edge. In the vicinity of the center of the interferogram, the carrier frequency is so small that the demodulated phase in this region is not reliable. This is a disadvantage

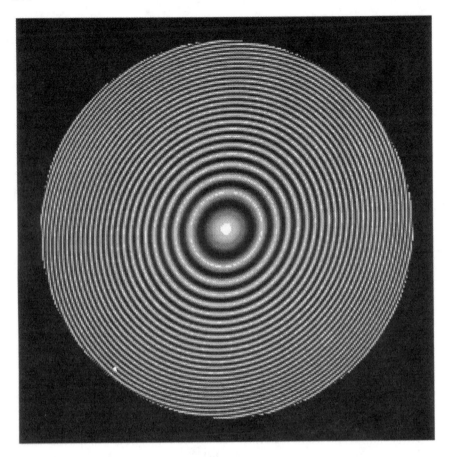

Figure 8.21 Interferogram with a circular carrier.

of this method. To reduce this problem, the circular carrier frequency should be as large as possible, provided the Nyquist limit is not exceeded.

8.4 PHASE DEMODULATION WITH A CIRCULAR CARRIER

The phase demodulation of the interferogram (hologram reconstruction) can be performed using either an on-axis spherical or a tilted spherical wavefront. These two methods, although quite similar, have some small but important differences, as will be described.

8.4.1 Phase Demodulation with a Spherical Reference Wavefront

The demodulation using an on-axis spherical wavefront with almost the same curvature as that used to introduce the circular carrier is illustrated in Fig. 8.22. Then this spherical reference wavefront may be written as

$$r(x, y) = \exp[i(kD_r(x^2 + y^2))] = \exp(i(kD_r S^2)) \tag{8.54}$$

where $S^2 = x^2 + y^2$, and the curvature of this wavefront is close to that of the original spherical wavefront that produced the hologram (circular carrier). In other words, the value of the coefficient D_r for the reference beam must be as close as possible to the value of the coefficient D for the spherical beam introducing the circular carrier.

The product between the interferogram irradiance $s(x, y)$ in Eq. 8.51 and the illuminating wavefront amplitude $r(x, y)$ is

$$s(x, y)r(x, y) = a \exp[i(kD_r S^2)]$$
$$+ \frac{b}{2} \exp[ik((D + D_r)S^2 - W(x, y))]$$
$$+ \frac{b}{2} \exp[-ik((D - D_r)S^2 - W(x, y))] \tag{8.55}$$

The first term on the right is the zero-order beam corresponding to the illuminating spherical wavefront. Its spatial frequency is zero at the center and increases with the square of S toward the edge of the pupil.

$$f_r(x, y) = \frac{2D_r S}{\lambda} \tag{8.56}$$

The second term is of order -1. It is the conjugate wavefront with deformations opposite to those of the wavefront under test. Its curvature is about twice the reference wavefront curvature, and its spatial frequency is

$$f_{-1}(x, y) = \frac{2(D + D_r)S}{\lambda} - \frac{1}{\lambda}\frac{\partial W(x, y)}{\partial S} \tag{8.57}$$

The third term is the first order of diffraction and represents the reconstructed wavefront, with only a slight difference in curvature, and its spatial frequency is

$$f_{+1}(x, y) = \frac{2(D - D_r)S}{\lambda} - \frac{1}{\lambda}\frac{\partial W(x, y)}{\partial S} \tag{8.58}$$

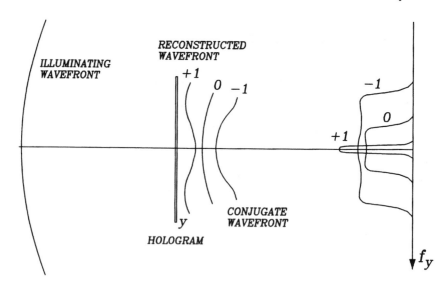

Figure 8.22   Phase demodulation in an interferogram with a circular carrier using a spherical reference wavefront.

The Fourier spectra of these three beams are concentric and overlap each other. However, the wavefront to be measured ($+1$) can be nearly isolated because of the different diameters of these spectra. Equation 8.55 can also be written

$$s(x, y)r(x, y) = z_C(x, y) + iz_S(x, y)$$
$$= s(x, y)\cos(kD_rS^2) + is(x, y)\sin(kD_rS^2) \tag{8.59}$$

We see that an interferogram with a circular carrier can be phase demodulated by multiplying the signal by these cosine and sine functions with a quadratic phase close to that used to introduce the circular carrier.

Using a two-dimensional digital low pass filter, the first two terms in Eq. 8.55 are eliminated, obtaining

$$\bar{z}_C(x, y) + i\bar{z}_S(x, y) = \frac{b}{2}\exp[-ik((D - D_r)S^2 - W(x, y))]$$
$$= \frac{b}{2}\cos[k((D - D_r)S^2 - W(x, y))]$$
$$- i\frac{b}{2}\sin[k((D - D_r)S^2 - W(x, y))] \tag{8.60}$$

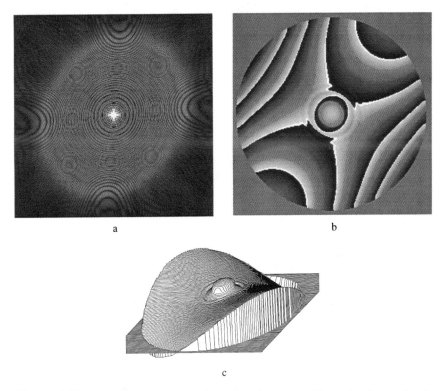

Figure 8.23 Phase demodulation of the interferogram with a circular carrier, in Fig. 8.21, using a demodulating spherical wavefront. (a) Spectrum; (b) phase map; (c) unwrapped phase. The small circular center deformation is due to the imperfect order isolation at the center of the circular fringes.

Thus, the wavefront under reconstruction is given by

$$k[(D - D_r)S^2 - W(x, y)] = -\tan^{-1} \frac{\bar{z}_S(x, y)}{\bar{z}_C(x, y)} \tag{8.61}$$

An example of phase demodulation using a circular carrier is presented in Fig. 8.23.

8.4.2 Phase Demodulation with a Tilted Plane Reference Wavefront

The phase demodulation method described by Moore and Mendoza-Santoyo (1995) is basically a modification of that of Kreis (1986a, 1986b) for the Fourier method. We will consider here a circular carrier, but we will see that the method is more general and also applies to interferograms with systems of closed fringes.

To understand how the demodulation can be made with closed fringes, let us consider the interference along one diameter in an interferogram with a circular carrier. In Fig. 8.24a we have a flat wavefront interfering with a spherical wavefront. In Fig. 8.24b the spherical wavefront has been replaced by a discontinuous wavefront in which the sign of the left half has been reversed. Both pairs of wavefronts produce the same interferogram, with the same signal as in Fig. 8.24c. In the first case the phase increases monotonically from the center to the edges. In the second case the phase increases monotonically from the left to the right. If we assume that what we have is the second case, we can phase demodulate in the standard manner, multiplying by the functions sine and cosine and low pass filtering these two functions. However, to obtain the correct result we must reverse the sign of the left half of the wavefront.

Using now the holographic analogy, let us consider the interferogram with a circular carrier, illuminated with a tilted plane wavefront as illustrated in Fig. 8.25. This illuminating tilted plane reference wavefront may be written as

$$r(x, y) = \exp[i(2\pi f_r x)]$$

$$= \cos(2\pi f_r x) + i \sin(2\pi f_r x) \tag{8.62}$$

The product of the interferogram irradiance $s(x, y)$ in Eq. 8.50 and the illuminating wavefront amplitude $r(x, y)$ is

$$s(x, y) r(x, y) = a \exp[i(2\pi f_r x)]$$

$$+ \frac{b}{2} \exp\{i[2\pi f_r x + k(DS^2 - W(x, y))]\}$$

$$+ \frac{b}{2} \exp\{-i[-2\pi f_r x + k(DS^2 - W(x, y))]\} \tag{8.63}$$

The first term is the tilted flat wavefront (zero order). The second term is the conjugate wavefront, and the last term is the reconstructed wavefront to be measured. The wavefront to be measured and the conjugate wavefront differ only in the sign of the deformations with respect to the reference plane.

The Fourier spectrum of expression 8.63 is illustrated in Fig. 8.26. We see that these three spots are concentric but shifted laterally with respect to the axis.

If we use a rectangular low pass filter ($N \times 1$ convolution mask) as shown on the right-hand side of Fig. 8.26, we can see that we are isolating the reconstructed wavefront for the positive y half-plane and the conjugate wavefront for the negative y half-plane. The conjugate wavefront is equal in magnitude to the

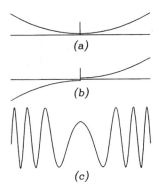

Figure 8.24 Interfering wavefronts. (a) A flat wavefront and a spherical wavefront; (b) a flat wavefront and a discontinuous wavefront with two spherical portions; (c) signal for both cases.

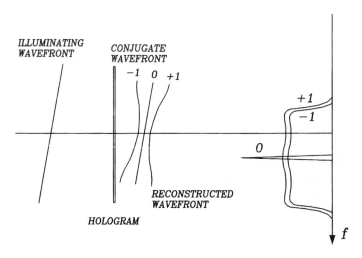

Figure 8.25 Phase demodulation in an interferogram with a circular carrier using a tilted plane reference wavefront.

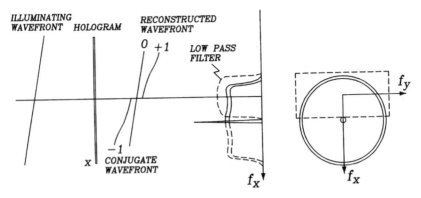

Figure 8.26 Fourier spectrum produced by an interferogram with a circular carrier (Gabor hologram) when illuminated with a tilted flat reference wavefront.

reconstructed wavefront, but with the opposite sign. Thus, we obtain the wavefront under measurement by just changing the sign of the retrieved wavefront deformations for the negative half-plane. It is easy to understand that there are singularities in the vicinity of the points where the slope of the fringes is zero.

We can also write expression 8.63 as

$$s(x, y)r(x, y) = z_C(x, y) + i z_S(x, y)$$
$$= s(x, y)\cos(2\pi f_r x) + i s(x, y)\sin(2\pi f_r x) \tag{8.64}$$

Again we see that the phase demodulation of an interferogram with a circular carrier can be done by multiplying the signal by the functions cosine and sine with a reference frequency. This reference frequency is chosen near the maximum spatial frequency in the interferogram.

Using two-dimensional digital low pass filtering, the first two terms in Eq. 8.63 are eliminated, obtaining

$$\bar{z}_C(x, y) + i\bar{z}_S(x, y) = \frac{b}{2}\exp\{-i[-2\pi f_r x + k(DS^2 - W(x, y))]\}$$
$$= \frac{b}{2}\cos[-2\pi f_r x + k(DS^2 - W(x, y))]$$
$$- i\frac{b}{2}\sin[-2\pi f_r x + k(DS^2 - W(x, y))] \tag{8.65}$$

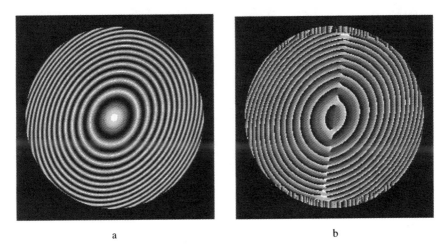

a   b

Figure 8.27 Phase map of demodulated interferogram with a circular carrier. (a) Interferogram; (b) retrieved phase. A reference frequency near the highest value in the interferogram was used. Notice the imperfect demodulation near the upper and lower edges, where noise with the reference demodulating freqency appears.

Thus, the retrieved wavefront is given by

$$[-2\pi f_r x + k(DS^2 - W(x, y))] = -\tan^{-1} \frac{\bar{z}_S(x, y)}{\bar{z}_C(x, y)} \qquad (8.66)$$

which, as we know, gives us the wavefront to be measured by changing the sign of the phase for negative values of y.

An example of phase demodulation using a circular carrier and a reconstructing tilted plane wavefront is presented in Fig. 8.27.

8.5 FOURIER TRANSFORM PHASE DEMODULATION WITH A LINEAR CARRIER

The wavefront deformations in an interferogram with a linear carrier may also be calculated with a procedure using Fourier transforms. This method was originally proposed by Takeda *et al.* (1982) using one-dimensional Fourier transforms along one scanning line. Later, Macy (1983) applied Takeda's method to extend it to two dimensions by adding the information from many scanning lines, obtaining slices of the two-dimensional phase. Then Bone *et al.* (1986) extended Macy's work to full two dimensions by using two-dimensional Fourier transforms. They also suggested techniques to reduce phase errors introduced by the finite boundaries.

To describe the procedure, let us assume that we calculate the Fourier transform of the interferogram with a large tilt. The minimum magnitude of this tilt from a geometrical point of view is the same as that used in direct interferometry. However, even if this tilt is increased more, the images with orders -1 and $+1$ still partially overlap the light with zero order. The reason is that diffraction effects due to the finite size of the aperture produce rings around the three Fourier images. The presence of these rings makes it impossible to completely separate the three images so that the zero-order image may be isolated. These diffraction rings due to the finite boundary of the interferogram may be substantially reduced by either of the two mechanisms described in Sec. 8.1.4.

Figure 8.28 shows the result of applying a two-dimensional Hamming window to an interferogram and its effect on its Fourier transform.

Another important way to avoid the presence of high spatial frequency noise in the Fourier images is to subtract irradiance irregularities in the continuum. These may be easily subtracted by measuring the irradiance in a pupil without interference fringes and then subtracting the irregularities from the interference pattern. This continuum may be measured in many ways, as described by Roddier and Roddier (1987). They also described several ways to eliminate the effects of turbulence in the interferogram.

Once the interference pattern has been cleaned and the fringes extended outside of the pupil or the Hamming filter has been applied, a fast Fourier transform (see Chap. 2) is used to obtain the Fourier space images. Once the three Fourier spots are clear and separated from each other, a circular boundary is selected, as shown in Fig. 8.29, around one of the first-order images. Then all irradiance values outside this circular boundary are multiplied by zero to isolate the selected image.

After the desired image is isolated, its center is shifted to the origin and its Fourier transform is obtained. The result is the wavefront under test.

To mathematically describe this procedure, let us write the expression for the signal in the form

$$s(x, y) = g(x, y) + h(x, y) \exp[i2\pi f x] + h^*(x, y) \exp[-i(2\pi f x)] \quad (8.67)$$

where * denotes a complex conjugate and f_c is the carrier spatial frequency. The variable $s(x, y)$ is the signal in the interferogram after the irradiance irregularities have been subtracted and the Hamming filter has been applied or the fringes have been extrapolated outside of the pupil. We have written all variables with lowercase letters so that the Fourier transforms are represented with capital letters and $h(x, y)$ has been defined by

$$h(x, y) = 0.5b(x, y) \exp[-ikW(x, y)] \quad (8.68)$$

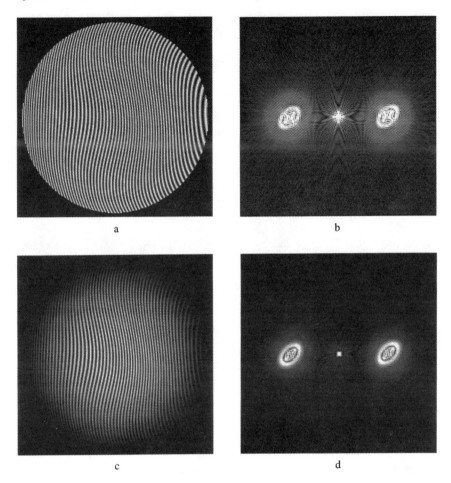

Figure 8.28 Interferogram and its Fourier transform, before and after applying a Hamming filter. (a) Interferogram and (b) its Fourier transform. (c) The same interferogram after applying Hamming function and (d) its Fourier transform.

If we take the Fourier transform of the signal $s(x, y)$, using some Fourier transform properties, we may write

$$S(f_x, f_y) = G(f_x, f_y) + H(f_x - f_0, f_y) + H^*(f_x + f_0, f_y) \qquad (8.69)$$

where the coordinates in the Fourier plane are f_x and f_y.

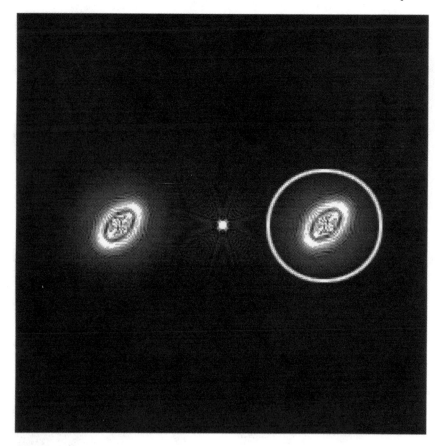

Figure 8.29 Isolating desired spectrum spot in interferogram in Fourier method.

A low pass filter function may be used to isolate the desired term, for example, the Hamming filter, thus obtaining

$$S(f_x, f_y) = H(f_x - f_0, f_y) \qquad (8.70)$$

Shifting this function to the origin in the Fourier plane we have

$$S(f_x, f_y) = H(f_x, f_y) \qquad (8.71)$$

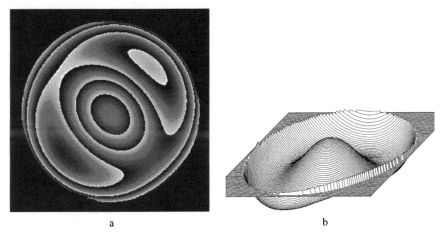

a   b

Figure 8.30 Phase demodulation of interferogram in Fig. 8.26 using the Fourier transform method. (a) Phase map; (b) wavefront deformations after phase unwrapping.

Now, taking the inverse Fourier transform of this term we obtain

$$h(x, y) = 0.5b(x, y) \exp[-ikW(x, y)] \tag{8.72}$$

Hence the wavefront deformation is given by

$$W(x, y) = -\frac{1}{k} \tan^{-1} \frac{\text{Im}\{h(x, y)\}}{\text{Re}\{h(x, y)\}} \tag{8.73}$$

As an example, the wavefront obtained from the interferogram in Fig. 8.21 is illustrated in Fig. 8.30.

Reviews on the Fourier method had been published by Takeda (1989) and Kujawinska and Wójciak (1991a). Kujawinska and Wójciak (1991a, 1991c) also described practical details for the implementations of Fourier demodulation. Simova and Stoev (1993) applied this technique to holographic moiré fringe patterns.

8.5.1 Sources of Error in the Fourier Transform Method

The Fourier transform method has some advantages but also some important limitations compared to other phase modulation methods for interferograms with a linear carrier. There are several factors that may introduce errors into the phase

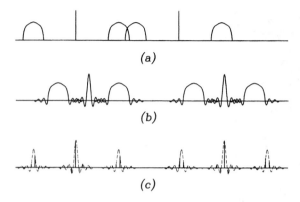

Figure 8.31 Graphical illustration of errors due to the discrete nature of the fast Fourier transform. (a) Aliasing; (b) energy leakage; (c) picket fence.

calculated with the Fourier transform method, as pointed out in detail by several researchers, for example, by Kujawinska and Wójciak (1991a, 1991c), Takeda (1987, 1989), Nugent (1985), Green *et al.* (1988), Frankowski *et al.* (1989), and Malcolm *et al.* (1989). Schmit *et al.* (1992) studied these errors for the one-dimensional case.

The main errors are inherent to the discrete nature of the fast Fourier transform. The continuous Fourier transform cannot be evaluated; instead, the discrete fast Fourier transform is used. Thus, the following are some of the possible sources of phase errors.

1. *Aliasing*. If the sampling frequency is not high enough, as in Fig. 8.31a, the Nyquist limit is exceeded and some nonexisting spatial frequencies may appear in the computed wavefront.
2. *Picket fence*. This error is produced by the discrete calculation of the fast Fourier transform. We see in Fig. 8.31c that not all frequency components appear in the calculated discrete Fourier transform. It is easy to understand that after filtering and taking the inverse Fourier transform, some wavefront spatial frequencies may disappear in the calculated wavefront.
3. *Energy leakage*. This is the most important source of phase errors in the Fourier method. As we pointed out before, if the tilt is not high enough, and the pupil is finite, the side ripples of the Fourier transforms of each order interfere with each other, as in Fig. 8.31b. This effect may cause serious phase errors in the retrieved wavefront due to leakage of energy

of some spatial frequencies into adjacent spatial frequencies. Increasing the tilt, using window functions like the Hamming filter, or extrapolating fringes outside of the pupil limits reduce this error.

4. *Multiple reflections or spurious fringes in the interferogram.* Multiple reflections or spurious fringes inside and outside the interferogram pupil distort the signal, introducing harmonic components. In this case the minimum frequency of the linear carrier is three times that required by expression 8.3, as pointed out in Sec. 8.1.3. The reason is that the harmonic components cannot be filtered out if their spatial frequency is lower than the maximum fringe frequency in the interferogram. Then the proper low pass filtering should be performed.

5. *Light detector nonlinearity.* Nugent (1985) showed that if the light detector has a nonlinear response to the light irradiance, the harmonics due to this nonlinearity produce phase errors.

6. *Random noise.* Bone *et al.* (1986) showed that the expected rms phase error is

$$\delta\phi_{\text{rms}} = \pi\sqrt{\frac{\alpha}{2}}\sigma m \qquad (8.74)$$

where $\alpha = n/N$ is the ratio of the number of spectral sample points n within the filter passband to the number of sample points N, σ is the rms value of the noise, and m is the mean modulation amplitude.

7. *Quantization errors.* Frankowski *et al.* (1989) proved that quantization noise cannot contribute to phase errors. The error for 6 bits is smaller than one-thousandth of a wavelength.

A comparison between phase shifting interferometry and the Fourier transform method from the viewpoint of their noise characteristics has been reported by Takeda (1987).

8.5.2 Spatial Carrier Frequency, Spectrum Width, and Interferogram Domain Determination

The magnitude of the spatial carrier frequency, the filter width, and the interferogram domain limits are three important parameters that have to be determined with the highest possible precision. They can be obtained in an automatic manner, as described by Kujawinska (1993), but they can also be obtained using operator-assisted methods.

As pointed out before, to measure and then remove the spatial carrier (tilt) from the interferogram, Takeda *et al.* (1982), Macy (1983), and Lai and Yatagai

(1994) performed a lateral translation of the Fourier transform of the interferogram. However, the magnitude of the translation must be previously determined and it cannot be exactly made, since the Fourier transform is calculated at discrete spatial frequency values. As a result, we are bound to obtain a residual tilt in the calculated interferogram, but this linear term can then be removed in the final result.

The filter width determination is another problem that must be solved. Takeda and Mutoh (1983) suggested that the limits of the Fourier band to be filtered and preserved are the maximum and minimum local fringe spatial frequencies. This is true for large wavefront deformations, where we can neglect diffraction effects. Kujawinska *et al.* (1990) suggested another method to determine both the carrier frequency and the spectrum width. The carrier frequency is determined by locating the maximum value of the Fourier transform. The filter width is determined by isolating the area in the frequency space, where Fourier transform values above a certain threshold are found.

The simplest (but not the most precise) way to find the filter width and location is with operator intervention, by observing on the computer screen the image of the two-dimensional Fourier transform and manually selecting there a circle around the first order, visually estimating its location and size.

8.6 FOURIER TRANSFORM PHASE DEMODULATION WITH A CIRCULAR CARRIER

We have seen in Sec. 8.4.1 that an interferogram with a circular carrier can be demodulated, following the holographic analogy, by using a reconstructing tilted flat wavefront without a linear carrier. This method can also be used to demodulate with the Fourier transform. In this case the flat reconstructing wavefront does not need to be tilted, as illustrated in Fig. 8.32. This method of demodulating with closed fringes has been described by Kreis (1986a, 1986b).

If all frequencies greater than or equal to zero are filtered out as shown in Fig. 8.33, we isolate the reconstructed wavefront for the positive y half-plane and the conjugate wavefront for the negative y half-plane. The wavefront to be measured is obtained if the sign of the phase for positive values of y is changed.

Kreis (1986a, 1986b) showed that this method can be extended to the demodulation of fringe patterns with closed fringes, not necessarily with a circular carrier. Then the fringe pattern has to be processed with two orthogonal rectangular filters, as shown in Fig. 8.34.

The problem of analyzing an interferogram with closed fringes as well as the problem of recording in a single interferogram information about two events by the use of crossed fringes was studied by Pirga and Kujawinska (1995, 1996).

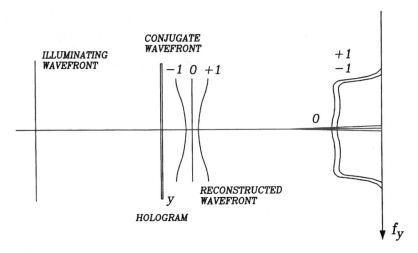

Figure 8.32 Demodulation of an interferogram with a circular carrier (Gabor hologram) with a flat reference wavefront.

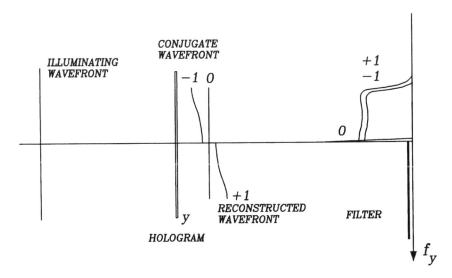

Figure 8.33 Spatial frequencies in an interferogram with a circular carrier (Gabor hologram) when illuminated with a flat reference wavefront, after filtering out all positive spatial frequencies f_y.

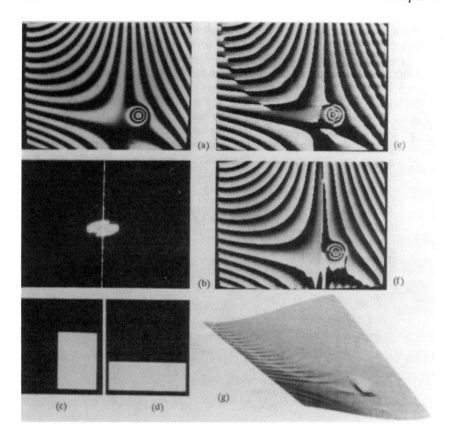

Figure 8.34 Demodulation of an interferogram with closed fringes with a flat reference wavefront. (a) Interferogram; (b) spectrum; (c) and (d) phase maps; (e) filters; and (f) calculated phase. (From Kreis 1986a.)

REFERENCES

Bone D. J., H.-A. Bachor, and R. J. Sandeman, "Fringe-Pattern Analysis Using a 2-D Fourier Transform," *Appl. Opt.*, **25**, 1653–1660 (1986).

Burton D. R. and M. J. Lalor, "Managing Some of the Problems of Fourier Fringe Analysis," *Proc. SPIE*, **1163**, 149–160 (1989).

Chan P. H., P. J. Bryanston-Cross, and S. C. Parker, "Spatial Phase Stepping Method of Fringe Pattern Analysis," *Opt. and Lasers in Eng.*, **23**, 343–356 (1995).

Choudry A. and M. Kujawinska, "Fourier Transform Method for the Automated Analysis of Fringe Pattern," *Proc. SPIE*, **1135**, 113–118 (1989).

Dörband B., W. Wiedmann, U. Wegmann, W. Kübler, and K. R. Freischlad, "Software Concept for the New Zeiss Interferometer," *Proc. SPIE*, **1332**, 664–672 (1990).

Frankowski G., I. Stobbe, W. Tischer, and F. Schillke, "Investigation of Surface Shapes Using a Carrier Frequency Based Analysis System," *Proc. SPIE*, **1121**, 89–100 (1989).

Freischlad K., M. Küchel, K. H. Schuster, U. Wegmann, and W. Kaiser, "Real-Time Wavefront Measurement with Lambda/10 Fringe Spacing for the Optical Shop," *Proc. SPIE*, **1332**, 18–24 (1990a).

Freischlad K., M. Küchel, W. Wiedmann, W. Kaiser, and M. Mayer, "High Precision Interferometric Testing of Spherical Mirrors with Long Radius of Curvature," *Proc. SPIE*, **1332**, 8–17 (1990b).

Green R. J., J. G. Walker, and D. W. Robinson, "Investigation of the Fourier Transform Method of Fringe Pattern Analysis," *Opt. and Lasers in Eng.*, **8**, 29–44 (1988).

Hatsuzawa T., "Optimization of Fringe Spacing in a Digital Flatness Test," *Appl. Opt.*, **24**, 2456–2459 (1985).

Horman M. H., "An Application of Wavefront Reconstruction to Interferometry," *Appl. Opt.*, **4**, 333–336 (1965).

Ichioka Y. and M. Inuiya, "Direct Phase Detecting System," *Appl. Opt.*, **11**, 1507–1514 (1972).

Kreis T., "Digital Holographic Interference-Phase Measurement Using the Fourier Transform Method," *J. Opt. Soc. Am. A*, **3**, 847–855 (1986a).

Kreis T., "Fourier Transform Evaluation of Holographic Interference Patterns," *Proc. SPIE*, **814**, 365–371 (1986b).

Küchel M., "The New Zeiss Interferometer," *Proc. SPIE*, **1332**, 655–663 (1990).

Küchel M., "Methods and Apparatus for Phase Evaluation of Pattern Images Used in Optical Measurement," U.S. Patent 5,361,312 (1994).

Küchel M., Personal communication (1997).

Kujawinska M., "Spatial Phase Measurement Methods," in *Interferogram Analysis*, D. W. Robinson and G. T. Reid, Eds., Institute of Physics Publ., Bristol and Philadelphia, p. 294, 1993.

Kujawinska M. and J. Wójciak, "High Accuracy Fourier Transform Fringe Pattern Analysis," *Opt. and Lasers in Eng.*, **14**, 325–339 (1991a).

Kujawinska M. and J. Wójciak, "Spatial-Carrier Phase Shifting Technique of Fringe Pattern Analysis," *Proc. SPIE*, **1508**, 61–67 (1991b).

Kujawinska M. and J. Wójciak, "Spatial Phase Shifting Techniques of Fringe Pattern Analysis in Photomechanics," *Proc. SPIE*, **1554**, 503–513 (1991c).

Kujawinska M., A. Spik, and J. Wójciak, "Fringe Pattern Analysis Using Fourier Transform Techniques," *Proc. SPIE*, **1121**, 130–135 (1989).

Kujawinska M., M. Salbut, and K. Patorski, "Three Channel Phase Stepped System for Moiré Interferometry," *Appl. Opt.*, **29**, 1633–1636 (1990).

Lai G. and T. Yatagai, "Use of the Fast Fourier Transform Method for Analysing Linear and Equispaced Fizeau Fringes," **33**, 5935–5940 (1994).

Macy W. W., Jr., "Two-Dimensional Fringe Pattern Analysis," *Appl. Opt.*, **22**, 3898–3901 (1983).

Malcolm A., D. R. Burton, and M. J. Lalor, "A Study of the Effects of Windowing on the Accuracy of Surface Measurements Obtained from the Fourier Analysis of

Fringe Patterns," in *Proc. FASIG, "Fringe Analysis 1989,"* Loughborough, UK, 1989.

Melozzi M., L. Pezzati, and A. Mazzoni, "Vibration-Insensitive Interferometer for On-Line Measurements," *Appl. Opt.,* 34, 5595–5601 (1995).

Mertz L., "Real Time Fringe Pattern Analysis," *Appl. Opt.,* 22, 1535 (1983).

Moore A. J. and F. Mendoza-Santoyo, "Phase Demodulation in the Space Domain Without a Fringe Carrier," *Opt. and Lasers in Eng.,* 23, 319–330 (1995).

Nugent K. A., "Interferogram Analysis Using an Accurate Fully Automatic Algorithm," *Appl. Opt.,* 24, 3101–3105 (1985).

Peng X., S. M. Shou, and Z. Gao, "An Automatic Demodulation Technique for a Non-Linear Carrier Fringe Pattern," *Optik,* 100, 11–14 (1995).

Pirga M. and M. Kujawinska, "Two Directional Spatial-Carrier Phase-Shifting Method for Analysis of Crossed and Closed Fringe Patterns," *Opt. Eng.,* 34, 2459–2466 (1995).

Pirga M. and M. Kujawinska, "Errors in Two Directional Spatial- Carrier Phase-Shifting Method," *Proc. SPIE,* 2544, 112–121 (1996).

Ransom P. L. and J. V. Kokal, "Interferogram Analysis by a Modified Sinusoid Fitting Technique," *Appl. Opt.,* 25, 4199 (1986).

Roddier C. and F. Roddier, "Interferogram Analysis Using Fourier Transform Techniques," *Appl. Opt.,* 26, 1668–1673 (1987).

Roddier C. and F. Roddier, "Wavefront Reconstruction Using Iterative Fourier Transforms," *Appl. Opt.,* 30, 1325–1327 (1991a).

Roddier C. and F. Roddier, "Reconstruction of the Hubble Space Telescope Mirror Figure from Out-of-Focus Stellar Images," *Appl. Opt.,* 30, 1325–1327 (1991b).

Schmit J., K. Creath, and M. Kujawinska, "Spatial and Temporal Phase-Measurement Techniques: A Comparison of Major Error Sources in One Dimension," *Proc. SPIE,* 1755, 202–211 (1992).

Servín M. and F. J. Cuevas, "A Novel Technique for Spatial Phase-Shifting Interferometry," *J. Mod. Opt.,* 42, 1853–1862 (1995).

Servín M. and R. Rodríguez-Vera, "Two-Dimensional Phase Locked Loop Demodulation of Interferogram," *J. Mod. Opt.,* 40, 2087–2094 (1993).

Servín M., D. Malacara, and F. J. Cuevas, "Direct Phase Detection of Modulated Ronchi Rulings Using a Phase Locked Loop," *Opt. Eng.,* 33, 1193–1199 (1994).

Servín M., R. Rodríguez-Vera, and D. Malacara, "Noisy Fringe Pattern Demodulation by an Iterative Phase Locked Loop," *Opt. and Lasers in Eng.,* 23, 355–366 (1995).

Shough D. M., O. Y. Kwon, and D. F. Leary, "High Speed Interferometric Measurement of Aerodynamic Phenomena," *Proc. SPIE,* 1221, 394–403 (1990).

Simova E. S. and K. N. Stoev, "Automated Fourier Transform Fringe- Pattern Analysis in Holographic Moiré," *Opt. Eng.,* 32, 2286–2294 (1993).

Takeda, M., "Temporal Versus Spatial Carrier Techniques for Heterodyne Interferometry," *Proc. SPIE,* 813, 329–330 (1987).

Takeda, M., "Spatial Carrier Heterodyne Techniques for Precision Interferometry and Profilometry: An Overview," *Proc. SPIE,* 1121, 73–88 (1989).

Takeda M. and K. Mutoh, "Fourier Transform Profilometry for the Automatic Measurement of 3-D Object Shapes," *Appl. Opt.,* 22, 3977–3982 (1983).

Takeda M. and Q.-S. Ru, "Computer-Based Highly Sensitive Electron-Wave Interferometry," *Appl. Opt.*, **24**, 3068–3071 (1985).

Takeda M. and Z. Tung, "Subfringe Holographic Interferometry by Computer-Based Spatial-Carrier Fringe-Pattern Analysis," *J. Opt. (Paris)*, **16**, 127–131 (1985).

Takeda M., H. Ina, and S. Kobayashi, "Fourier-Transform Method of Fringe-Pattern Analysis for Computer-Based Topography and Interferometry," *J. Opt. Soc. Am.*, **72**, 156–160 (1982).

Toyooka S., "Phase Demodulation of Interference Fringes with Spatial Carrier," *Proc. SPIE*, **1121**, 162–165 (1990).

Toyooka S. and Y. Iwaasa, "Automatic Profilometry of 3-D Diffuse Objects by Spatial Phase Detection," *Appl. Opt.*, **25**, 1630–1633 (1986).

Toyooka S. and M. Tominaga, "Spatial Fringe Scanning for Optical Phase Measurement," *Opt. Commun.*, **51**, 68–70 (1984).

Toyooka S., K., K. Ohashi, K. Yamada, and K. Kobayashi, "Real-Time Fringe Processing by Hybrid Analog-Digital System," *Proc. SPIE*, **813**, 33–35 (1987).

Vlad V. I. and D. Malacara, "Direct Spatial Reconstruction of Optical Phase from Phase-Modulated Images," in *Progress in Optics, Vol. 3*, E. Wolf, ed., Elsevier-Holland, 1994.

Womack K. H., "Interferometric Phase Measurement Using Spatial Synchronous Detection," *Opt. Eng.*, **23**, 391–395 (1984).

Interferogram Analysis with Moiré Methods

9.1 MOIRÉ TECHNIQUES

When two slightly different periodic structures are superimposed, a moiré fringe pattern appears (Sciammarella 1982; Reid 1984; Patorski 1988). Traditionally, moiré patterns have been analyzed from a geometrical point of view, but alternative methods have also been used. Some typical applications of moiré techniques were described in Chap. 1. Moiré methods are studied in this chapter as a tool for the analysis of interferograms.

The superposition of periodic structures to form the moiré patterns can be performed in two different basic ways.

1. Multiplication of the irradiances of the two images. This process may be implemented, for example, by superimposing the slides of two images. This is the most common method. The irradiance transmission of the combination is equal to the product of the two transmittances. Thus, the contrast in the moiré pattern is smaller than the contrast in either of the two images. There is an interesting holographic interpretation of the multiplicative moiré that will be described in this chapter.

2. Addition or subtraction of the irradiances of the two images. This method is less common (Rosenblum *et al.* 1991) than the multiplicative method, because it is more difficult to implement in practice. The advantage of this method is that since the two images (irradiances) are additively su-

perimposed, the contrast in the moiré image is higher than in the multi-
plicatively superimposed images.

9.2 MOIRÉ PATTERN FORMED BY TWO INTERFEROGRAMS WITH A LINEAR CARRIER

To mathematically analyze the moiré fringes from a geometrical point of view
using the multiplicative method, let us consider a transparency with phase-
modulated structure, which may be an interferogram with a linear carrier (tilt),
whose transmittance, assuming the maximum contrast, may be described by

$$T(x, y) = 1 + \cos(kx \sin\theta - kW(x, y)) \tag{9.1}$$

where $W(x, y)$ represents the wavefront deformations with respect to a close
reference sphere (frequently a plane) and the angle θ introduces the linear carrier
by means of a wavefront tilt about the x axis.

Let us now superimpose this interferogram to be evaluated on another reference interferogram with an irradiance transmittance given by

$$T_r(x, y) = 1 + \cos\left(\frac{2\pi}{d_r}x - kW_r(x, y) + \phi\right) \tag{9.2}$$

where $W_r(x, y)$ is any possible aspheric deformation of the wavefront producing
this interferogram, with respect to the same reference sphere used to measure
$W(x, y)$; d_r is the vertex spatial period of the reference linear carrier; and ϕ is
its phase at the origin. The transmittance of the combination is the product of
these two individual transmittances.

Thus, if the moiré pattern is produced by the multiplicative method, the
transmitted signal $s(x, y)$ is

$$s(x, y) = \{1 + \cos[k(x \sin\theta - W(x, y))]\}$$
$$\times \left[1 + \cos\left(\frac{2\pi}{d_r}x - kW_r(x, y) + \phi\right)\right] \tag{9.3}$$

from which we may obtain

$$s(x, y) = 1 + \cos[k(x \sin\theta - W(x, y))]\cos\left(\frac{2\pi}{d_r}x - kW_r(x, y) + \phi\right)$$
$$+ \cos[k(x \sin\theta - W(x, y))] + \cos\left(\frac{2\pi}{d_r}x - kW_r(x, y) + \phi\right) \tag{9.4}$$

Let us now use the trigonometric identity

$$\cos \alpha \cos \beta = \frac{1}{2}\cos(\alpha + \beta) + \frac{1}{2}\cos(\alpha - \beta) \tag{9.5}$$

thus obtaining

$$
\begin{aligned}
s(x, y) = 1 &+ \frac{1}{2}\cos\left[\left(k\sin\theta - \frac{2\pi}{d_r}\right)x - \phi - k(W(x, y) - W_r(x, y))\right] \\
&+ \frac{1}{2}\cos\left[\left(k\sin\theta + \frac{2\pi}{d_r}\right)x + \phi - k(W(x, y) + W_r(x, y))\right] \\
&+ \cos k[x\sin\theta - W(x, y)] + \cos\left(\frac{2\pi}{d_r}x - kW_r(x, y) + \phi\right) \tag{9.6}
\end{aligned}
$$

It is important to note that although each of the cosine functions may have positive as well as negative values, the total signal function has only positive values.

This result is general, for spherical as well as for aspheric wavefronts, but now we will consider separately the case of a reference interferogram with tilt fringes and that of a reference aspheric interferogram.

9.2.1 Moiré with Interferograms of Spherical Wavefronts

When the wavefront that produced the interferogram to be evaluated is nearly spherical, the reference interferogram must be ideally perfect, which means, as pointed out before, that it is formed by straight, parallel, and equidistant fringes. Thus, we assume that the reference wavefront is spherical and $W_r(x, y)$ is equal to zero, transforming expression 9.6 into

$$
\begin{aligned}
s(x, y) = 1 &+ \frac{1}{2}\cos\left[\left(k\sin\theta - \frac{2\pi}{d_r}\right)x - \phi - k(W(x, y))\right] \\
&+ \frac{1}{2}\cos\left[\left(k\sin\theta + \frac{2\pi}{d_r}\right)x + \phi - k(W(x, y))\right] \\
&+ \cos k[x\sin\theta - W(x, y)] + \cos\left(\frac{2\pi}{d_r}x + \phi\right) \tag{9.7}
\end{aligned}
$$

The first term on the right-hand side of Eq. 9.7 is a constant and thus has zero spatial frequency. Since ϕ is a constant, we see that the spatial frequency $f_2(x, y)$, along the x coordinate, of the second term is

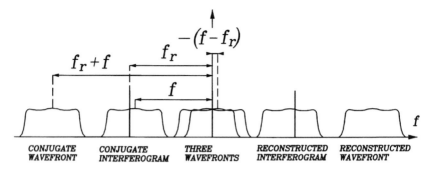

Figure 9.1 Fourier spectrum with the spatial frequencies of the moiré pattern.

$$f_2(x, y) = \pm \left(f - f_r - \frac{1}{\lambda} \frac{\partial W(x, y)}{\partial x} \right) \tag{9.8}$$

the spatial frequency $f_3(x, y)$, along x, of the third term is

$$f_3(x, y) = \pm \left(f + f_r - \frac{1}{\lambda} \frac{\partial W(x, y)}{\partial x} \right) \tag{9.9}$$

and the spatial frequency $f_4(x, y)$, along x, of the fourth term is

$$f_4(x, y) = \pm \left(f - \frac{1}{\lambda} \frac{\partial W(x, y)}{\partial x} \right) \tag{9.10}$$

where the interferogram carrier frequency f and the reference carrier frequency f_r are given by

$$f = \frac{\sin \theta}{\lambda} \quad \text{and} \quad f_r = \frac{1}{d_r} \tag{9.11}$$

Finally, the frequency of the fifth term is the reference frequency. Figure 9.1 shows the Fourier spectrum with the spatial frequency distribution of this moiré pattern.

Equation 9.7 represents the resulting irradiance pattern, but when observing moiré patterns the high frequency components need to be filtered out by any of several possible methods, for example, by defocusing or digital filtering. It is important to notice that the low pass filtering reduces the contrast of the pattern. Let us assume that the carrier frequencies f and f_r are close to each other.

We also impose the condition that the central frequency lobes in Fig. 9.1 are sufficiently separated from their neighbors that they can be isolated. Thus, the carrier spatial frequency of the interferogram, along the x coordinate, must have a value such that

$$f > \frac{2}{\lambda} \left(\frac{\partial W(x, y)}{\partial x} \right)_{max} \tag{9.12}$$

for all points inside the pattern.

If we use a low pass filter that cuts out all spatial frequencies higher than $f/2$, leaving only the central lobes in Fig. 9.1, we get

$$s(x, y) = 1 + \frac{1}{2} \cos \left[\left(k \sin \theta - \frac{2\pi}{d_r} \right) x + k W(x, y) + \phi \right] \tag{9.13}$$

which is the signal or irradiance of the interferogram, without any tilt (if $f = f_r$). From this result we may derive two important conclusions:

1. The moiré between the interferogram with a large tilt and the linear ruling modifies the carrier frequency (or removes it for $f_r = f$). It is interesting to note that the minimum allowed linear carrier to be able to remove this carrier with the moiré effect is twice the value required to phase demodulate the interferogram with a linear carrier using the methods in Chap. 7.
2. The phase of the final interferogram after the low pass filter may be changed if the constant phase ϕ of the linear ruling is changed. This effect has been used in some phase shifting schemes (Dorrio *et al.* 1995a, 1995b, 1996).

Figure 9.2a shows an example of an aberrated spherical interferogram. The reference interferogram is an interferogram of a perfect wavefront with tilt, as shown in Fig. 9.2b.

The resulting moiré pattern looks like that shown in Fig. 9.3a. Figure 9.3b shows the moiré image after low pass filtering.

The magnification or minification and hence the spatial frequency of the reference ruling can be modified to change the moiré pattern appearance. Two possible ways of doing this are illustrated in Fig. 9.4. In Fig. 9.4a the two transparencies are placed one over the other, with a short distance between them. The apparent magnification is changed by moving the reference ruling a small distance along the optical axis, changing the separation between the two transparencies.

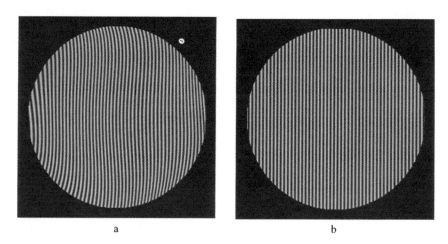

Figure 9.2 (a) Interferogram of an aberrated spherical wavefront with a linear carrier. (b) Interferogram of a perfect spherical wavefront with a linear carrier.

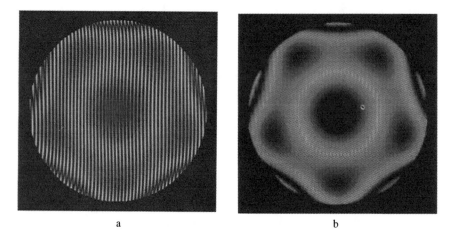

Figure 9.3 (a) Moiré pattern formed by interferograms of a spherical wavefront with a linear carrier, one of them aberrated. (b) Moiré image after low pass filtering. The histogram has been adjusted to compensate for the reduction in the contrast due to the low pass filtering.

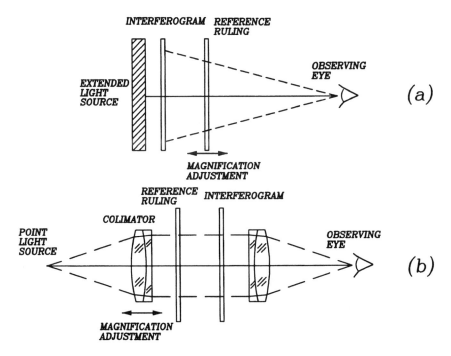

Figure 9.4 Optical arrangement used to observe the moiré pattern between an interferogram with a linear carrier and a linear ruling with adjustable linear carrier frequency.

In Fig. 9.4b the interferogram is placed at an integer multiple of the Rayleigh magnification of the reference ruling, so that an autoimage of the ruling is located close to the interferogram. Then the magnification is modified by moving the collimator along the optical axis to make the light beam slightly convergent or divergent.

When a ruling with a linear carrier is used as a reference, the magnification change can be a useful tool to visually remove the linear carrier or to change its magnitude. If the interferogram has a high frequency linear carrier, the spatial carrier (tilt) of the observed interferogram may be modified at will by moving the collimator along the axis. If the linear ruling is rotated, a spatial carrier (tilt) component in the y direction as well as in the x direction is introduced. We pointed out before that a lateral movement of the reference linear ruling introduces a constant phase shift (piston term). These effects can be used for teaching or demonstration purposes.

9.2.2 Moiré with Interferograms of Aspheric Wavefronts

When two perfect aspheric interferograms are superimposed, a moiré pattern formed by straight and parallel lines is observed. If the two interferograms are slightly different, the moiré fringes represent the difference between the two wavefronts producing a null test.

The general expression 9.5 must now be used. The first term on the right in Eq. 9.6 has zero spatial frequency. The spatial frequency in the x direction of the second term, $f_2(x, y)$, is

$$f_2(x, y) = \pm \left(f - f_r - \frac{1}{\lambda} \frac{\partial (W(x, y) - W_r(x, y))}{\partial x} \right) \tag{9.14}$$

the spatial frequency in the x direction of the third term, $f_3(x, y)$, is

$$f_3(x, y) = \pm \left(f + f_r - \frac{1}{\lambda} \frac{\partial (W(x, y) + W_r(x, y))}{\partial x} \right) \tag{9.15}$$

the spatial frequency in the x direction of the fourth term, $f_4(x, y)$, is

$$f_4(x, y) = \pm \left(f - \frac{1}{\lambda} \frac{\partial W(x, y)}{\partial x} \right) \tag{9.16}$$

and finally, the frequency of the fifth term is

$$f_5 = f_r - \frac{1}{\lambda} \frac{\partial W_r(x, y)}{\partial x} \tag{9.17}$$

The Fourier spectrum for this case, when an aspheric interferogram forms the moiré pattern with a reference aspheric interferogram, is shown in Fig. 9.5.

As pointed out earlier, when observing moiré patterns the high frequency components are filtered out. Let us assume that the frequencies f and f_r are close to each other. We use a low pass filter that cuts out all spatial frequencies equal to or higher than the width of the central lobes. To be able to isolate the lowest frequency terms we impose the condition that

$$f > \frac{1}{\lambda} \left(\frac{\partial (2W(x, y) - W_r(x, y))}{\partial x} \right)_{\max} \tag{9.18}$$

We may find

$$s(x, y) = 1 + \frac{1}{2} \cos \left[\left(k \sin \theta - \frac{2\pi}{d_r} \right) x - k(W(x, y) - W_r(x, y)) \right] \tag{9.19}$$

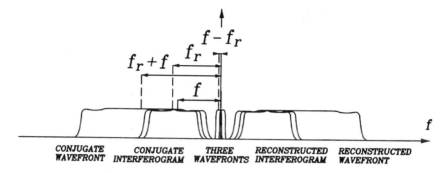

Figure 9.5 Fourier spectrum with the spatial frequencies of the moiré pattern when an aspheric reference is used.

Figure 9.6a shows an interferogram with spherical aberration plus some other high order aberrations. Figure 9.6b shows an interferogram with pure spherical aberration, to be used as a reference.

Then the transmittance of the combination is as in Fig. 9.7a. Figure 9.7b shows the low pass filtered moiré for two aspheric wavefronts.

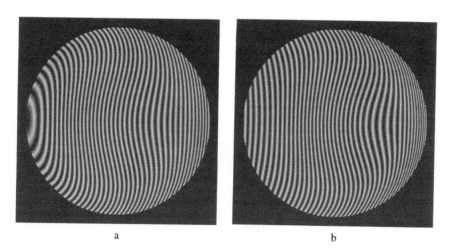

Figure 9.6 (a) Interferogram of an aberrated aspheric wavefront with a linear carrier; (b) interferogram of perfect aspheric wavefront with a linear carrier.

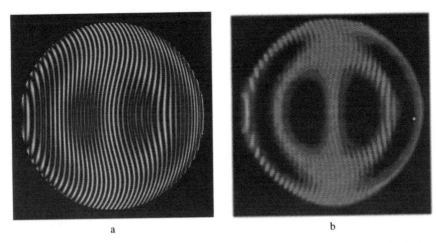

a   b

Figure 9.7 (a) Moiré pattern produced by the superposition of two aspheric interferograms, one of them aberrated, and (b) low pass filtered moiré after contrast enhancement.

If the wavefront under test is equal to the reference wavefront, we obtain a pattern of straight, parallel, and equidistant lines if the linear carriers of the two interferograms are different, as in any null test.

9.3 MOIRÉ PATTERN FORMED BY TWO INTERFEROGRAMS WITH A CIRCULAR CARRIER

Let us now study the moiré patterns between an interferogram with a circular carrier (defocusing) and an interferogram of a perfect wavefront with defocusing (circular ruling). All expressions will now be written in polar coordinates, (S, θ), defined as in Chap. 4, Sec. 4.3. The first image is an aberrated interferogram with a circular carrier (defocusing), whose transmittance may be written as

$$T(S, \theta) = 1 + \cos[k(DS^2 - W(S, \theta))] \tag{9.20}$$

where $W(S, \theta)$ is the wavefront deformation and kDS^2 is the radial spatial phase of the circular carrier.

Let us now superimpose on this interferogram another reference interferogram of a nonaberrated aspheric interferogram. This interferogram has perfect circular symmetry, but it can be decentered in the positive direction of x a small distance a, with an irradiance transmittance given by

$$T_r(S, \theta) = 1 + \cos\{k[D_r((x-a)^2 + y^2) - W_r(S, \theta)]\}$$
$$= 1 + \cos[k(D_r(S^2 + a^2 - 2ax) - W_r(S, \theta))] \qquad (9.21)$$

where $W_r(S, \theta)$ is the aspheric wavefront deformation of the reference interferogram and $kD_r S^2$ is the radial spatial phase of the reference circular ruling.

The transmittance of the combination is the product of these two individual transmittances, given by $s(S, \theta)$,

$$s(S, \theta) = \{1 + \cos[k(DS^2 - W(S, \theta))]\}$$
$$\times \{1 + \cos[k(D_r(S^2 + a^2 - 2ax) - W_r(S, \theta))]\} \qquad (9.22)$$

from which we can obtain

$$s(S, \theta) = 1$$
$$+ \cos[k(DS^2 - W(S, \theta))]$$
$$\times \cos[k(D_r(S^2 + a^2 - 2ax) - W_r(S, \theta))]$$
$$+ \cos[k(DS^2 - W(S, \theta))]$$
$$+ \cos[k(D_r(S^2 + a^2 - 2ax) - W_r(S, \theta))] \qquad (9.23)$$

Using Eq. 9.5, we obtain

$$s(S, \theta) = 1 + \frac{1}{2}\cos\{k[(D - D_r)S^2 - D_r(a^2 - 2ax) - (W(S, \theta) - W_r(S, \theta))]\}$$
$$+ \frac{1}{2}\cos\{k[(D + D_r)S^2 + D_r(a^2 - 2ax) - (W(S, \theta) + W_r(S, \theta))]\}$$
$$+ \cos[k(DS^2 - W(S, \theta))]$$
$$+ \cos[k(D_r(S^2 + a^2 - 2ax) - W_r(S, \theta))] \qquad (9.24)$$

This result is valid for both spherical and aspheric reference interferograms. Now we will consider these two cases separately.

9.3.1 Moiré with Interferograms of Spherical Wavefronts

If the wavefront that produced the interferogram to be evaluated is nearly spherical, the reference interferogram must be one of a spherical wavefront with defocusing, similar to a Fresnel zone plate or a Gabor plate. Then, if the reference wavefront is spherical and $W_r(x, y)$ is equal to zero, expression 9.24 becomes

$$s(S, \theta) = 1 + \frac{1}{2} \cos[k((D - D_r)S^2 - D_r(a^2 - 2ax) - W(S, \theta))]$$

$$+ \frac{1}{2} \cos[k((D + D_r)S^2 + D_r(a^2 - 2ax) - W(S, \theta))]$$

$$+ \cos[k(DS^2 - W(S, \theta))] + \cos[k(D_r(S^2 + a^2 - 2ax))] \quad (9.25)$$

Considering that the reference pattern is centered ($a = 0$), the first term on the right in Eq. 9.24 has zero spatial frequency. The radial spatial frequency of the second term, $f_2(S, \theta)$, is

$$f_2(S, \theta) = f(S) - f_r(S) - \frac{1}{\lambda} \frac{\partial W(S, \theta)}{\partial S} \tag{9.26}$$

the radial spatial frequency of the third term, $f_3(S, \theta)$, is

$$f_3(S, \theta) = f(S) + f_r(S) - \frac{1}{\lambda} \frac{\partial W(S, \theta)}{\partial S} \tag{9.27}$$

and the radial spatial frequency of the fourth term, $f_4(x, y)$, is

$$f_4(S, \theta) = f(S) - \frac{1}{\lambda} \frac{\partial W(S, \theta)}{\partial S} \tag{9.28}$$

where

$$f(S) = 2kDS \quad \text{and} \quad f_r(S) = 2kD_r S \tag{9.29}$$

Finally, the frequency of the fifth term is the reference frequency $f_r(S)$.

Equation 9.25 represents the resulting irradiance pattern, but when observing moiré patterns the high frequency components are filtered out by any of many possible methods, for example, by defocusing. Let us assume that the values of the linear carriers of both interferograms are close to each other. We also assume that the lowest frequency terms can be isolated, by requiring that the minimum radial frequency in the interferogram is such that

$$f > \frac{2}{\lambda} \left(\frac{\partial W(S, \theta)}{\partial S} \right)_{\text{max}} \tag{9.30}$$

for all points inside the moiré pattern.

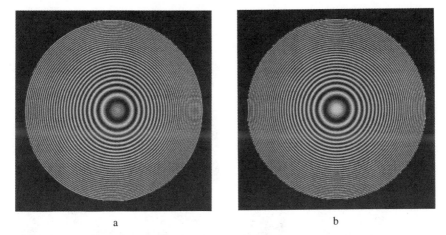

a   b

Figure 9.8 (a) Interferogram of an aberrated spherical wavefront with a circular carrier. (b) Reference interferogram of a perfect spherical wavefront with a circular carrier.

If we use a low pass filter that cuts out all spatial frequencies equal to or higher than the reference frequency $f_r(S)$, the second term is eliminated because its frequency is more than twice the carrier frequency. After the low pass filtering process we may find

$$s(S, \theta) = 1 + \frac{1}{2}\cos[k((D - D_r)S^2 - D_r(a^2 - 2ax) - W(S, \theta))] \qquad (9.31)$$

which is the interferogram with the spherical reference wavefront (defocus magnitude) modified or made flat (defocus removed) when $D = D_r$. Also, a tilt is added with the decentration a.

An example of an interferogram of this type is shown in Fig. 9.8a, and the reference interferogram with a perfect wavefront and a circular carrier is illustrated in Fig. 9.8b.

The moiré pattern obtained by the superposition of these two structures is illustrated in Fig. 9.9a. The low pass filtered moiré is in Fig. 9.9b.

9.3.2 Moiré with Interferograms of Aspheric Wavefronts

If the wavefront to be evaluated is aspheric as in Fig. 9.10(a), the reference interferogram can also be aspheric as in Fig. 9.10(b). Then $W_r(x, y)$ is not equal to zero and the general expression 9.24 must be used. In this case we have a null test for aspheric surfaces.

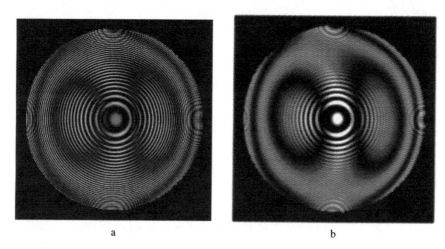

Figure 9.9 (a) Moiré image produced by interferograms with spherical wavefronts, one of them aberrated, with a circular carrier. (b) Filtered moiré after contrast enhancement.

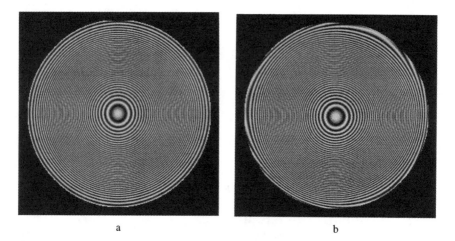

Figure 9.10 (a) Interferogram of an aberrated aspherical wavefront with a circular carrier. (b) Interferogram of a perfect aspheric wavefront with a circular carrier.

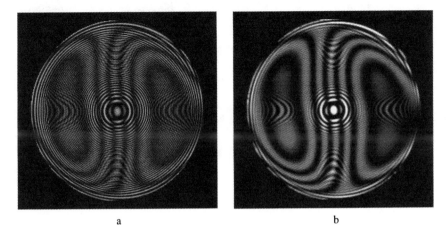

a   b

Figure 9.11 (a) Moiré pattern produced by interferograms of aspheric wavefronts, one of them aberrated, with a circular carrier. (b) Filtered moiré image after contrast enhancement.

The moiré pattern produced by these two interferograms is shown in Fig. 9.11a. The low pass filtered moiré image is in Fig. 9.11b.

9.4 SUMMARY OF MOIRÉ EFFECTS

Moiré methods are useful tools to detect aberrations in interferograms as well as for teaching demonstrations of the effect of tilt and defocusing in interferograms.

As described in Sec. 2.2, the apparent magnification of the reference ruling can be changed. This effect is useful in linear as well as in circular rulings. Table 9.1 summarizes the main operations that can be performed with moiré patterns of interferograms, by modifying the axial position (magnification) or the lateral position of the reference ruling.

Table 9.1 Effect Produced by a Displacement of the Reference Pattern

	Reference ruling displacement	
Reference ruling	Lateral displacement	Axial displacement (magnification)
Linear	Piston term (phase)	Tilt (linear carrier)
Circular	Tilt (linear carrier)	Focus (circ. carrier)

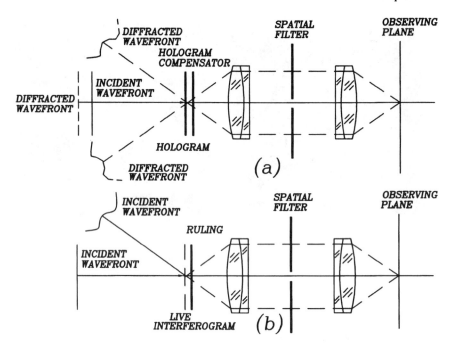

Figure 9.12 Moiré patterns between an interferogram and a ruling, (a) with a recorded interferogram and (b) with a "live" interferogram.

9.5 HOLOGRAPHIC INTERPRETATION OF MOIRÉ PATTERNS

The holographic point of view of the interferograms studied in Chap. 8 can also be used to interpret the moiré patterns of interferograms. To illustrate this model, let us consider the case of a linear reference ruling.

Let us assume that the linear ruling is illuminated with a plane wavefront impinging perpendicularly on this ruling as in Fig. 9.12a. After this ruling there will be three diffracted beams illuminating the hologram. Each of these flat wavefronts will generate its own three wavefronts—the zero-order, the wavefront under reconstruction, and the conjugate wavefront—after passing through the hologram. So after the hologram there will be a total of nine wavefronts, as illustrated in Fig. 9.13.

The lowest and the uppermost wavefronts in this figure are the wavefront under reconstruction and the conjugate wavefront. They correspond to the exp $[-iz]$ and exp $[+iz]$ components of the cosine function in the fourth term on the right in Eq. 9.7.

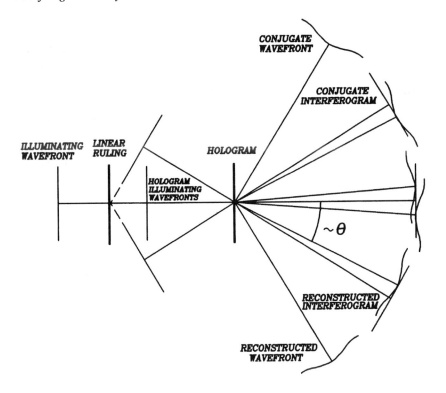

Figure 9.13 Holographic interpretation of moiré patterns. Generation of nine wavefronts.

Then we have a reconstructed image of the interferogram and a reconstructed image of the conjugate interferogram corresponding to the second and the last terms.

Near the optical axis we have, almost overlapping, the reconstructed wavefront, its conjugate, and a flat wavefront. They come from the third term and the constant term.

To conclude this chapter, let us mention an important conclusion that may be derived from the theory just described and in particular from Eq. 9.24. If two interferograms are formed by the interference between a flat reference wavefront and a distorted wavefront, different in each case, then the moiré pattern formed by these two interferograms is identical to the interferogram that would be obtained by the interference of the two distorted wavefronts. In other words, the moiré pattern of two interferograms represents the difference between the wavefront distortions (aberrations) in these two interferograms. Thus, any aberration common to the two interferograms will be canceled out.

REFERENCES

Dorrío B. V., A. F. Doval, C. López, R. Soto, J. Blanco-García, J. L. Fernández, and M. Pérez Amor, "Fizeau Phase-Measuring Interferometry Using the Moiré Effect," *Appl. Opt.*, **34**, 3639–3643 (1995a).

Dorrío B. V., J. Blanco-García, A. F. Doval, C. López, R. Soto, J. Bugarín, J. L. Fernández, and M. Pérez Amor, "Surface Evaluation Combining the Moiré Effect and Phase-Stepping Techniques in Fizeau Interferometry," *Proc. SPIE*, **2730**, 346–349 (1995b).

Dorrío B. V., J. Blanco-García, C. López, A. F. Doval, R. Soto, J. L. Fernández, and M. Pérez-Amor, "Phase Error Calculation in a Fizeau Interferometer by Fourier Expansion of the Intensity Profile," *Appl. Opt.*, **35**, 61–64 (1996).

Patorski K., "Moiré Methods in Interferometry," *Opt. and Lasers in Eng.*, **8**, 147–170 (1988).

Reid G. T., "Moiré Fringes in Metrology," *Opt. and Lasers in Eng.*, **5**, 63–93 (1984).

Rosenblum W. M., D. K. O'Leary, and W. J. Blaker, "Computerised Moiré Analysis of Progressive Addition Lenses," *Optom. Vision Sci.*, **69**, 936–940 (1992).

Sciammarella C. A., "The Moiré Method. A Review," *Exp. Mech.*, **22**, 418–433 (1982).

10

Wavefront Slope and Curvature Tests

10.1 WAVEFRONT DETERMINATION BY SLOPE SENSING

Wavefront slopes can be measured by using testing methods to measure the transverse ray aberrations in the x and y directions, which are directly related to the partial derivatives of the wavefront under analysis. Many of these tests use screens. Two typical examples are the Hartmann and Ronchi tests. The Hartmann test (Ghozeil 1992) uses a screen with holes or strips lying perpendicular to the propagation direction of the wavefront under test as shown in Fig. 10.1.

A screen with an array of circular holes is placed over a concave reflecting surface under test. Each of the narrow beams of light reflected on each hole returns to an observation screen called the Hartmann plate. Here we measure the deviation on the Hartmann plate of the reflected light beams with respect to the ideal positions. These deviations are the transverse aberrations TA_x and TA_y, measured along the x and y axes, respectively.

In the Ronchi test (Cornejo 1992), the screen is a ruling placed near the point of convergence of the returning aberrated wavefront, as shown in Fig. 10.2.

An imaging optical system is used to observe the projected shadows of the ruling lines over the surface under test. This imaging system may be the eye in qualitative tests, but it may be a lens in quantitative tests. By measuring the fringe deformations in the projected shadows, the transverse aberration in the

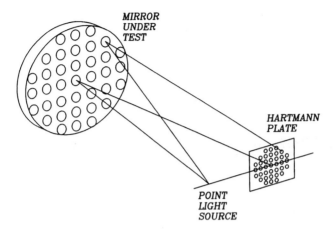

Figure 10.1 Optical arrangement in the Hartmann test.

direction perpendicular to the ruling lines is easily computed. If the ruling lines are along the y axis, the transverse aberration TA_x is measured. If the ruling lines are along the x axis, the transverse aberration TA_y is measured. In other words, two different measurements with two orthogonal ruling orientations are needed to measure the two components of the transverse aberration. Hartmann and Ronchi tests were compared by Welsh *et al.* (1995).

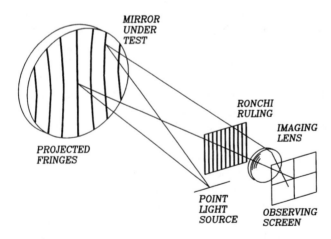

Figure 10.2 Optical arrangement in the Ronchi test.

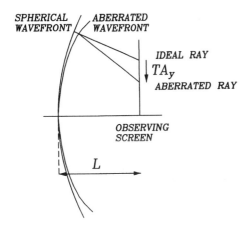

Figure 10.3   Measuring the transverse aberration in an aberrated wavefront.

Another system that measures the wavefront slopes is the lateral shear interferometer (Mantravadi 1992) described in Chap. 1, when the lateral shear is small compared with the period of the maximum spatial frequency to be detected in the wavefront deformations. Under these conditions the lateral shear interferometer is identical to the Ronchi test.

Thus, in these tests one measures the transverse aberrations at an observation plane located at a distance L from the wavefront being measured, as shown in Fig. 10.3.

These transverse aberrations are related to the wavefront slopes in the x and y directions as

$$\frac{\partial W(x, y)}{\partial x} = -\frac{TA_x}{L - W(x, y)} \approx -\frac{TA_x}{L} \tag{10.1}$$

and

$$\frac{\partial W(x, y)}{\partial y} = -\frac{TA_y}{L - W(x, y)} \approx -\frac{TA_y}{L} \tag{10.2}$$

Thus, to obtain the shape of the testing wavefront one must use an integration procedure. There are many ways to perform this integration, as will now be described.

One method is to use the trapezoidal rule, as suggested by Ghozeil (1992), which may be mathematically expressed as

$$W(x, y) = -\frac{1}{L} \int_0^x TA_x dx$$

$$= \frac{1}{L} \sum_2^N \left(\frac{TA_{x(n)} + TA_{x(n-1)}}{2} \right) (x_n - x_{n-1}) \tag{10.3}$$

Another method is to first interpolate the transverse aberration discrete measurements by means of a two-dimensional polynomial fitting and then perform the integration analytically as described by Cornejo (1992).

A third method is a least squares solution to the integration problem. This integration procedure has the advantage of being path-independent and robust to noise.

Some other wavefront measuring techniques analyze the wavefront by curvature sensing (Roddier 1990). This method is based on the fact that the irradiance of an aberrated wavefront changes along its propagation direction according to the transport equation applied to the irradiance beam.

10.2 SCREENS USED IN WAVEFRONT SLOPE MEASUREMENTS

As described earlier, the two most common methods to measure the ray transverse aberration are the Ronchi and Hartmann tests. It was also pointed out that two orthogonal Ronchi images are needed to fully determine the aberration of the wavefront. In contrast, only one Hartmann testing screen is needed to collect all the data regarding the wavefront aberration. The main advantage of using screen or ruling tests along with a CCD camera is that the measuring dynamic range of the tested wavefront is increased. That is, the fact that the slope of the testing wavefront is sensed instead of the wavefront itself allows us to increase the magnitude of the wavefront aberration that can be tested. This is why the most popular methods for testing large optics such as telescopes' primary mirrors use screens. So for a given number of pixels of a CCD camera it is possible to measure more waves of aberration using screen tests than with a standard interferometer, which measures the wave aberrations in a more direct way.

10.2.1 Wavefront Slope Analysis with Linear Gratings

As mentioned earlier, a linear grating fringe pattern is easier to analyze using standard carrier fringe detecting procedures such as the Fourier method, the synchronous method, or the spatial phase shifting method. These techniques have already been discussed in this book.

We may start with a simplified mathematical model for the transmittance of a linear grating (Ronchi rulings are normally made of binary, not sinusoidal, transmittance, but for mathematical simplicity we consider a sinusoidal ruling):

$$T_x(x, y) = \frac{1 + \cos(\omega_0 x)}{2} \tag{10.4}$$

The linear ruling is placed at the plane where the aberrated wavefront is to be measured. If we place a light detector at a distance L from the plate, due to the wavefront aberrations we will obtain a distorted irradiance pattern that will be approximately given by

$$I_x(x, y) = \frac{1}{2} + \frac{1}{2} \cos \left(\omega_0 x + \omega_0 L \frac{\partial W(x, y)}{\partial y} \right) \tag{10.5}$$

Then the irradiance $I_x(x, y)$ will be a distorted version of the transmittance $T_x(x, y)$. The shadow of the ruling, when illuminated with a wavefront with spherical aberration, produces a shadow over a CCD video array, as shown in Fig. 10.4.

As pointed out earlier, if there is no rotational symmetry, it is necessary to detect two orthogonal shadow patterns to completely describe the gradient field of the wavefront under test. The second linear ruling is located at the same testing plane but with its strip lines oriented orthogonally to those of the first ruling. That is,

$$T_y(x, y) = \frac{1 + \cos(\omega_0 y)}{2} \tag{10.6}$$

The lines in this transparency are perpendicular to those of the first one.

Thus the distorted image of the Ronchi ruling at the data collecting plane will be given by

$$I_y(x, y) = \frac{1}{2} + \frac{1}{2} \cos \left(\omega_0 y + \omega_0 L \frac{\partial W(x, y)}{\partial x} \right) \tag{10.7}$$

We may use any of the carrier fringe methods described in this book to demodulate these two Ronchigrams (Eqs. 10.5 and 10.7).

Once the detected and unwrapped phase of the ruling's shadows has been obtained, one needs to integrate the resulting gradient field. To integrate this phase gradient one may use path-independent integration such as least squares

Figure 10.4 Typical Ronchi pattern with spherical aberration.

integration. The least squares integration of the gradient field may be stated as the function that minimizes the quadratic merit function

$$
U(\hat{W}) = \sum_{(x,y) \in L} \left[\hat{W}(x_{i+1}, y_j) - \hat{W}(x_i, y_j) - \left(\frac{\partial W(x, y)}{\partial x} \right)_{x=x_i, y=y_j} \right]^2
$$

$$
+ \sum_{(x,y) \in L} \left[\hat{W}(x_i, y_{j+1}) - \hat{W}(x_i, y_j) - \left(\frac{\partial W(x, y)}{\partial y} \right)_{x=x_i, y=y_j} \right]^2
$$

$$(10.8)$$

where the hat function $\hat{W}$ is the estimated wavefront and we have approximated the derivatives of the searched phase along the x and y axes as first-order differences of the estimated wavefront. The least squares estimator may be obtained from U by simple gradient descent applied to all pixels, as

$$\hat{W}^{k+1}(x, y) = \hat{W}^k(x, y) - \tau \frac{\partial U(\hat{W})}{\partial \hat{W}(x, y)} \tag{10.9}$$

or by using a faster algorithm such as the conjugate gradient or transform method (Fried 1977; Hunt 1979; Ghiglia and Romero 1994).

10.2.2 Moiré Deflectometry and Talbot Interferometry

In moiré deflectometry or Talbot interferometry, as described in Chap. 1, the observing plane is located at the first Talbot autoimage of the ruling. Thus, the distance d_T is made equal to the Rayleigh distance L_R given by

$$L_R = \frac{2d^2}{\lambda} \tag{10.10}$$

The resulting deflectograms or Talbot interferograms may be analyzed in the same way as the Ronchigrams.

10.2.3 Lateral Shear Interferometry

In lateral shear interferometry, as described in Chap. 1, the interference pattern is formed with two mutually laterally displaced copies of the wavefront under analysis, as shown in Fig. 10.5.

The mathematical form of the irradiance of a lateral shear fringe pattern may be written as

$$
\begin{aligned}
I_x(x, y) &= \frac{1}{2} + \frac{1}{2} \cos[k(W(x - S, y) - W(x, y))] \\
&= \frac{1}{2} + \frac{1}{2} \cos[k \Delta_x W(x, y)]
\end{aligned} \tag{10.11}
$$

where $k = 2\pi/\lambda$ and S is the lateral shear. As Eq. 10.11 shows, one also needs the orthogonally displaced interferogram to completely describe the wavefront under analysis. The orthogonal interferogram may be written as

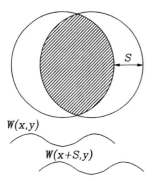

$W(x,y)$

$W(x+S,y)$

Figure 10.5 A shear interferometer produces the interference pattern between two copies of the laterally sheared wavefront.

$$I_y(x, y) = \frac{1}{2} + \frac{1}{2} \cos[k(W(x, y - T) - W(x, y))]$$

$$= \frac{1}{2} + \frac{1}{2} \cos[k\Delta_y W(x, y)] \tag{10.12}$$

where T is the lateral shear orthogonal to S. The fringe patterns in Eqs. 10.11 and 10.12 may be transformed into carrier frequency interferograms by introducing a large and known amount of defocusing into the testing wavefront (Mantravadi 1992). Having linear carrier fringe patterns, one may proceed to their demodulation using standard techniques of fringe carrier analysis as seen in this book.

The demodulated and unwrapped difference wavefront may be integrated using the path-independent integration procedure presented here. Assume that we have already estimated and unwrapped the interesting phase of the two orthogonally sheared interferograms. Using this information, the least squares wavefront reconstruction may be stated as the minimizer of the merit function

$$U(\hat{W}) = \sum_{(x,y)\in L_x} [\hat{W}(x - S, y) - \hat{W}(x, y) - \Delta_x W(x, y)]^2$$

$$+ \sum_{(x,y)\in L_y} [\hat{W}(x, y - T) - \hat{W}(x, y) - \Delta_y W(x, y)]^2$$

$$= \sum_{(x,y)\in L_x} U_x^2(x, y) + \sum_{(x,y)\in L_y} U_y^2(x, y) \tag{10.13}$$

where the "hat" function represents the estimated wavefront and L_x and L_y are two-dimensional lattices containing valid phase data in the x and y shear directions. However, the minimization problem stated in Eq. 10.13 is not well posed, because the matrix that results from setting the gradient of U equal to zero is not invertible. Fortunately, we may apply classical regularization to this inverse problem and find the expected smooth solution of the problem (Thikonov 1963). In classical regularization theory, the regularizer consists of a linear combination of the squared magnitudes of derivatives of the estimated wavefront inside the domain of interest. In particular, one may use a discrete approximation to the Laplacian to obtain the second-order potentials:

$$R_x(x_i, y_j) = \hat{W}(x_{i-1}, y_j) - 2\hat{W}(x_i, y_j) + \hat{W}(x_{i+1}, y_j)$$

$$R_y(x_i, y_j) = \hat{W}(x_i, y_{j-1}) - 2\hat{W}(x_i, y_j) + \hat{W}(x_i, y_{j+1}) \tag{10.14}$$

Therefore, the regularized merit function becomes

$$U(\hat{W}) = \sum_{(x,y)\in L_x} U_x^2(x, y) + \sum_{(x,y)\in L_y} U_y^2(x, y)$$

$$+ \lambda \sum_{(x,y)\in \text{Pupil}} [R_x^2(x, y) + R_y^2(x, y)] \tag{10.15}$$

where Pupil refers to the two-dimensional lattice inside the pupil of the wavefront being tested. The estimated wavefront obtained using these second-order potentials as regularizers makes the solution behave like a thin metallic plate attached to the observations by linear springs. The regularizing potentials discourage large changes in the estimated wavefront among neighboring pixels. As a consequence, the searched-for solution will be relatively smooth. The parameter λ controls the amount of smoothness of the estimated wavefront. If the observations have a negligible amount of noise, then the parameter λ may be set to a small value (~ 0.1); if the observations are noisy, then λ may be set to a higher value (in the range of 0.5–11.0) to filter out some noise. It should be remarked that the use of regularizing potentials in this case is a must, even for noise-free observations, to yield a stable solution of the least squares integration for lateral displacements greater than two pixels. As analyzed in Servín *et al.* (1996), this is because the inverse operator that performs the least squares integration has poles in the frequency domain.

The estimated wavefront may be calculated using simple gradient descent as

$$\hat{W}^{k+1}(x, y) = \hat{W}^k(x, y) - \tau \frac{\partial U(\hat{W})}{\partial \hat{W}(x, y)} \tag{10.16}$$

applied to all pixels, where τ is the convergence rate. This optimizing method is not very fast. One normally uses faster algorithms such as the conjugate gradient.

10.2.4 Hartmann Test

The Hartmann test is a well-known technique for testing large optical components (Ghozeil 1992). The Hartmann technique samples the wavefront under analysis using a screen of uniformly spaced holes situated at the pupil plane,

$$
\text{HS}(x, y) = \sum_{n=-N/2}^{N/2} \sum_{m=-N/2}^{N/2} h(x - nd, y - md) \tag{10.17}
$$

where $\text{HS}(x, y)$ is the Hartmann screen and $h(x, y)$ are the small holes which are uniformly spaced in the Hartmann screen. Finally, d is the space between holes of the screen. A typical Hartmann screen is shown in Fig. 10.6.

The collimated rays of light that pass through the screen holes (Eq. 10.17) are then captured by a photographic plate at some distance L from the screen. The uniformly spaced array of holes at the instrument's pupil is then distorted at the photographic plane by the spherical aberration of the wavefront under test. The screen deformations are then proportional to the slope of the aspheric wavefront, which is

$$
H(x, y) = \left[\sum_{(n,m)=-N/2}^{N/2} h'\left(x - nd - L\frac{\partial W(x, y)}{\partial x}, y - md - L\frac{\partial W(x, y)}{\partial x} \right) \right]
$$
$$
\times P(x, y) \tag{10.18}
$$

where $H(x, y)$ is the Hartmanngram obtained at a distance L from the Hartmann screen. The function $h'(x, y)$ is the image of the screen's hole $h(x, y)$ as projected at the Hartmanngram plane. Finally, $P(x, y)$ is the pupil of the wavefront being tested. As Eq. 10.18 shows, only one Hartmanngram is needed to fully estimate the wavefront's gradient. The frequency content of the estimated wavefront will be limited by the sampling theorem to the period d of the screen holes. Figure 10.7 shows a Hartmanngram of a 62 cm paraboloidal mirror.

Traditionally these Hartmanngrams (the distorted image of the screen at the photographic plate's plane) are analyzed by measuring the centroid of the spot images $h'(x, y)$ generated by the screen holes $h(x, y)$. The deviations of these centroids from their uniformly spaced positions (unaberrated positions) are recorded. As Eq. 10.18 shows, these deviations are proportional to the aspheric aberration's slope. The centroid coordinates give a two-dimensional discrete field of the wavefront gradient that has to be integrated and interpolated over

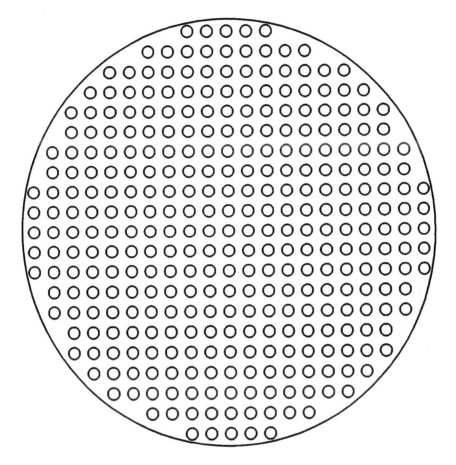

Figure 10.6 Typical Hartmann screen used in the Hartmann screen test.

regions without data. The wavefront's gradient field is normally integrated by using the trapezoidal rule (Ghozeil 1992). The trapezoidal rule follows several independent integration paths and their average outcome. In this way one may approach a path-independent integration. Using this integration procedure, the wavefront is known only at the hole's position. Although this integration technique may give good wavefront estimates, the manual hole position estimation is a time-consuming process. Finally, a polynomial or spline wavefront fitting is necessary to estimate the wavefront's values at places other than the discrete points where the gradient data are collected. If a two-dimensional polynomial for the wavefront's gradient is proposed, it is then fitted by least squares to the slope

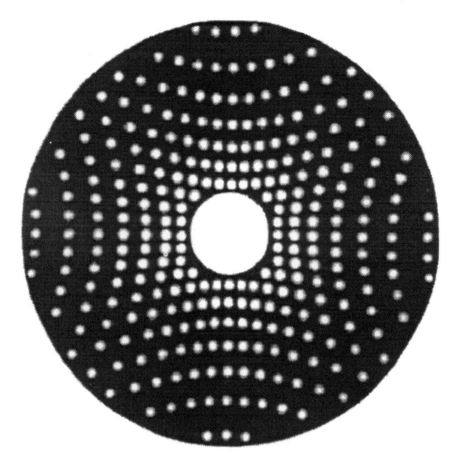

Figure 10.7   Hartmanngram of a 62 cm paraboloidal primary mirror.

data. This polynomial must contain every possible type of wavefront aberration; otherwise, some unexpected features (especially at the edges) of the wavefront may be filtered out. On the other hand, if one uses a high degree polynomial (in order to avoid filtering out any wavefront aberration), the estimated continuous wavefront may oscillate wildly in regions where no data are collected.

10.3 WAVEFRONT CURVATURE SENSING

The observation of defocused stellar images, known as the star test, has been used for many years as a sensitive method to detect small wavefront deformations. The principle of this method is based on the fact that the illumination in a defocused image is not homogeneous if the wavefront has deformations. These deformations may be interpreted as variations in the local curvature of the wavefront. If the focus is shortened, the light energy will be concentrated at a shorter focus and vice versa. An obvious consequence is that the illumination at two observation planes located symmetrically with respect to the focus has different illumination densities. However, this test remained for a long time only a qualitative visual test.

10.3.1 The Laplacian and Local Average Curvatures

Roddier (1988) and Roddier *et al.* (1988) proposed a quantitative wavefront evaluation method indirectly based on the star test principle, which measures wavefront local curvatures. The local curvatures c_x and c_y of a nearly flat wavefront in the x and y directions are given by the second partial derivatives of this wavefront, as

$$c_x = \frac{\partial^2 W(x, y)}{\partial x^2} \quad \text{and} \quad c_y = \frac{\partial^2 W(x, y)}{\partial y^2} \tag{10.19}$$

Hence, the Laplacian, defined by

$$\nabla^2 W(x, y) = 2\rho(x, y) = \frac{\partial^2 W(x, y)}{\partial x^2} + \frac{\partial^2 W(x, y)}{\partial y^2} \tag{10.20}$$

is twice the value of the average local curvature $\rho(x, y)$. This whole expression is known as the Poisson equation. To solve the Poisson equation to obtain the wavefront deformations $W(x, y)$, we need to have (1) the average local curvature distribution $\rho(x, y)$, which is a scalar field since no direction is involved, as in the wavefront slopes, and (2) the radial wavefront slopes at the edge of the circular pupil, to be used as Neumann boundary conditions.

As described by Roddier *et al.* (1988), the simplest method to solve the Poisson equation once the Laplacian has been determined is the Jacobi iteration algorithm. Noll (1978) showed that Jacobi's method is essentially the same as that derived by Hudgin (1977) to find the wavefront from slope measurements. Equivalent iterative Fourier methods to obtain the wavefront without having to directly solve the Poisson equation are described in Sec. 10.3.4.

10.3.2 Irradiance Transport Equation

Let us consider a light beam propagating with an average direction along the z axis after passing through a diffracting aperture (pupil) on the xy plane. The irradiance and the wavefront shape both change continuously along the trajectory. As proved by Teague (1983), the wave disturbance $u(x, y, z)$ at a point (x, y, z) can be found with good accuracy, even with a diffracting aperture with sharp edges, by using the Huygens-Fresnel diffraction theory if a paraxial approximation is taken. This approximation considers the Huygens wavelets as emitted in a narrow cone, with a parabolic approximation for the wavefront shape of each wavelet. This can be considered a geometrical optics approximation. Teague (1983) and Steibl (1984) showed that if a wide diffracting aperture much larger than the wavelength is assumed, the disturbance at any plane with any value of z may be found with the differential equation

$$\nabla^2 u(x, y, z) + 2k^2 u(x, y, z) + 2ik \frac{\partial u(x, y, z)}{\partial z} = 0 \tag{10.21}$$

where $k = 2\pi/\lambda$. We may consider a solution to this equation of the form

$$u(x, y, z) = I^{1/2}(x, y, z) \exp(ikW(x, y, z)) \tag{10.22}$$

where $I(x, y, z)$ is the irradiance. If we substitute this disturbance expression into the differential equation, we can obtain, after some algebraic steps, a complex function that should be made equal to zero. Then, equating real and imaginary parts to zero, we obtain

$$\frac{\partial W}{\partial z} = 1 + \frac{1}{4k^2 I} \nabla^2 I - \frac{1}{2} \nabla W \cdot \nabla W - \frac{1}{8k^2 I^2} \nabla I \cdot \nabla I \tag{10.23}$$

and

$$\frac{\partial I}{\partial z} = -\nabla I \cdot \nabla W - I \nabla^2 W \tag{10.24}$$

where the (x, y, z) dependence has been omitted for notational simplicity and the Laplacian ∇^2 and the gradient operators ∇ work only on the lateral coordinates x and y.

Expression 10.23 is the phase transport equation, which can be used to find the wavefront shape at any point along the trajectory. Expression 10.24 is the irradiance transport equation. Ichikawa *et al.* (1988) reported an experimental demonstration of phase retrieval based on this equation.

Following an interesting discussion by Ichikawa *et al.* (1988), one may note in the irradiance transport equation the following interpretation for each term:

1. The gradient $\nabla W(x, y, z)$ represents the direction and magnitude of the local tilt of the wavefront, and $\nabla I(x, y, z)$ is the direction in which the irradiance value changes with maximum speed. Thus their scalar product $\nabla I(x, y, z) \cdot \nabla W(x, y, z)$ is the irradiance variation along the optical axis **z** due to the local wavefront tilt. Ichikawa *et al.* (1988) call this a *prism term*.

2. The term $I(x, y, z)\nabla^2 W(x, y, z)$ can be interpreted as the irradiance along the z axis caused by the local wavefront average curvature. Ichikawa *et al.* (1988) called this a *lens term*.

In sum, these terms describe the variation of the beam irradiance caused by the wavefront deformations as it propagates along the z axis. This means that the transport equation is a geometrical optics approximation, valid away from sharp apertures and as long as the aperture is large enough compared with the wavelength. To gain even more insight into the nature of this equation, it can be rewritten as

$$-\frac{\partial I(x, y, z)}{\partial z} = \nabla \cdot [I(x, y, z)\nabla W(x, y, z)] \tag{10.25}$$

and recalling that ∇W is a vector representing the wavefront local slope, we can easily see that the transport equation represents the law of light energy conservation, which is analogous to the law of mass or charge conservation, frequently expressed as

$$-\frac{\partial \rho}{\partial t} = \nabla \cdot (\rho v) \tag{10.26}$$

with ρ and v being the mass or charge density and the flow velocity, respectively.

10.3.3 Laplacian Determination with Irradiance Transport Equation

Roddier *et al.* (1990) used the transport equation to measure the wavefront. Let $P(x, y)$ be the transmittance of the pupil, which is equal to 1 inside the pupil and 0 outside. Furthermore, we assume that the illumination at the pupil's plane is uniform and equal to a constant I_0 inside the pupil. Hence, the irradiance gradient $\nabla I(x, y, 0) = 0$ everywhere except at the pupil's edge, where

$$\nabla I(x, y, 0) = -I_0 n \delta_c \tag{10.27}$$

where δ_c is a Dirac distribution around the pupil's edge and n is a unit vector perpendicular to the edge and pointing outward. Substituting this gradient into the irradiance transport equation, one obtains

$$\left(\frac{\partial I(x, y, z)}{\partial z}\right)_{z=0} = -I_0 \cdot \left(\frac{\partial W(x, y, z)}{\partial n}\right)_{z=0} \delta_c - I_0 P(x, y)\nabla^2 W(x, y, z) \tag{10.28}$$

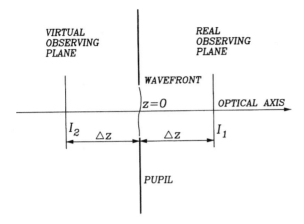

Figure 10.8 Irradiance measured in two planes symmetrically placed with respect to the pupil.

where the derivative on the right-hand side is the wavefront derivative in the outward direction perpendicular to the pupil's edge. Curvature sensing consists in taking the difference between the illumination observed in two planes symmetrically located with respect to the diffracting stop, as shown in Fig. 10.8.

Thus, the measured irradiances at these two planes are

$$I_1(x, y, \Delta z) = I_0 + \left(\frac{\partial I(x, y, z)}{\partial z} \right)_{z=0} \Delta z$$

and   (10.29)

$$I_2(x, y, -\Delta z) = I_0 - \left(\frac{\partial I(x, y, z)}{\partial z} \right)_{z=0} \Delta z$$

When the wavefront is perfectly flat at the pupil, the Laplacian at all points inside the pupil and the radial slope at the edge of the pupil are both zero. Then, $I_2(x, y, -\Delta z)$ is equal to $I_1(x, y, \Delta z)$. Having obtained these data, one may form the so-called sensor signal as

$$s(x, y, \Delta z) = \frac{I_1 - I_2}{I_1 + I_2} = \frac{1}{I_0} \left(\frac{\partial I(x, y, z)}{\partial z} \right)_{z=0} \Delta z$$ (10.30)

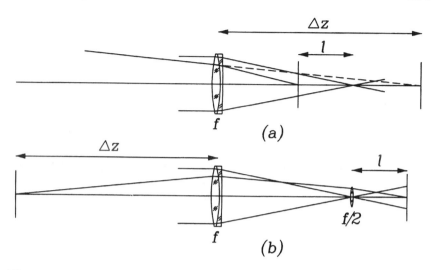

Figure 10.9 Two conjugate planes, one plane before refraction on the optical system, at a distance Δz from the pupil, and the second plane after refraction, at a distance l from the focus of the system. (a) With the first plane at the back of the pupil and the second plane inside the focus. (b) With the first plane in front of the pupil and the second plane outside the focus, using an auxiliary small lens of focal length $f/2$.

Substituting Eq. 10.28 into Eq. 10.30 yields

$$\frac{I_1 - I_2}{I_1 + I_2} = \left(\frac{\partial W(x, y)}{\partial n} \delta_c - P(x, y) \nabla^2 W(x, y) \right) \Delta z \qquad (10.31)$$

Thus with the irradiance I_1 and I_2 in two planes symmetrically located with respect to the pupil ($z = 0$), we obtain the left-hand term of this expression. This gives us the Laplacian of $W(x, y)$ (average local curvature) for all points inside the aperture and the wavefront slope $\partial W/\partial n$ around the pupil's edge $P(x, y)$ as a Neumann boundary condition to be used when solving Poisson's equation.

The two planes on which the irradiance has to be measured are symmetrically located with respect to the diffracting pupil. In other words, one plane is real because it is located after the pupil, but the other plane is virtual because it is located in front of the pupil. In practice this problem has an easy solution because the diffracting aperture is the pupil of the lens to be evaluated, typically a telescope objective.

As we see in Fig. 10.9, a plane at a distance l inside the focus is conjugate to a plane at a distance Δz after the pupil. On the other hand, if a small lens with focal length $f/2$ is placed at the focus of the objective, a plane at a distance l

outside the objective focus is conjugate to a plane at a distance Δz in front of the pupil. In both cases the distance Δz and the distance l are related by

$$\Delta z = \frac{f(f-1)}{l} \tag{10.32}$$

Roddier and Roddier (1991) pointed out that the small lens of length $f/2$ is not necessary if l is small compared with f. We must take into account that one defocused image is rotated $180°$ with respect to the other and also any possible difference in the magnification of the two images. The important consideration is that the subtracted and added irradiances in the two measured images must correspond to the same point (x, y) on the pupil.

The measurements of the irradiance have to be made close enough to the pupil that the diffraction effects are negligible and the geometric approximation remains valid. Let us assume that the wavefront to be measured has some corrugations and deformations of scale r_0 (maximum spatial period). With the diffraction grating equation we see that these corrugations spread out the light over a narrow cone with an angular diameter $\alpha = \lambda/r_0$. Thus, the illumination in the plane of observation can be considered a blurred pupil image. Let us now impose the condition that the maximum allowed blurring at a distance Δz is equal to $r_0/2$. With this condition it is possible to show that the geometrical optics approximation implied in the transport irradiance equation is valid only if Δz is sufficiently small that the following condition is satisfied:

$$\Delta z \ll \frac{r_0^2}{2\lambda} \tag{10.33}$$

It is interesting to see that this distance Δz is one-fourth of the Rayleigh distance in Talbot autoimaging, as described in Chap. 1. This result is to be expected, since then the shadow of the grating is geometrical. If the light angular diameter spread α is known—for example, if this is equal to the atmospheric light seen in a telescope—we may also write

$$\Delta z \ll \frac{\lambda}{2\alpha^2} \tag{10.34}$$

When measuring in the converging beam, this condition implies that the defocusing distance l should be large enough that

$$l \gg \frac{f}{1 + r_0^2/2\lambda f} \tag{10.35}$$

In conclusion, the minimum defocusing distance depends on the maximum spatial frequency of the wavefront corrugation we want to measure. This frequency also determines the density of sampling points to be used to measure the irradiance in the defocused image.

10.3.4 Wavefront Determination with Iterative Fourier Transforms

Hardy and MacGovern (1987) measured slope differences to obtain the curvatures, and from them the Poisson equation is solved to obtain the wavefront. The curvature in the x direction is taken as the difference between two adjacent tilts in this direction, and in the same manner the curvature along the y axis is obtained. Then the average of these curvatures is calculated. They used the Hudgin (1977) algorithm to obtain this solution.

Roddier and Roddier (1991b) and Roddier *et al.* (1990) reported a method to obtain the wavefront deformations $W(x, y)$ from a knowledge of the Laplacian, by solving the Poisson equation using iterative Fourier transforms. To understand this method, let us take the Fourier transform of the Laplacian of the wavefront:

$$F\{\nabla^2(x, y)\} = F\left\{\frac{\partial^2 W(x, y)}{\partial x^2}\right\} + F\left\{\frac{\partial^2 W(x, y)}{\partial y^2}\right\} \tag{10.36}$$

On the other hand, from the derivative theorem in Sec. 2.3.4, we have

$$F\left\{\frac{\partial W(x, y)}{\partial x}\right\} = i2\pi f_x F\{W(x, y)\} \tag{10.37}$$

and similarly for the partial derivative with respect to y. In an identical manner we may also write

$$F\left\{\frac{\partial^2 W(x, y)}{\partial x^2}\right\} = i2\pi f_x F\left\{\frac{\partial W(x, y)}{\partial x}\right\} = -4\pi^2 f_x^2 F\{W(x, y)\} \tag{10.38}$$

Thus, it is easy to prove that

$$F\{\nabla^2(x, y)\} = -4\pi^2 F\{W(x, y)\}(f_x^2 + f_y^2) \tag{10.39}$$

Hence, in the Fourier domain the Fourier transform of the Laplacian operator translates into a multiplication of the Fourier transform of the wavefront $W(x, y)$ by $f_x^2 + f_y^2$.

The wavefront may be calculated if measurements of the slopes along x and y are available, as in the case of the Hartmann and Ronchi tests, with the expression

$$W(x, y) = -\frac{i}{2\pi} F^{-1} \left\{ \frac{f_x F\left\{\frac{\partial W(x, y)}{\partial x}\right\} + f_y F\left\{\frac{\partial W(x, y)}{\partial y}\right\}}{f_x^2 + f_y^2} \right\} \tag{10.40}$$

This simple approach works for a wavefront without any limiting pupil. In practice, however, the Laplacian is multiplied by the pupil function, to take into account its finite size. Thus, its Fourier transform is convolved with the Fourier transform of the pupil function. As a result, this procedure does not give correct results. An apodization in the Fourier space, that is, a filtering of the frequencies produced by the pupil boundaries, is needed, as in the Gershberg algorithm, described earlier in this book, to extrapolate the fringes outside of the pupil. The division by $f_x^2 + f_y^2$ produces this filtering. As a result of this filtering, just as in the Gershberg algorithm, after taking the inverse Fourier transform the wavefront extension is not restricted to the internal region of the pupil but extends outside the initial boundary.

The complete procedure to find the wavefront is thus an iterative process, as described in Fig. 10.10.

We can also retrieve the wavefront by taking the Fourier transform of the wavefront Laplacian, dividing it by $f_x^2 + f_y^2$, and taking the inverse Fourier transform, as follows:

$$W(x, y) = -\frac{1}{4\pi^2} F^{-1} \left\{ \frac{F\{\nabla^2 W(x, y)\}}{f_x^2 + f_y^2} \right\} \qquad (10.41)$$

An iterative algorithm, quite similar to the one just described, based on this expression has been also proposed by Roddier and Roddier (1991b), as shown in Fig. 10.11. The Laplacian is measured with the method before described with two defocused images. The Neumann boundary conditions are taken by setting the radial slope equal to zero within a narrow band surrounding the pupil. To better understand this boundary condition we may consider the wavefront curvature on the edge of the pupil as the difference between the slopes on each side of the pupil's edge. If the outer slope is set to zero, the curvature has to be equal to the inner slope. In other words, the edge radial slope is not arbitrarily separated from the inner curvature if this external slope is made equal to zero.

10.3.5 Wavefront Determination with Defocused Images

If the defocusing distance cannot be made large enough, the geometrical optics approximation assumed by the irradiance transport equation is not satisfied. In that case diffraction effects are important as in the classical star test and the method described in the preceding section cannot be applied. Thus, a different iterative method must be used.

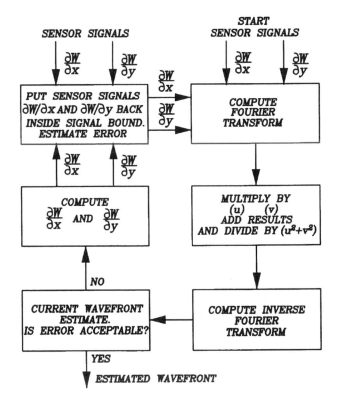

Figure 10.10 Iterative Fourier transform algorithm used to find the wavefront from the measured slopes. (After Roddier and Roddier, 1991b).

Gershberg and Saxton (1972) described an algorithm using a single defocused image, with the following steps:

1. An arbitrary guess of the wavefront deformations (phase and pupil transmission) is taken. The pupil transmission is frequently equal to 1, and the phase can be anything.
2. The defocused image (amplitude and phase) in the observation plane is computed with a fast Fourier transform.
3. The calculated amplitude is replaced by the observed amplitude (square root of the observed intensity). Then, the phase is calculated.
4. An inverse Fourier transform gives a new estimate of the incoming wavefront amplitude and phase (deformations).

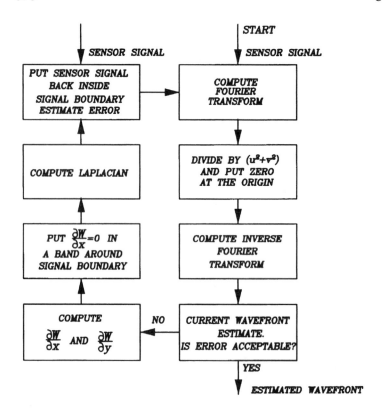

Figure 10.11 Iterative Fourier transform algorithm used to find the wavefront from measurement of the Laplacian. (After Roddier and Roddier, 1991b).

5. The calculated input amplitude is replaced by the known input amplitude (pupil transmission), keeping the calculated phase.

These steps are iterated until a reasonably small difference between measured and calculated amplitudes is obtained. This algorithm quickly converges at the beginning but then tends to stagnate.

Based on the work by Fienup and Wackermann (1986) and by Misell (1973a, 1973b), an improved method that converges more easily, using two defocused images, was described by Roddier and Roddier (1991a). This method was used to test the defective primary mirror of the Hubble telescope.

10.4 CONCLUSION

In this chapter we have presented the most important techniques for testing an optical wavefront by estimating the slope and curvature changes as it propagates along the experimental setup.

We have seen that the main advantage of the screen and curvature methods (especially if one is using a low resolution CCD camera to capture the interesting data) is the wider measuring dynamic range of the experimental arrangement. That is, these methods allow us to measure aberrations produced by many more fringes than standard interferometric methods such as temporal phase shifting. This increase of measuring range comes at the price of a proportional reduction in sensitivity. While commercial phase shifting interferometers may have a sensitivity as high as $\lambda/100$, a slope and curvature test may reach typically $\lambda/10$ accuracy.

A great advantage of the curvature sensing over all other testing methods analyzed in this book is its capacity to measure large optics in situ. That is, there is no need of any special experimental arrangement other than the optics where the lenses or mirrors are used.

REFERENCES

Cornejo A., "The Ronchi Test," in *Optical Shop Testing*, D. Malacara, Ed., John Wiley & Sons, New York, 1992.

Fienup J. R. and C. C. Wackermann, "Phase-Retrieval Stagnation Problems and Solutions," *J. Opt. Soc. Am. A*, **3**, 1897 (1986).

Fischer D. J., "Vector Formulation for Ronchi Shear Surface Fitting," *Proc. SPIE*, **1755**, 228–238 (1992).

Freischlad K., "Wavefront Integration from Difference Data," *Proc. SPIE*, **1755**, 212–218 (1992).

Freischlad K. and C. L. Koliopoulos, "Wavefront Reconstruction from Noisy Slope or Difference Data Using the Discrete Fourier Transform," *Proc. SPIE*, **551**, 74–80 (1985).

Fried D. L., "Least-Squares Fitting of a Wave-Front Distortion Estimate to an Array of Phase-Difference Measurements," *J. Opt. Soc. Am.*, **67**, 370–375 (1977).

Gershberg R. W. and W. O. Saxton, "A Practical Algorithm for the Determination of Phase from Image and Diffraction Plane Pictures," *Optik*, **35**, 237 (1972).

Ghiglia D. C. and L. A. Romero, "Robust Two Dimensional Weighted and Unweighted Phase Unwrapping That Uses Fast Transforms and Iterative Methods," *J. Opt. Soc. Am. A*, **11**, 107–117 (1994).

Ghozeil I., "Hartmann and Other Screen Tests," in *Optical Shop Testing*, D. Malacara, Ed., John Wiley & Sons, New York, 1992.

Hardy J. W. and A. J. MacGovern, "Shearing Interferometry: A Flexible Technique for Wavefront Measuring," *Proc. SPIE*, **816**, 180–195 (1987).

Horman M. H., "An Application of Wavefront Reconstruction to Interferometry," *Appl. Opt.*, **4**, 333–336 (1965).

Hudgin R. H., "Wave-Front Reconstruction for Compensated Imaging," *J. Opt. Soc. Am.*, **67**, 375–378 (1977).

Hunt B. R., "Matrix Formulation of the Reconstruction of Phase Values from Phase Differences," *J. Opt. Soc. Am.*, **69**, 393–399 (1979).

Ichikawa K., A. W. Lohmann, and M. Takeda, "Phase Retrieval Based on the Irradiance Transport Equation and the Fourier Transport Method: Experiments," *Appl. Opt.*, **27**, 3433–3436 (1988).

Mantravadi M. V., "Lateral Shearing Interferometers," in *Optical Shop Testing*, D. Malacara, Ed., John Wiley & Sons, New York, 1992.

Misell D. L., "An Examination of an Iterative Method for the Solution of the Phase Problem in Optics and Electron Optics. I: Test Calculations," *J. Phys. D. (Appl. Phys.)*, **6**, 2200 (1973a).

Misell D. L., "An Examination of an Iterative Method for the Solution of the Phase Problem in Optics and Electron Optics. II: Sources of Error," *J. Phys. D. (Appl. Phys.)*, **6**, 2217 (1973b).

Noll R. J., "Phase Estimates from Slope-Type Wave-Front Sensors," *J. Opt. Soc. Am.*, **68**, 139–140 (1978).

Roddier C. and F. Roddier, "Reconstruction of the Hubble Space Telescope Mirror Figure from Out-of Focus Stellar Images," *Proc. SPIE*, **1494**, 11–17 (1991a).

Roddier F. and C. Roddier, "Wavefront Reconstruction Using Iterative Fourier Transforms," *Appl. Opt.*, **30**, 1325–1327 (1991b).

Roddier C., F. Roddier, A. Stockton, and A. Pickles, "Testing of Telescope Optics: A New Approach," *Proc. SPIE*, **1236**, 756–766 (1990).

Roddier F., "Curvature Sensing and Compensation: A New Concept in Adaptive Optics," *Appl. Opt.*, **27**, 1223–1225 (1988).

Roddier F., "Wavefront Sensing and the Irradiance Transport Equation," *Appl. Opt.*, **29**, 1402–1403 (1990).

Roddier F., C. Roddier, and N. Roddier, "Curvature Sensing: A New Wavefront Sensing Method," *Proc. SPIE*, **976**, 203–209 (1988).

Servín M., D. Malacara, and J. L. Marroquín, "Wave-Front Recovery from Two Orthogonal Sheared Interferograms," *Appl. Opt.*, **35**, 4343–4348 (1996).

Steibl N., "Phase Imaging by the Transport Equation of Intensity," *Opt. Commun.*, **49**, 6–10 (1984).

Teague M. R., "Deterministic Phase Retrieval: A Green's Function Solution," *J. Opt. Soc. Am.*, **73**, 1434–1441 (1983).

Thikonov A. N., "Solution of Incorrectly Formulated Problems and the Regularization Method," *Sov. Math. Dokl.*, **4**, 1035–1038 (1963).

Welsh B. M., B. L. Ellerbroek, M. C. Roggemann, and T. L. Pennington, "Fundamental Performance Comparison of a Hartmann and a Shearing Interferometer Wave-Front Sensor," *Appl. Opt.*, **34**, 4186–4195 (1995).

Yatagai T., ''Fringe Scanning Ronchi Test for Aspherical Surfaces,'' *Appl. Opt.*, **23**, 3676–3679 (1984).

Yatagai T. and T. Kanou, ''Aspherical Surface Testing with Shearing Interferometer Using Fringe Scanning Detection Method,'' *Opt. Eng.*, **23**, 357–360 (1984).

11

Phase Unwrapping

11.1 THE PHASE UNWRAPPING PROBLEM

Optical interferometers may be used to measure a wide range of physical quantities. The interesting data item supplied by the interferometer is a fringe pattern, which is a cosinusoidal function phase modulated by the wavefront distortions being measured. As shown in Chap. 1, a fringe pattern or interferogram may be modeled by the expression

$$s(x, y) = a(x, y) + b(x, y) \cos(\phi(x, y)) \tag{11.1}$$

where $a(x, y)$ is a slowly varying background illumination, $b(x, y)$ is the amplitude modulation, which also is a low frequency signal, and $\phi(x, y)$ is the phase being measured. The purpose of computer-aided fringe analysis is to automatically detect the two-dimensional phase variation $\phi(x, y)$ that occurs over the interferogram due to the spatial change in the corresponding physical variable. The continuous interferogram is then imaged over a CCD video camera and digitized using a video frame grabber for further analysis in a digital computer.

There are a number of techniques to measure the interesting spatial phase variation of $\phi(x, y)$. Among them we can mention phase shifting interferometry, which requires at least three phase-shifted interferograms. The phase shift among the interferograms must be known all over the interferogram. In this case one may estimate the modulating phase at each resolvable image pixel. Phase

Table 11.1 Phase and Range of Values
According to Values in the Numerator (sin ϕ)
and the Denominator (cos ϕ) in the Expression
for tan ϕ

sin ϕ	cos ϕ	Adjusted phase	Range of values of ϕ
> 0	> 0	ϕ	$0-\pi/2$
> 0	< 0	$\pi - \phi$	$\pi/2-\pi$
< 0	> 0	$\phi + \pi$	$\pi-3\pi/2$
< 0	< 0	$2\pi - \phi$	$3\pi/2-2\pi$
> 0	0	$\pi/2$	$\pi/2$
0	< 0	π	π
< 0	0	$3\pi/2$	$3\pi/2$
0	≥ 0	0	0

shifting interferometry is the preferred technique whenever the atmospheric turbulence and mechanical conditions of the interferometer remain constant over
the time required to obtain the three phase-shifted interferograms. When the
above requirements are not fulfilled, one may analyze just one interferogram,
if carrier fringes are introduced to the fringe pattern, to obtain a spatial carrier
frequency interferogram. One may then analyze this interferogram using well-
known techniques such as the Fourier transform technique, the spatial carrier
demodulation technique, the spatial phase shifting technique, or the phase-locked
loop technique among others.

Except for the phase-locked loop technique, which does not introduce any
phase wrapping, all methods give the detected phase wrapped. The Carré method
wraps the phase modulo π, but all other methods wrap the phase modulo 2π, due
to the arctangent function involved in the phase estimation process, as illustrated
in Table 11.1.

An example of a phase map is given in Fig. 11.1, where the 2π dynamic
range is represented in gray levels. Black represents the phase value of $-\pi$, and
white the value of π. All other gray levels represent intermediate and linearly
mapped phase values.

The relationship between the wrapped phase and the unwrapped phase may
be stated as

$$\phi(x_i, y_j) = \phi_W(x_i, y_j) + 2\pi m(x_i, y_j), \qquad 1 \leq i \leq N; \ 1 \leq j \leq M \quad (11.2)$$

where $\phi_W(x, y)$ is the wrapped phase, $\phi(x, y)$ is the unwrapped phase, and
$m(x, y)$ is an integer-valued number called the field number.

Figure 11.1 Wrapped phase data mapped to gray levels for display purposes.

The unwrapping problem is trivial for phase maps calculated from good quality fringe data when the following two conditions are satisfied:

1. The signal is free of noise.
2. The Nyquist condition is not violated, which means that the absolute value of the phase difference between any two consecutive phase samples (pixels) is less than π.

Figure 11.2 represents the phase wrapping of a one-dimensional function. The lower zigzag curve is the wrapped function, and the upper curve passing

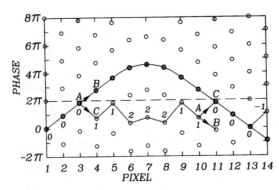

Figure 11.2 Phase unwrapping in one direction without noise and the appropriate Nyquist limited sampling frequency.

through the thick small circles is the unwrapped function. When unwrapping, several of the phase values should be shifted by an integer multiple of 2π to any of the small circles. The vertical distance between the circles is 2π. At pixel numbers 3 and 10 in Fig. 11.2 the phase step from pixel A to pixel B is smaller than π and the phase step from pixel A to pixel C is larger than π since the Nyquist condition is fulfilled.

Unwrapping is thus a simple matter of adding or subtracting 2π offsets at each discontinuity encountered in the phase data (Macy 1983; Bone 1991) or integrating the wrapped phase differences along a given coordinate (Itoh 1982; Ghiglia *et al.* 1987; Ghiglia and Romero 1994).

The unwrapping procedure consists in finding the correct field number for each phase measurement. In Fig. 11.2 the field numbers $m(x)$ for each pixel are marked near the wrapped value. Taking $m(x_1) = 0$, we can easily see that this field number has only three possibilities at each pixel, mathematically expressed by (Kreis 1986)

$$
\begin{aligned}
m(x_1) &= 0 \\
m(x_1) &= m(x_{i-1}) & \text{if } |\phi(x_i) - \phi(x_{i-1})| < \pi \\
m(x_1) &= m(x_{i-1}) + 1 & \text{if } |\phi(x_i) - \phi(x_{i-1})| \le -\pi \\
m(x_1) &= m(x_{i-1}) - 1 & \text{if } |\phi(x_i) - \phi(x_{i-1})| \ge \pi \\
& & i = 1, 2, \ldots, N
\end{aligned}
\tag{11.3}
$$

Kreis (1986) also described a method for unwrapping in two dimensions. Unwrapping becomes more difficult when the absolute phase difference between adjacent pixels at points other than discontinuities in the arctan function is

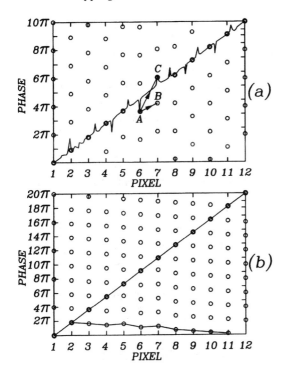

Figure 11.3 Phase unwrapping (a) in the presence of noise and (b) with over-sampling.

greater than π. These discontinuities may be introduced, as shown in Fig. 11.3, by (1) high frequency, high amplitude noise, (2) discontinuous phase jumps, or (3) regional undersampling in the fringe pattern.

Ghiglia *et al.* (1987) considered unwrapping the phase by isolating these erroneous discontinuities before starting the unwrapping process. Erroneous discontinuities or phase inconsistencies are detected when the sum of the wrapped phase differences around a square path of size L is different from zero. Inconsistencies generate phase errors (unexpected phase jumps) which propagate along the unwrapping direction. As a consequence, the unwrapping process becomes path-dependent, i.e., one may obtain different unwrapped phase fields depending on the unwrapping direction chosen.

An important step toward obtaining a robust path-independent phase unwrapper was given by Ghiglia and Romero (1994), who applied the ideas of Fried (1977) and Hudgin (1977) of least squares integration of phase gradients (Noll 1978; Hunt 1979; Takajo and Takahashi 1988) to the unwrapping problem. The phase gradient needed by Ghiglia and Romero (1994) is obtained as wrapped phase differences along the x and y directions. This wrapped gradient field is then least squares integrated to obtain the searched-for continuous phase. More recently Marroquín and Rivera (1995) extended the technique of least squares integration of the wrapped phase gradient by adding a regularization term to the sought-after solution in the form of a norm of potentials. Using this technique it is possible to filter out some noise in the unwrapped phase as well as to interpolate the solution over regions of invalid phase data (such as holes) with well-defined behavior inside them.

One drawback of the least squares integration (Ghiglia and Romero 1994) or its regularized extension (Marroquín and Rivera 1995) stems from the assumption that the phase difference between adjacent pixels is less than π in absolute value. That is, these techniques take the wrapped differences of the wrapped phase as if it were a true gradient field. Unfortunately this is not the case when severely noisy phase maps are being unwrapped. The phase gradient obtained here is actually wrapped in regions of high phase noise and high phase gradients. The use of the least squares unwrapping technique in very noisy phase maps leads to unwrapping errors due to reduction of the dynamic range in the unwrapped phase.

11.2 UNWRAPPING CONSISTENT PHASE MAPS

In this section we analyze two simple unwrapping techniques that apply to consistent phase maps. The first one unwraps full-field wrapped phase data. The second one deals with the unwrapping problem of consistent data within an arbitrary simply connected region.

11.2.1 Unwrapping Full-Field Consistent Phase Maps

The phase unwrapping technique shown in this section is one of the simplest methods to unwrap a good or nearly consistent (small phase noise) smooth phase map. The technique consists of integrating phase differences along a scanning path as shown in Fig. 11.4.

Let us assume that the full-field phase map is given by $\phi_W(x, y)$ in a regular two-dimensional lattice L of size $N \times N$ pixels. We may unwrap this phase map by unwrapping the first row ($y = 0$) of it and then taking the last value of it as the initial condition to unwrap along the following row of the phase map in the positive direction. We may do this by using the following formula along the first row:

$$\phi(x_{i+1}, y_0) = \phi(x_i, y_0) + V(\phi_w(x_{i+1}, y_0) - \phi(x_i, y_0)); \qquad 1 \leq i \leq N$$

$$(11.4)$$

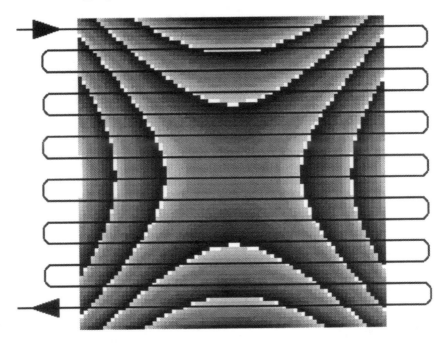

Figure 11.4 Scanning path followed by the proposed full-field phase unwrapper.

where $V(x) = [x - 2\pi \operatorname{int}(x/\pi)]^2$ valid in the interval $(-\pi, +\pi)$ is the wrapping function. This function is equal to $V(x) = \tan^{-1}[\sin(x)/\cos(x)]$ in the same range. In Eq. 11.4 we may use as the initial condition

$$\phi(x_0, y_0) = \phi_0 \qquad\qquad (11.5)$$

Having unwrapped along the first row, we may use the last unwrapped phase value as the initial condition to unwrap the following row $(j = 1)$ in the backward direction, that is,

$$\phi(x_{i-1}, y_1) = \phi(x_i, y_1) + V(\phi_w(x_{i-1}, y_1) - \phi(x_i, y_1)); \qquad 1 \le i \le N \qquad (11.6)$$

For the backward unwrapping direction (Eq. 11.6) we must use as the initial condition

$$\phi(x_{N-1}, y_1) = \phi(x_{N-1}, y_0) + V(\phi_w(x_{N-1}, y_1) - \phi(x_{N-1}, y_0)) \qquad (11.7)$$

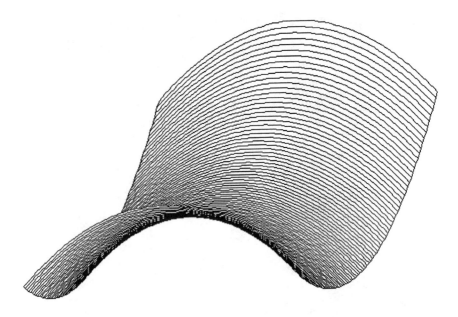

Figure 11.5 Unwrapped full-field phase data using the sequential technique.

The unwrapping then proceeds to the next row ($j = 2$) in the forward direction as

$$\phi(x_{i+1}, y_2) = \phi(x_i, y_2) + V(\phi_w(x_{i+1}, y_2) - \phi(x_i, y_2)); \qquad 1 \leq i \leq N$$
$$(11.8)$$

and the initial condition as

$$\phi(x_0, y_2) = \phi(x_0, y_1) + V(\phi_w(x_0, y_2) - \phi(x_0, y_1)) \qquad (11.9)$$

The scanning procedure just described is followed until the full-field phase map is unwrapped. The phase surface obtained using this sequential procedure is shown in Fig. 11.5.

11.2.2 Unwrapping Consistent Phase Maps Within a Simply Connected Region

If we do not have a full-field phase map, that is, if the shape of the consistent phase map is bounded by an arbitrary simply connected region such as the one shown in Fig. 11.6, the above-described algorithm (Eqs. 11.4-11.9) cannot be used.

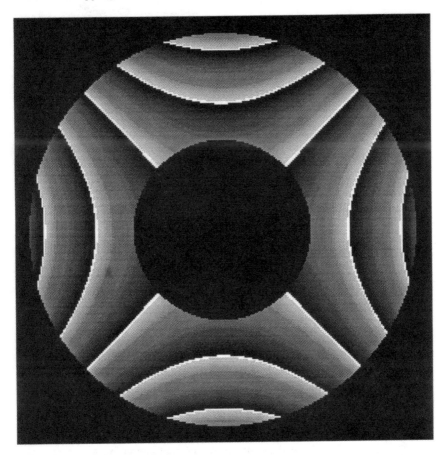

Figure 11.6 An example of a simply connected region containing valid phase data.

For this situation one may use the following algorithm to unwrap a consistent phase map. To start, define and set to zero an indicator function $\sigma(x, y)$ inside the domain D of valid phase data (as shown in Fig. 11.6). Then choose a seed or starting point inside D and assign to it an arbitrary phase value $\phi(x, y) = \phi_0$. Mark the visited site as unwrapped, i.e., set $\sigma(x, y)$ equal to 1. Now that the seed pixel phase is defined, we may proceed with the unwrapping process as follows:

1. Choose a pixel (x, y) inside D (at random or in any prescribed order).
2. Test if the visited site (x, y) inside D is already unwrapped.

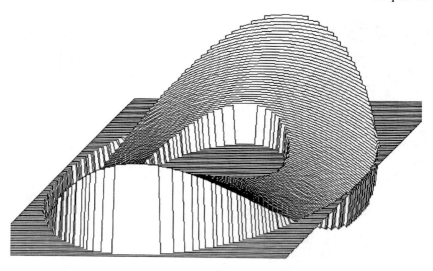

Figure 11.7 Noise-free phase unwrapped using the algorithm given in Sec. 11.2.2.

   a. If the selected site is marked as unwrapped [$\sigma(x, y) = 1$], return to the first statement.

   b. If the visited site is wrapped [for example, $\sigma(x, y) = 0$], then test for any adjacent (x', y') unwrapped pixel.

   c. If no adjacent pixel has already been unwrapped, return to the first statement.

   d. If an adjacent pixel (x', y') is found unwrapped, take its phase value $\phi(x', y')$ and use it to unwrap the current site (x, y) as

$$\phi(x, y) = \phi(x', y') + V(\phi_w(x, y) - \phi(x', y')) \qquad (11.10)$$

      where $V(.)$ is the wrapping function defined before.

5. Mark the selected site as unwrapped [$\sigma(x, y) = 1$].

6. Return to the first statement until all the pixels in D are unwrapped.

The algorithm just described will unwrap any simply connected bounded region D having valid and consistent wrapped phase data as shown in Fig. 11.7.

11.3 UNWRAPPING NOISY PHASE MAPS

One still may use the above-described algorithm to unwrap inconsistent phase maps corrupted by a small amount of noise. This may be done by marking the inconsistent wrapped phase pixels and excluding them from the unwrapping process as forbidden regions. Inconsistencies are present when multiples of 2π

radians cannot be added to each wrapped phase sample over a two-dimensional grid to eliminate all adjacent phase differences greater than π radians in magnitude. Marking the inconsistent pixels is not practical as the noise increases, given that the number of inconsistent marked pixels may grow very fast. For that reason we will not get into the details of such techniques.

Although many algorithms have been proposed for phase unwrapping in the presence of noise, here we are going to discuss only two that we feel are the most important ones for unwrapping inconsistent phase maps of smooth continuous functions. These algorithms are the least squares integration of wrapped phase differences (Ghiglia and Romero 1994) and the regularized phase-tracking unwrapper. This means that we are not going to discuss some important algorithms and techniques (Huntley 1989, 1994; Huntley *et al.* 1993; Bryanston-Cross and Quan (1993); Ströbel 1996) that can handle phase maps of noisy or discontinuous functions because we feel that these techniques fall outside the scope of this book.

11.3.1 Unwrapping by Means of Least Squares Integration

The least squares technique was first introduced by Ghiglia and Romero (1994) to unwrap inconsistent phase maps. To apply this method, start by estimating the wrapped phase gradient along the x and y directions. That is,

$$\phi_y(x_i, y_j) = V(\phi_w(x_i, y_j) - \phi_w(x_i, y_{i-1}))$$
$$\phi_x(x_i, y_j) = V(\phi_w(x_i, y_j) - \phi_w(x_{i-1}, y_j)) \tag{11.11}$$

Having an oversampled phase map, the phase differences in Eq. 11.11 will be everywhere in the range $(-\pi, +\pi)$. In other words, the estimated gradient will be unwrapped. Now we may integrate the phase gradient in a consistent way by means of a least squares integration. The integrated or searched for continuous phase will be the one which minimizes the cost function

$$U(\phi) = \sum_{i=2}^{N} \sum_{j=2}^{M} [\phi(x_i, y_i) - \phi(x_{i-1}, y_j) - \phi_x(x_i, y_j)]^2$$
$$+ \sum_{i=2}^{N} \sum_{j=2}^{M} [\phi(x_i, y_j) - \phi(x_i, y_{j-1}) - \phi_y(x_i, y_j)]^2 \tag{11.12}$$

This expression applies whenever we have a full-field wrapped phase. Let us assume that we have valid phase data only inside a two-dimensional region marked by an indicator function $\sigma(x, y)$. That is, we will have valid phase data

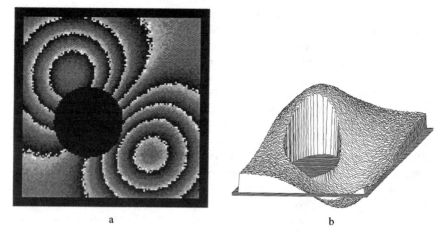

Figure 11.8 (a) Computer-generated noisy phase map. (b) Unwrapped phase using least squares integration of wrapped differences.

for $\sigma(x, y) = 1$ and invalid phase data for $\sigma(x, y) = 0$. We then may modify our cost function to include the indicator function as follows:

$$U = \sum_{i=2}^{N} \sum_{j=2}^{M} [\phi(x_i, y_j) - \phi(x_{i-1}, y_j) - \phi_x(x_i, y_j)]^2 \sigma(x_i, y_j)\sigma(x_{i-1}, y_j)$$

$$+ \sum_{i=2}^{N} \sum_{j=2}^{M} [\phi(x_i, y_j) - \phi(x_i, y_{j-1}) - \phi_y(x_i, y_j)]^2 \sigma(x_i, y_j)\sigma(x_i, y_{j-1})$$

$$(11.13)$$

The estimated unwrapped phase $\phi(x, y)$ may be found, for example, by using simple gradient descent at all pixels, as

$$\phi^{k+1}(x, y) = \phi^k(x, y) - \tau \frac{\partial U}{\partial \phi(x, y)} \qquad (11.14)$$

where k is the iteration number and τ is the convergence rate of the gradient search system (typically around $\tau = 0.1$). There are faster algorithms to obtain the searched for unwrapped phase; among them we can mention conjugate gradient techniques or transform methods (Ghiglia and Romero 1994).

Consider the noisy phase map of Fig. 11.8a. The wrapped phase $\phi_W(x, y)$ in this map is obtained as the sum of two Gaussians with different signs. Fig-

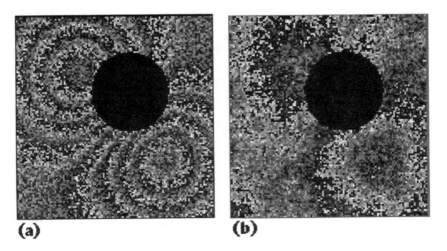

(a) **(b)**

Figure 11.9 (a) Highly noisy phase map. (b) Phase map obtained after unwrapping and then wrapping again for comparison purposes. We can see that the technique fails to recover the full dynamic range of the modulating phase. That is so because the wrapped first-order difference is a bad estimator of the true phase gradient in such a noisy phase map.

ure 11.8b shows the unwrapped phase map obtained using the least squares integration technique developed by Ghiglia and Romero (1994).

Figure 11.9 shows a highly noisy phase map and the phase after unwrapping and then wrapping again for comparison purposes. This phase was obtained with the least squares integration technique of wrapped differences (Ghiglia and Romero 1994), applied to the same phase map but with more noise added. We note that the method is not as successful as with less noise. There is a substantial decrease in the phase dynamic range.

11.3.2 The Regularized Phase Tracking Unwrapper

From Eq. 11.2 we can see that the unwrapping inverse problem is ill-posed, that is, the $m(x, y)$ field is not uniquely determined by the observations. This means that the unwrapping problem cannot be solved unless additional (prior) information about the expected unwrapped phase $\phi(x, y)$ is provided. Smoothness is a typical piece of prior information that constrains the search space of unwrapped functions, and this information may be incorporated into the unwrapping algorithm by using regularization theory (Marroquín and Rivera 1995).

To regularize the phase unwrapping problem it is necessary to find a suitable merit function that uses at least two terms which contribute to constrain the searched unwrapped field. These terms are related to

1. The fidelity between the estimated function and the observations.
2. Prior knowledge about the spatial behavior of the unwrapped phase.

It is then assumed that the searched phase function is the one which minimizes this merit function.

In classical regularization one uses a pixelwise error between the searched function and the observed data and the norm of a differential operator over the searched function as regularizer. In the proposed regularized phase tracking (RPT) technique, however, we assume that in a small region of the image one may consider the data smooth enough that it can be modeled by a plane. This plane must be close to the observed phase map in the wrapped space (item 1, above). A phase plane such as this must adapt itself to every region in the phase map so that its local slope changes continuously in the two-dimensional space. Using these motivations we postulate that the estimated fringe pattern's phase $\phi(x, y)$ must minimize the following merit function at each site (x, y) containing valid phase data:

$$U_{x,y}(\phi, \omega_x, \omega_y) = \sum_{(\epsilon,\eta)\epsilon(N_{x,y}\cap L)} \{V(\phi_W(\epsilon, \eta) - \phi_e(x, y, \epsilon, \eta))]$$

$$+ \lambda[\phi(\epsilon, \eta) - \phi_e(x, y, \epsilon, \eta)]^2 \sigma(\epsilon, \eta)\} \qquad (11.15)$$

and

$$\phi_e(x, y, \epsilon, \eta) = \phi(x, y) + \omega_x(x, y)(x - \epsilon) + \omega_y(x, y)(y - \eta) \qquad (11.16)$$

The functions $\phi_W(x, y)$, and $\phi(x, y)$ are the wrapped and the searched unwrapped phase, respectively, estimated at pixel (x, y). L is the two-dimensional domain that has valid wrapped phase data. $N_{x,y}$ is a small neighborhood around the coordinate (x, y). As explained below, the function $\sigma(\epsilon, \eta)$ is an indicator field which equals 1 if the site (ϵ, η) has already been unwrapped and 0 otherwise. We can see from Eq. 11.16 that we are approximating the local behavior of the unwrapped phase by a plane whose parameters $\phi(x, y)$, $\omega_x(x, y)$, and $\omega_y(x, y)$ are determined in such a way that the merit function $U_{x,y}(\phi, \omega_x, \omega_y)$ at each site (x, y) in L is minimized.

The first term in Eq. 11.15 attempts to keep the local phase model close to the observed phase map in a least squares sense within the neighborhood

$N_{x,y}$ (item 1, above). The second term enforces our assumption of smoothness and continuity of the unwrapped phase (item 2, above) using only previously unwrapped pixels marked by $\sigma(x, y)$. We can see that the second term will contribute a small amount to the value of the merit function $U_{x,y}(\phi, \omega_x, \omega_y)$ only for smooth unwrapped phase functions. Note also that the local phase plane is adapted simultaneously to the observed data [in the wrapped space using the wrapping operator $V(x)$] and to the continuous unwrapped phase marked by $\sigma(x, y)$.

To unwrap the phase map $\phi_W(x, y)$ we need to find the minimum of the merit function $U_{x,y}(\phi, \omega_x, \omega_y)$ (Eq. 11.15) with respect to the fields $\phi(x, y)$, $\omega_x(x, y)$, and $\omega_y(x, y)$. To this end we propose to find a minimum of $U_{x,y}(\phi, \omega_x, \omega_y)$ according to the sequential unwrapping algorithm described below.

The proposed unwrapping strategy in L is calculated as follows. To start, set the indicator function to zero [$m(x, y) = 0$ in L]. Then choose a seed or starting point inside L to begin the unwrapping process. Optimize the chosen site for $U_{x,y}(\phi, \omega_x, \omega_y)$ by adapting the triad $\phi_0(x, y), \omega_x(x, y), \omega_y(x, y)$ until a minimum is reached, and mark the visited site as unwrapped, i.e., set $\sigma(x, y)$ equal to 1. Now that the seed pixel is unwrapped, we carry out the unwrapping process as follows:

1. Choose a pixel inside L (at random or in any prescribed order).
2. Test if the visited site is already unwrapped.
   a. If the selected site is marked as unwrapped [i.e., $\sigma(x, y) = 1$], return to the first statement.
   b. If the visited site is wrapped [i.e., $\sigma(x, y) = 0$], then test for any adjacent unwrapped pixel (x', y').
   c. If no adjacent pixel (x', y') has already been unwrapped, return to the first statement.
   d. If an adjacent pixel (x', y') is found unwrapped, take its optimized triad $(\phi, \omega_x, \omega_y)$ and use it as the initial condition to minimize the merit function $U_{x,y}(\phi, \omega_x, \omega_y)$ (Eq. 11.15) at the chosen site (x, y).
5. Once the minimum for $U_{x,y}(\phi, \omega_x, \omega_y)$ in (x, y) is reached, mark the selected site as unwrapped [i.e., $\sigma(x, y) = 1$].
6. Return to the first statement until all the pixels in L are unwrapped.

An intuitive way of regarding this iteration is as a "crystal growing" (CG) process in which new molecules (planes) are added to the bulk in the particular orientation (slope) which minimizes the local crystal energy given the geometric orientation of the adjacent and previously positioned molecules.

We may use simple gradient descent to optimize $U_{x,y}$, moving the triad $(\phi, \omega_x, \omega_y)$ according to

$$\phi^{k+1}(x, y) = \phi^k(x, y) - \tau \frac{\partial U_{x,y}(\phi, \omega_x, \omega_y)}{\partial \phi(x, y)} \tag{11.17a}$$

$$\omega_x^{k+1}(x, y) = \omega_x^k(x, y) - \tau \frac{\partial U_{x,y}(\phi, \omega_x, \omega_y)}{\partial \omega_x(x, y)} \tag{11.17b}$$

$$\omega_y^{k+1}(x, y) = \omega_y^k(x, y) - \tau \frac{\partial U_{x,y}(\phi, \omega_x, \omega_y)}{\partial \omega_y(x, y)} \tag{11.17c}$$

where τ is the convergence rate of the gradient search system. As mentioned before, the initial condition for Eqs. 11.17 is chosen from any adjacent unwrapped pixel. In practice, the τ parameter in Eq. 11.17a may be multiplied by about 10 to accelerate the convergence rate of the gradient search.

The first global phase estimation just described is usually very close to the actual unwrapped phase; if needed, one may perform additional global iterations to improve the phase estimation process. The additional iterations may be performed using again Eqs. 11.17, but now taking as initial condition the last estimated values at the same site (x, y) [not the ones at a neighborhood site (x', y') as in the first global CG iteration]. Note that for the additional global phase estimations, the indicator function $\sigma(x, y)$ in Eq. 11.15 is now everywhere equal to 1; therefore, one may scan the lattice in any desired order whenever all the sites are visited at each global iteration. In practice, only one or two additional global iterations are needed to reach a stable minimum of $U_{x,y}(\phi, \omega_x, \omega_y)$ at each site (x, y) in the two-dimensional lattice L

One may argue that the first term in Eq. 11.15 may suffice to unwrap the observed phase map. But the simplified system was found to give good results only for small phase noise (between -0.2π and 0.2π). For higher amounts of phase noise (between -0.7π and 0.7π), the second term (the regularizing plane over the unwrapped phase) makes a substantial improvement in the noise robustness of the RPT system.

The parameter λ and the size of the neighborhood $N_{x,y}$ are related to the unwrapped phase bandwidth and to the robustness of the RPT algorithm. For example, for a very low frequency highly inconsistent phase map, the size of $N_{x,y}$ should be large so the RPT system can properly track the smooth unwrapped phase in such a noisy field. Once the size of $N_{x,y}$ is chosen, the value of the λ parameter in Eq. 11.5 is not very critical. A value of $\lambda = 2$ was used for all the results herein presented. The computational speed of the RPT technique is related to the size of the neighborhood $N_{x,y}$ as well as to the size of the

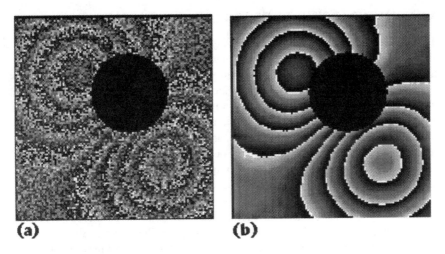

(a)   **(b)**

Figure 11.10   (a) A highly noisy phase map (also shown in Fig. 11.9a). (b) Phase obtained using the regularized phase tracking (RPT) technique, shown after unwrapping and then wrapping again for comparison purposes. We can see that the RPT technique works better than the least squares technique (Fig. 11.9b) for severe phase noise.

lattice L. In the experiments presented, the size of $N_{x,y}$ ranged from a 5×5 pixels to 11×11 pixels. Having reasonably good phase maps, a neighborhood $N_{x,y}$ of 3×3 pixels may be enough and the RPT system will give very fast and reliable results.

As in a crystal growing process, the size of the neighborhood $N_{x,y}$ in the RPT technique is very critical. If it succeeds, the RPT system will move the whole unwrapping system to the right attractor. If the CG algorithm reaches a wrong attractor, the RPT system will give a wrong result. In these circumstances one should try another neighborhood $N_{x,y}$ for the RPT system and compute the solution again.

Figure 11.10b shows the phase obtained from the noisy phase map of Fig. 11.10a after unwrapping using the RPT unwrapper and then wrapping again for comparison purposes. One can appreciate the capacity of the RPT system to remove noise while preserving almost unchanged the original phase dynamic range. The noise introduced in Fig. 11.10a may roughly be considered as the maximum noise tolerated by the proposed RPT unwrapper. Note also how the unwrapped phase was almost unaffected near the image boundaries despite the large amount of noise.

11.4 UNWRAPPING SUBSAMPLED PHASE MAPS

Testing of aspheric wavefronts is nowadays routinely carried out in the optical shop by the use of commercial interferometers. The testing of deep aspherics is limited by the aberrations of the interferometer's imaging optics as well as the spatial resolution of the CCD video camera used to gather the interferometric data. The CCD video arrays come typically with 256×256 or 512×512 image pixels. The number of CCD pixels limits the highest recordable frequency over the CCD array to π rad/pixel. As seen in Chap. 2, this maximum recordable frequency is called the Nyquist limit of the sampling system. The detected phase map of an interferogram having frequencies higher than the Nyquist limit contains false fringes and is said to be aliased. Another factor to take into account is the fact that CCD detector elements have a finite size, which can be almost as large as the pixel separation. In this case the contrast of the sub-Nyquist sampled image is strongly reduced, as described in Chap. 2 and illustrated in Fig. 2.12. Thus, aliasing fringes cannot be observed with these kinds of detectors unless a CCD detector with small detector elements is used. Aliasing fringes are quite useful for unwrapping sub-Nyquist sampled phase maps, as will now be described.

Subsampled phase maps cannot be unwrapped using standard techniques such as the ones presented so far. One can nevertheless still unwrap such an under-sampled phase map, and we have mainly two situations to deal with:

1. One has enough knowledge about the wavefront being tested to null test the wavefront under analysis (Greivenkamp 1987; Servín and Malacara 1996a).
2. The expected wavefront is smooth. Then one may introduce this prior knowledge into the unwrapping process (Greivenkamp 1987; Servín and Malacara 1996b).

Both cases are discussed in this section. The main requirement for application of the techniques presented here is a CCD camera that has detectors much smaller than the spatial separation among them (Greivenkamp 1987), as pointed out before. An alternative is to use a mask with small holes over the CCD array to reduce the light-sensitive area of the CCD pixels. This requirement allows us to have a strong signal even for thin interferogram fringes (see Fig. 2.12c). To obtain the undersampled phase map, one may use any well-known phase stepping interferometric technique using phase-shifted undersampled interferograms.

To illustrate the principle of sub-Nyquist phase unwrapping in one dimension, Fig. 11.11 shows the unwrapped phase in a wavefront produced by an optical system with spherical aberration.

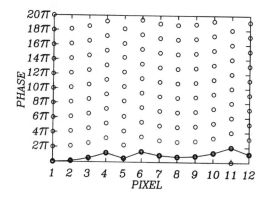

Figure 11.11 Wrapped phase for a wavefront with spherical aberration, with sub-Nyquist sampling.

The correct unwrapping result is shown in Fig. 11.12a. However, if no previous knowledge about the wavefront shape is available, the result in Fig. 11.12b would be obtained.

The undersampled interferogram may be imaged directly over the CCD video array with the aid of an optical interferometer as demonstrated in Chap. 1. If the CCD sampling rate is x_s over the x direction and y_s over the y direction and the diameter of the light-sensitive area of the CCD is d, we may write the mathematical expression for the sampling operation over the interferogram irradiance (Eq. 11.1) as (Greivenkamp 1987)

$$S[s(x, y)] = \left[s(x, y) ** \text{circ} \left(\frac{\rho}{d} \right) \right] \text{comb} \left(\frac{x}{x_s}, \frac{y}{y_s} \right), \qquad \rho = (x^2 + y^2)^{1/2}$$

$$(11.18)$$

where the function $S[s(x, y)]$ is the sampling operator over the irradiance given by Eq. 11.1. The symbol $**$ indicates a two-dimensional convolution. The term $\text{circ}(\rho/d)$ is the circular size of the CCD detector. The comb function is an array of delta functions with the same spacing as the CCD pixels. The phase map of the sampled interferogram in Eq. 11.18 may be obtained using, for example, three phase-shifted interferograms as

$$s_1(x, y) = a(x, y) + b(x, y) \cos(\phi(x, y) + \alpha)$$

$$s_2(x, y) = a(x, y) + b(x, y) \cos(\phi(x, y))$$

$$s_3(x, y) = a(x, y) + b(x, y) \cos(\phi(x, y) - \alpha)$$

$$(11.19)$$

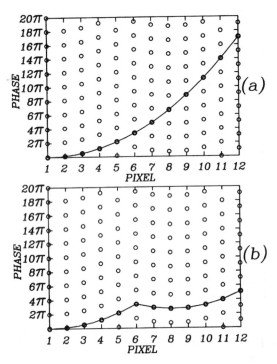

Figure 11.12 Unwrapped phase for a wavefront with spherical aberration with sub-Nyquist sampling. (a) Correct phase; (b) phase obtained if no previous knowledge is available.

where α is the phase shift. Using well-known formulas we can find the subsampled wrapped phase map as

$$\phi_w(x, y) = \tan^{-1}\left[\frac{1 - \cos \alpha}{\sin \alpha}\left(\frac{S[s_1(x, y)] - S[s_3(x, y)]}{2S[s_1(x, y)] - S[s_2(x, y)] - S[s_3(x, y)]}\right)\right]$$

$$\times \sigma(x, y) \tag{11.20}$$

where $\sigma(x, y)$ is an indicator function which equals 1 if one has valid phase data and 0 otherwise. As Eq. 11.20 shows, the obtained phase is a modulo 2π of the true undersampled phase due to the arctangent function involved in the phase detection process. Figure 11.13 shows an example of a subsampled phase map of pure spherical aberration.

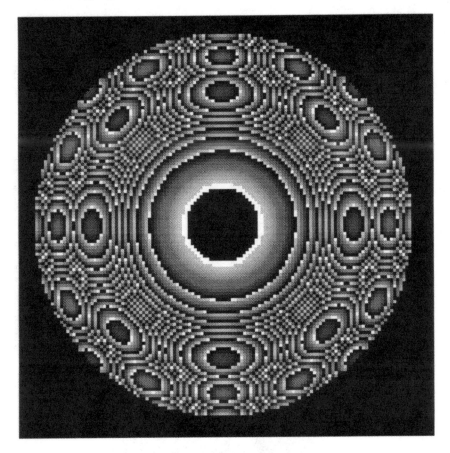

Figure 11.13 Subsampled phase map corresponding to pure spherical aberration.

11.4.1 Null Fringe Analysis of Subsampled Phase Maps Using a Computer-Stored Compensator

As mentioned above, one way to deal with deep aspheric wavefronts is to use an optical, diffractive, or software compensator. Optical or diffractive compensators reduce the number of aberration fringes so they can be analyzed without aliasing. To construct the compensator one must have a good knowledge of the testing wavefront up to a few aberration fringes. The ''remaining'' aberration fringes constitute the error between the expected or ideal wavefront and the actual one from the test optics. In this way one may analyze the remaining uncompensated fringes using standard fringe analysis techniques. Fortunately, in optical shop testing one typically has a good knowledge about the kind and amount of aberration expected (in the final stages of the manufacturing process) at the testing plane. This knowledge allows us to construct the proper optical or diffractive compensator.

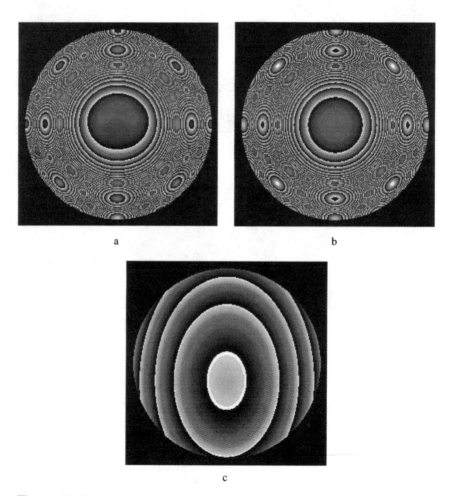

Figure 11.14   (a) Subsampled phase map obtained using Eq. 11.19. (b) Ideal or expected subsampled phase map. (c) Phase error between the two phase maps according to Eq. 11.21.

In this section we deal with another kind of compensator called a "software" compensator (Servín and Malacara 1996a). The software compensator does not need to be constructed (as an optical or diffractive compensator), it only needs to be calculated numerically inside the computer. This software compensator, however, does need a specially constructed CCD video array with a small light detector size d with respect to the spatial separation (x_s, y_s) of the pixels (see Eq. 11.18).

Assuming that the expected or ideal wavefront $\phi_i(x, y)$ differs from the detected phase $\phi_w(x, y)$ by a few wavelengths, we may form an over-sampled wrapped wavefront error $\Delta\phi_w(x, y)$ as

$$\Delta\phi_w(x, y) = \tan^{-1}\{\tan[\phi_w(x, y) - \phi_i(x, y)]\}\sigma(x, y) \qquad (11.21)$$

We may then unwrap the wavefront error $\Delta\phi_w(x, y)$ by using standard unwrapping techniques. To obtain the unwrapped testing wavefront, the unwrapped error and the ideal wavefront are added:

$$\phi(x, y) = [\phi_i(x, y) + \Delta\phi(x, y)]\sigma(x, y) \qquad (11.22)$$

where $\Delta\phi(x, y)$ is the unwrapped phase error. As mentioned before, the limitation of the technique presented in this section resides in the fact that the error wavefront (Eq. 11.21) must be oversampled. This requirement is the same as the one when a holographic or diffractive compensator is used. That is, the wavefront being tested must be close enough to the expected ideal wavefront to obtain a compensated interferogram having spatial frequencies below the Nyquist upper bound over the CCD array. In summary, the problem of building an optical or holographic compensator is replaced herein by the construction of a special-purpose CCD video array or the construction of a mask of small holes in contact with the CCD array. The considerable benefit of this approach is that once the CCD mask or the specially built CCD array is available, the need to build special-purpose diffractive or holographic compensators disappears. The use of this technique is shown in Fig. 11.14. Figure 11.14a shows a subsampled phase map that is being analyzed. This phase map is then "compared," using Eq. 11.21, with the expected one shown in Fig. 11.14b. Their phase difference (the phase error between them) is shown in Fig. 11.14c.

As in the case when an optical compensator is used, the positioning of the CCD array used to collect the interference irradiance is very critical. A mispositioning of the compensator, or in this case the CCD array, may give erroneous measurements.

11.4.2 Unwrapping Smooth Continuous Subsampled Phase Maps

In the last subsection we discussed the problem of unwrapping undersampled phase maps. The method is based on having a good enough prior knowledge of the kind and amount of aberrations to perform a null test on the detected phase map.

In this section we generalize the problem of unwrapping undersampled phase maps to smooth wavefronts. That is, the only prior knowledge about the wavefront being analyzed is its smoothness. This is far less restrictive than the null testing technique presented in the last section. Analysis of interferometric data beyond the Nyquist frequency was first proposed by Greivenkamp (1987), who assumed that the wavefront being tested is smooth up to the first or second derivative. Greivenkamp's approach to unwrapping subsampled phase maps consists of adding multiples of 2π each time a discontinuity in the phase map is found. The number of 2π's added is determined by the smoothness condition imposed on the wavefront in its first or second derivative along the unwrapping direction. Although Greivenkamp's approach is robust against noise, its weakness resides in the fact that it is a path-dependent phase unwrapper.

In this section we present a method (Servín and Malacara 1996) which overcomes the path dependency of the Greivenkamp approach while preserving its noise robustness. In this case an estimation of the local wrapped curvature (or wrapped Laplacian) of the subsampled phase map $\phi_w(x, y)$ (Eq. 11.20) is used to unwrap the interesting deep aspheric wavefront. Once we have the local wrapped curvature along the x and y directions we may use least squares integration to obtain the unwrapped continuous wavefront. The local wrapped curvature is obtained as

$$
\begin{aligned}
L_x(x_i, y_j) &= V(\phi_w(x_{i-1}, y_j) - 2\phi_w(x_i, y_j) + \phi_w(x_{i+1}, y_j)) \\
L_y(x_i, y_j) &= V(\phi_w(x_i, y_{j-1}) - 2\phi_w(x_i, y_j) + \phi_w(x_i, y_{i+1}))
\end{aligned}
\tag{11.23}
$$

If the absolute value of the discrete wrapped Laplacian given by Eqs. 11.23 is less than π, its value will be nonwrapped. Then we may obtain the unwrapped phase $\phi(x, y)$ as the function which minimizes the quadratic merit function (least squares)

$$
U = \sum_{(x,y)\epsilon\sigma(x,y)} U_x(x, y)^2 + U_y(x, y)^2
\tag{11.24}
$$

where $\sigma(x, y)$ is an indicator or mask function which equals 1 if we have valid phase data and 0 otherwise. The functions $U_x(x, y)$ and $U_y(x, y)$ are given by

$$U_x(x_i, y_j) = L_x(x_i, y_j) - [\phi(x_{i-1}, y_j) - 2\phi(x_i, y_j) + \phi(x_{i+1}, y_j)]$$
$$U_y(x_i, y_j) = L_y(x_i, y_j) - [\phi(x_i, y_{j-1}) - 2\phi(x_i, y_j) + \phi(x_i, y_{j+1})]$$ (11.25)

The minimum of the cost function given by Eq. 11.24 is obtained when its partial with respect to $\phi(x, y)$ equals zero. Therefore the set of linear equations that must be solved is

$$\frac{\partial U}{\partial \phi(x, y)} = U_x(x_{i-1}, y_j) - 2U_x(x_i, y_j) + U_x(x_{i+1}, y_j)$$

$$+ U_y(x_i, y_{j-1}) - 2U_y(x_i, y_j) + U_y(x_i, y_{j+1})$$ (11.26)

Several methods may be used to solve this system of linear equations; among others, there is the simple gradient descent,

$$\phi^{k+1}(x, y) = \phi^k(x, y) - \eta \frac{\partial U}{\partial \phi(x, y)}$$ (11.27)

where the parameter η is the rate of convergence of the gradient search. The simple gradient descent is quite slow for this application. Instead, we have used the conjugate gradient to speed up the computing time.

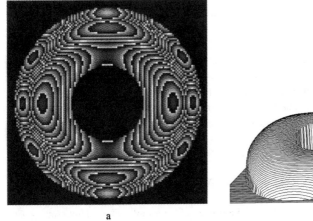
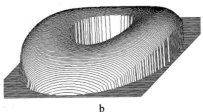

a   b

Figure 11.15 (a) Subsampled phase map of a wavefront with a central obstruction. (b) Wire mesh of the unwrapped phase map according to the least squares integration of wrapped phase curvature presented in this section.

A result that shows the use of the unwrapping technique presented in this section is presented in Fig. 11.15. Figure 11.15a shows a subsampled phase map, and Fig. 11.15b shows the unwrapped phase in wire mesh.

11.5 CONCLUSION

We have analyzed some important techniques to unwrap phase maps of continuous and smooth functions. To start with, we presented two algorithms to unwrap good quality phase maps. The first of these applies only to full-field phase maps, but the second may be applied to a phase map bounded by an arbitrary single connected shape.

Afterward we presented the unwrapping technique that uses least squares integration of phase gradients to obtain the searched continuous phase. We also pointed out its main limitation, namely the estimation of the phase gradient as the wrapped difference of two consecutive pixels along the x and y directions. This gradient phase estimation works well only for relatively small phase noise. This is because a very noisy phase map may have differences between two adjacent pixels exceeding π or $-\pi$ radians.

Next, we presented the two-dimensional RPT phase unwrapping system which is capable of unwrapping severely degraded phase maps. The RPT technique is based on a two-dimensional regularized phase tracking system. This unwrapping system tracks the instantaneous phase and its gradient, adapting a plane to the estimated wrapped and unwrapped phases simultaneously. In other words, the system fits the best least squares tangent plane at each pixel in the wrapped and unwrapped phase spaces within a small neighborhood $N_{x,y}$. Once the least squares best plane is found at a given location, the constant term of this plane $\phi(x, y)$ gives the estimated unwrapped phase at the (x, y) location, and the slope (ω_x, ω_y) estimates the local frequency.

Finally we analyzed two techniques to deal with subsampled interferograms. For the null unwrapping technique we must know the wrapped wavefront up to a few wavelengths. The second technique is more general; the only prior assumption about the testing wavefront is smoothness up to its second derivative.

REFERENCES

Bone D. J., "Fourier Fringe Analysis: The Two Dimensional Phase Unwrapping Problem," *Appl. Opt.*, **30**, 3627–3632 (1991).

Bryanston-Cross P. J. and C. Quan, "Examples of Automatic Phase Unwrapping Applied to Interferometric and Photoelastic Images," in *Proceedings of the 2nd International Workshop on Automatic Processing of Fringe Patterns*, W. Jüptner and W. Osten, Eds., Akademie Verlag, Bremen, 1993.

Fried D. L., "Least-Squares Fitting a Wave-Front Distortion Estimate to an Array of Phase Difference Measurements," *J. Opt. Soc. Am.*, **67**, 370–375 (1977).

Ghiglia D. C. and L. A. Romero, ''Robust Two Dimensional Weighted and Unweighted Phase Unwrapping That Uses Fast Transforms and Iterative Methods,'' *J. Opt. Soc. Am. A*, **11**, 107–117 (1994).

Ghiglia D. C., G. A. Mastin, and L. A. Romero, ''Cellular Automata Method for Phase Unwrapping,'' *J. Opt. Soc. Am.*, **4**, 267–280 (1987).

Greivenkamp J. E., ''Sub-Nyquist Interferometry,'' *Appl. Opt.*, **26**, 5245–5258 (1987).

Hudgin R. H., ''Wave-Front Reconstruction for Compensated Imaging,'' *J. Opt. Soc. Am.*, **67**, 375–378 (1977).

Hunt B. R., ''Matrix Formulation of the Reconstruction of Phase Values from Phase Differences,'' *J. Opt. Soc. Am.*, **69**, 393–399 (1979).

Huntley J. M., ''Noise Immune Phase Unwrapping Algorithm,'' *Appl. Opt.*, **28**, 3268–3270 (1989).

Huntley J. M., ''Phase Unwrapping - Problems & Approaches,'' in *Proc. FASIG, Fringe Analysis'94*, York University, UK, p. 391, 1994.

Huntley J. M., R. Cusack, and H. Saldner, ''New Phase Unwrapping Algorithms,'' in *Proceedings of the 2nd International Workshop on Automatic Processing of Fringe Patterns*, W. Jüptner and W. Osten, Eds., Akademie Verlag, Bremen, p. 391, 1993.

Itoh K., ''Analysis of the Phase Unwrapping Algorithm,'' *Appl. Opt.*, **21**, 2470 (1982).

Kreis T., ''Digital Holographic Interference-Phase Measurement Using the Fourier-Transform Method,'' *J. Opt. Soc. Am. A*, **3**, 847–855 (1986).

Macy W., Jr., ''Two Dimensional Fringe Pattern Analysis,'' *Appl. Opt.*, **22**, 3898–3901 (1983).

Marroquín J. L. and M. Rivera, ''Quadratic Regularization Functionals for Phase Unwrapping,'' *J. Opt. Soc. Am. A*, **12**, 2393–2400 (1995).

Noll R. J., ''Phase Estimates from Slope-Type Wave-Front Sensors,'' *J. Opt. Soc. Am.*, **68**, 139–140 (1978).

Servín M. and D. Malacara, ''Sub-Nyquist Interferometry Using a Computer Stored Reference,'' *J. Mod. Opt.*, **43**, 1723–1729 (1996a).

Servín M. and D. Malacara, ''Path-Independent Phase Unwrapping of Subsampled Phase Maps,'' *Appl. Opt.*, **35**, 1643–1649 (1996b).

Ströbel B., ''Processing of Interferometric Phase Maps as Complex-Value Phasor Images,'' *Appl. Opt.*, **35**, 2192–2198 (1996).

Takajo H. and T. Takahashi, ''Least Squares Phase Estimation from Phase Differences,'' *J. Opt. Soc. Am. A*, **5**, 416–425 (1988).

12

Measurement of Aspheric Wavefronts

12.1 INTRODUCTION

The most common types of interferometers, with the exception of lateral or rotational shear interferometers, produce interference patterns in which the fringes are straight, equidistant, and parallel when the wavefront under test is perfect and spherical with the same radius of curvature as the reference wavefront. If the surface under test does not have a perfect shape, the fringes will not be straight and their separations will be variable. The deformations of the wavefront may be determined by a mathematical examination of the shape of the fringes. Since the fringe separations are not constant, in some places the fringes will be widely spaced, but in others the fringes will be too close together.

It is desirable that the spherical aberration of the wavefront under test be compensated in some way so that the fringes appear straight, parallel, and equidistant, for a perfect wavefront. This is called a null test and may be accomplished by means of some special configurations. These special configurations may be used to have a null test of a conic surface. Almost all of these surfaces have rotational symmetry.

12.2 ASPHERIC COMPENSATORS

An aspheric or null compensator is an optical element with spherical aberration designed to compensate the spherical aberration in an aspheric wavefront under

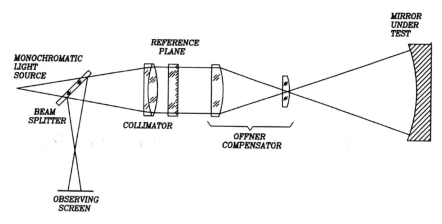

Figure 12.1 Offner compensator.

test. They have been described in detail in the literature (Offner and Malacara 1992). It is beyond the scope of this book to describe them in detail.

As a typical example, the well-known Offner compensator is illustrated in Fig. 12.1.

12.3 IMAGING THE PUPIL ON THE OBSERVATION PLANE

An aberrated wavefront continuously changes shape as it travels. Thus, if the optical system is not perfect, the interference pattern will also continuously change as the beam advances, as shown in Fig. 12.2. The change in shape of a traveling wavefront has been studied and calculated taking into account diffraction by Jóźwicki (1990). The errors of an instrument are represented by wavefront distortions on the pupil, and hence the interferogram should be taken at the pupil.

12.3.1 Imaging the Pupil Back on Itself

When testing a lens with any of the configurations described in Fig. 1.2 in Chap. 1, the wavefront travels twice through the lens, the second time after being reflected at the small mirror in front of the lens. If the aberration is small, the total wavefront deformation is twice the deformation introduced in a single pass through the lens. However, if the aberration is large, this is not so, because the wavefront changes in its travel from the lens to the mirror and back to the lens. Then the spot in the surface on which the defect is located is

WAVEFRONTS

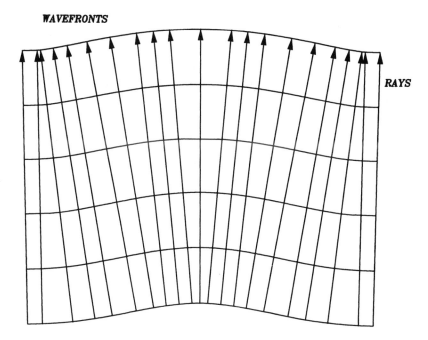

RAYS

Figure 12.2 Change in the shape of a wavefront as it travels.

not imaged back onto itself by the concave or convex mirror, and the ray will not pass through this defect the second time. Great confusion then results in regard to the interpretation of the interferogram, since the defect is not precisely duplicated by the double pass through the lens (Dyson 1959).

It may be shown that the image of the lens is formed at a distance S from the lens given by

$$S = \frac{2(F - r)^2}{2F - r} \tag{12.1}$$

where F is the focal length and r is the radius of curvature of the surface ($r > 0$ for a convex mirror and $r < 0$ for a concave mirror). We may see that the ideal mirror is convex and very close to the lens ($r \sim F$).

An appropriate optical configuration has to be used if the lens under test has a large aberration, in order to image its pupil back on itself. Any auxiliary lenses or mirrors must preserve the wavefront shape. Some examples of these arrangements are presented in Fig. 12.3 (Malacara and Menchaca 1985).

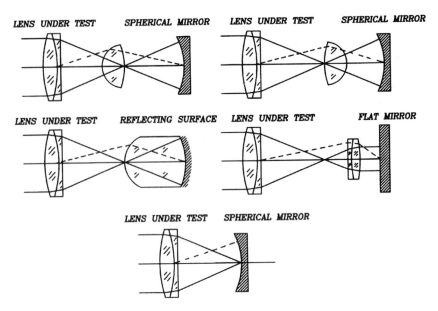

Figure 12.3 Some optical arrangements to test a lens, imaging its pupil back on itself.

For microscope objectives this solution is not satisfactory because the ideal place to observe the fringes is at the back focal plane. However, in this case the Dyson system illustrated in Fig. 12.4 is an ideal solution. It is interesting to point out that Dyson's system may be used to place the self-conjugate plane at the concave or convex surface while maintaining the concentricity of the surfaces.

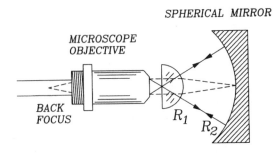

Figure 12.4 Dyson's system to test microscope objectives.

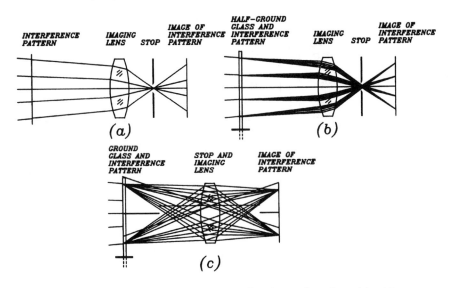

Figure 12.5 Imaging the interferogram on the observation plane (a) without any rotating ground glass, (b) with a rotating half-ground glass, and (c) with a rotating ground glass.

12.3.2 Imaging the Pupil on the Observation Screen

The second problem is to image the interference pattern on the detector, observation screen, or photographic plate. The imaging lens does not need to preserve the wavefront shape, since it is generally placed after the beam splitter and thus both interfering wavefronts pass through this lens. However, this lens has to be designed in such a way that the interference pattern is imaged without any distortion, assuming that the pupil of the system is at the closest image of the light source, as shown in Fig. 12.5a. A rotating ground glass in the plane of the interferogram might be useful sometimes to reduce the noise due to speckle and dust in the optical components. Ideally, this rotating glass should not be completely ground in order to reduce the loss of brightness and to keep the stop of the imaging lens at the original position, as shown in Fig. 12.5b. If the rotating glass is completely ground, the stop of the imaging lens should be shifted to the lens in order to use all available light, but then the lens must be designed with this new stop position taken into consideration, as in Fig. 12.5c.

When a distorted wavefront propagates in space, its shape is not preserved but changes continuously along its trajectory. From a geometrical point of view, that is, neglecting diffraction, only spherical and flat wavefronts keep their shape,

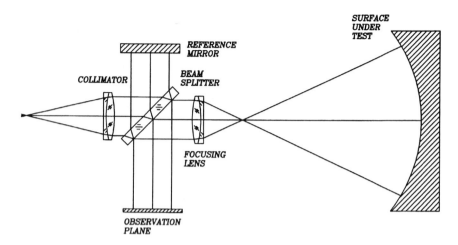

Figure 12.6 Conic mirror tested in a Twyman-Green interferometer.

changing only their radius of curvature. This is a well-known fact that should be taken into account in interferometry of wavefronts. As an example, let us consider the Twyman-Green interferometer shown in Fig. 12.6. A conic or spherical mirror is tested by means of this interferometer. If the mirror has a conic shape, the spherical aberration is compensated with a lens with the proper amount of spherical aberration of the opposite sign.

The wavefront reflected on the surface under test is combined at the beam splitter with a perfectly flat reference wavefront. The focusing lens has to be designed so that the returning wavefront is perfectly flat if the surface under test has no defects. If the surface has a distorted shape, the reflected wavefront is also distorted. Thus, the wavefront going out of the focusing lens and returning to the beam splitter will not be flat but distorted. However, the deformations in the wavefront going out of the focusing lens are not the same as the deformations at the surface under test.

12.3.3 Requirements on the Imaging Lens

To have an interference pattern that is directly related to the wavefront deformations on the surface under test, it has to be observed at a plane that is conjugate to this surface, as has been described many times in the literature by several authors, including Slomba and Figoski (1978), Jóźwicki (1989, 1990), Malacara and Menchaca (1985), Selberg (1987), and Malacara (1992). This is the purpose of the projection lens, which has to form an image of the surface under test on the observation plane. The following two requirements must be satisfied by this lens, as shown in Fig. 12.7:

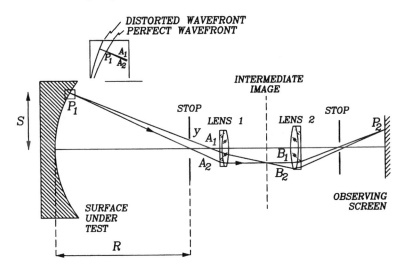

Figure 12.7 Optical system to image the pupil of the system under test on the observation plane.

1. The height of point P_2 above the optical axis should be strictly linear with the height of point P_1 above the optical axis. In other words, there should be no distortion. This ensures that a straight fringe on the surface under test will also be a straight fringe on the observation screen. This condition is not absolutely necessary if the fringe distortion is taken into account in the computer analysis of the fringes.
2. A point object P_1 must correspond to a point image P_2. By Fermat's principle, the optical path through A_1B_1 is then equal to the optical path through A_2B_1. Let us assume that a perfect surface sends the reflected ray from P_1 through A_1. A distorted wavefront sends a ray that passes through P_1 and toward A_2. Both rays then arrive together at point P_2. Then, since the optical paths are equal, any phase difference between the two rays at point P_1 is the same when they arrive at point P_2.

If these conditions are satisfied, the interferograms are identical. It must be noted that lens 2 need not produce a perfect wavefront, since both wavefronts are refracted on this lens and any deformation is introduced in both wavefronts to the same degree.

The imaging lens design must include the complete system with all lenses between the surface under test and the observation screen. The points where the light beams converge may be considered the stops of the lens system. So the system may have two or more virtual stops. There is an intermediate image, as we see in Fig. 12.7. However, the observation plane cannot be located at this position for two reasons: first because it is very unlikely that it has the required dimensions, and second because the system would be so asymmetrical that the distortion would be extremely large.

The complete system with lenses 1 and 2 is more symmetrical, making distortion correction easier. The stop diameter is given by the maximum transverse aberration at the stop. This maximum transverse aberration is a function of three factors: (1) the degree of asphericity of the surface under test, (2) the deformation error in this surface, and (3) the tilt between the wavefront under test and the reference wavefront. In general, this aperture is extremely small, even with large transverse aberrations.

Let us now analyze the degree of correction required for each of the five Seidel aberrations.

Spherical aberration. This aberration increases with the fourth power of the aperture. Thus, it does not need to be highly corrected, because the aperture is very small. A large amount of spherical aberration may be tolerated.

Coma. This aberration increases with the cube of the aperture in the tangential plane and with the square of the aperture in the sagittal plane. Thus, the correction of this aberration is more necessary than that of the spherical aberration, but the most important is the sagittal coma. If a large tilt is introduced in the interferogram, introducing straight fringes perpendicular to the tangential plane, the fringes in the vicinity of the this plane are affected by coma less than the fringes on the sagittal plane.

Petzval curvature. Ideally, the curvature of the surface under test must be taken into account, by curving the object plane by the same amount. The wavefront aberration due to this curvature increases with the square of the aperture. However, this aberration is not so important as long as the ray transverse aberration in the observation plane remains small, as we will see later.

Astigmatism. The wavefront aberration produced by this aberration, like the Petzval curvature, increases with the square of the aperture. Therefore, another criterion should be that the ray transverse should be small.

Distortion. This aberration, as explained earlier, may be ignored if the compensation is made in the computer analysis of the fringes. However, it is always easier to correct it on the lens. Again, the criterion is the magnitude of the ray transverse aberration.

The slope of the aberrated wavefront with respect to the ideal wavefront (reference wavefront) is

$$\frac{\partial W}{\partial S} = \frac{\Delta W}{\Delta S} \tag{12.2}$$

where ΔW is the change in the wavefront deformation if the height of point P_1 changes by an amount ΔS. Let us assume that the magnification of the whole lens system is m. Then the magnitude of the transverse ray aberration TA on the observation plane corresponds to an object height shift ΔS given by

$$m = \frac{\text{TA}}{\Delta S} \tag{12.3}$$

Thus, we may see that

$$\text{TA} = \frac{m \Delta W}{\partial W / \partial S} \tag{12.4a}$$

and to find the maximum allowable ray transverse aberration TA_{max}, we may see that if ΔW_{max} is the maximum permissible error in the wavefront measurement, the corresponding maximum value of this ray transverse aberration is

$$\text{TA}_{max} = \frac{m \Delta W_{max}}{\partial W / \partial S} \tag{12.4b}$$

If the minimum separation between two consecutive fringes on the surface under test is σ_1 and ΔW_{max} is a fraction $1/n$ of the wavelength ($\Delta W_{max} = \lambda / n$), we may write

$$\text{TA}_{max} = \frac{m \sigma_1}{n} \tag{12.5}$$

Hence, if the minimum separation between two consecutive fringes in the observation plane is σ_2, given by $\sigma_2 = m \sigma_1$, we may see that

$$\text{TA}_{max} = \frac{\sigma_2}{n} \tag{12.6}$$

which means that the maximum permissible transverse aberration in the projecting optical system is equal to a predetermined fraction of the minimum separation between the fringes in the observation plane.

When the interferogram is observed with a two-dimensional detector, a wavefront tilt or aberration may be introduced into the limit imposed by the detector. Then the maximum transverse aberration is approximately equal to the resolution power of the detector given by the separation between two consecutive pixels or detector elements.

The stop semiaperture y may be obtained by using the minimum fringe separation as follows:

$$y = \frac{\lambda R}{\sigma_1} = \frac{m \lambda R}{\sigma_2} \tag{12.7}$$

where R is the radius of curvature of the mirror under test, as shown in Fig. 12.7.

If the distortion aberration is not software-compensated in the computer analysis, the transverse aberration has to be measured from the Gaussian image position. Otherwise, it is measured from the center of gravity of the image. If the magnification of the system is much smaller than 1, the interferogram in the observation plane is very small and the requirement for a small transverse aberration may be quite strong.

The principles to be followed in the design of projecting lenses for interferometry have been described using the Twyman-Green instrument as an example, but they may be applied to Fizeau interferometers as well.

12.4 REFERENCE SPHERE DEFINITION IN WAVEFRONT MEASUREMENT

When digitizing an interferogram with a detector array, the sampling theorem requires the minimum local fringe spacing or period to be greater than twice the pixel separation. Thus, given a detector, there is a minimum fringe period that may be allowed. This minimum period, in turn, is set by the wavefront asphericity and the testing method. In this section we study the optimum defocusing and tilt for testing aspheric wavefronts with as much asphericity as possible in a non-null test configuration. We will determine, closely following Malacara-Hernández et al. (1996), which is the maximum asphericity that may be measured with various testing methods and the optimum tilt and defocusing in each of these methods.

A general expression for an aspheric wavefront deformation W for different focus shifts and only primary spherical aberration is

$$W = aS^2 + bS^4 \tag{12.8}$$

where a is the defocusing term and b is the primary spherical aberration coefficient. Figure 12.8 shows the values of the wavefront deformation W for three different focus settings to be described later.

The first derivative W' with respect to S is the radial slope of this wavefront, given by

$$W' = 2aS + 4bS^3 \tag{12.9}$$

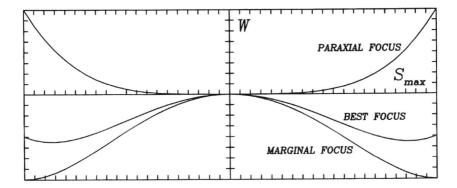

Figure 12.8 Aspheric wavefront deformations at the paraxial focus, best focus, and marginal focus, with primary spherical aberration.

The radial derivatives for the three focus positions are illustrated in Fig. 12.9. If we plot this wavefront slope W', any change in the focus or in the amount of tilt may be easily represented in this graph. As shown in Fig. 12.10, a tilt is a vertical displacement of the curve, and a change in focus is represented by a small rotation of the graph about the origin.

The wavefront can be measured with respect to many reference spheres by selecting the defocusing coefficient a. Here we will study the three main possibilities.

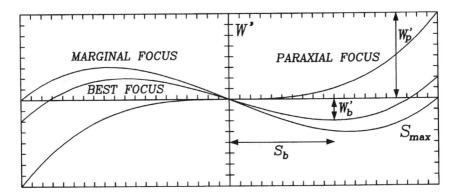

Figure 12.9 Wavefront radial slopes at the paraxial focus, best focus, and marginal focus for a wavefront with primary spherical aberration. The maximum radial slope for the best focus is at S_b and at the edge of the pupil.

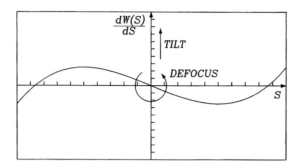

Figure 12.10 Tilt and defocus effect on the derivative of a wavefront. A defocus rotates the curve about the origin, and a tilt displaces the curve vertically.

12.4.1 Paraxial Focus

The paraxial focus is defined by a zero defocusing coefficient. Thus, the slope of the wavefront at the edge of the pupil, referred to the paraxial focus ($a = 0$), is

$$W'_p = 4bS^3_{max} \tag{12.10}$$

where S_{max} is the semidiameter of the wavefront.

12.4.2 Best Focus

The best focus is defined as the focus value that minimizes the absolute value of the maximum radial slope over the pupil. This maximum slope $W'_b = -W'(S_b)$ occurs at the edge S_{max} of the pupil and at some intermediate pupil radius S_b, but with opposite values. Opposite signs and the same magnitude for the radial slope means that the transverse aberrations $TA(S_b)$ and $TA(S_{max})$ are also equal in magnitude but with opposite signs. This is the condition for the waist of the caustic.

Hence, the optimum or best focus occurs when the center of the reference sphere is located at the waist of the caustic as illustrated in Fig. 12.11.

Thus, we may write

$$W'_b = -W'(S_b) = W'(S_{max}) \tag{12.11}$$

After some algebraic manipulation using first derivatives as well as the second derivative of W, it is possible to show that the ratio of the maximum wave-

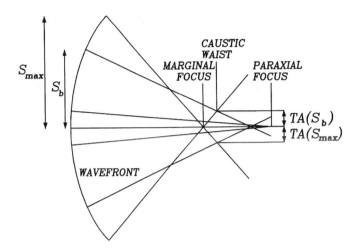

Figure 12.11 An aspheric wavefront and its caustic waist, showing the paraxial, marginal, and best focus.

front deformation at the paraxial focus to the best focus position is a constant given by

$$\frac{W_p'}{W_b'} = 4 \tag{12.12}$$

It remains to calculate the value of the defocusing term for the best focus position given the spherical aberration wavefront. Again, after some algebraic manipulation we may find

$$a = -\frac{3}{2}bS_{\max}^2 \tag{12.13}$$

12.4.3 Marginal Focus

The wavefront slope W_m' at the marginal focus and the edge of the pupil has to be zero. Thus,

$$W_m'(S_{\max}) = 2aS_{\max} + 4bS_{\max}^3 = 0 \tag{12.14}$$

Hence the defocusing coefficient at the marginal focus is

$$a = -2bS_{\max}^2 \tag{12.15}$$

and the first radial derivative of the wavefront at the marginal focus is then

$$W' = 4b(S^3 - S_{max}^2 S) \tag{12.16}$$

Then the maximum slope value of this wavefront deformation is given by equating the second radial derivative with respect to S to zero. Thus we obtain for the radial position S_m of this maximum deformation the value

$$S_m = \frac{S_{max}}{\sqrt{3}} \tag{12.17}$$

The ratio between the slope maxima at the paraxial and marginal foci may be shown to be

$$\frac{W'_p}{W'_m} = -2.6 \tag{12.18}$$

12.5 OPTIMUM TILT AND DEFOCUSING IN DIFFERENT INTERFEROGRAM ANALYSIS METHODS

The optimum tilt magnitude and reference sphere (defocusing) for the different interferogram analysis methods may now be estimated using these results.

The sampling theorem requires the minimum local fringe spacing or period to be greater than twice the pixel separation. Thus, given a detector, there is a minimum fringe period that may be allowed. This minimum period, in turn, is set by the wavefront asphericity and the testing method, as pointed out by Creath and Wyant (1987).

The fringe period s (or its fringe frequency f) in the interferogram is related to the wavefront slope by the relation

$$f = \frac{1}{s} = \frac{W'}{\lambda} \tag{12.19}$$

On the other hand, from geometrical optics, the slope W' of the wavefront is related to the ray transverse aberration by

$$W' = \frac{TA}{r} \tag{12.20}$$

where r is the radius of curvature of the reference wavefront. The condition to maximize the minimum fringe period is equivalent to minimizing the ray transverse aberration, which happens at the waist of the caustic. In other words, the best focus position is obtained when the center of the reference sphere is at the center of the waist of the caustic.

$$W'_p = 4 \frac{W_p}{S_{max}} \tag{12.21}$$

$$f_{max} = \frac{1}{S_{min}} = \frac{W'_P}{\lambda} = 4 \frac{W_p}{\lambda S_{max}} = 4 \frac{n_p}{S_{max}} \tag{12.22}$$

where n_p is the number of fringes at the paraxial focus, without any tilt.

$$\frac{S_{max}}{S_{min}} = 4N_p \leq \frac{p}{3} \tag{12.23}$$

$$p \leq 12 n_p \tag{12.24}$$

The minimum number of pixels p in one direction is equal to

$$p \geq 12 \frac{n_p}{\eta} \tag{12.25}$$

where η is the *relative minimum fringe period*. The larger the relative minimum fringe period, the greater the degree of asphericity that can be tested using that wavefront configuration.

12.5.1 Phase Shifting Techniques

In this case no tilt is necessary, but the focus may be adjusted with any value. Let us consider three focus possibilities:

Paraxial focus. In this case the minimum fringe period is defined as the unit ($\eta = 1$). The phase shifting method may be used, but to obtain the maximum asphericity capacity this focus setting is not the optimum.

Best focus. At the best focus we obtain the maximum possible local minimum fringe period ($\eta = 4$) of all configurations. Then this is the optimum focus for testing the maximum degree of asphericity.

Marginal focus. With this focus setting the local minimum fringe period η is equal to 2.6, better than the paraxial focus but worse than the best focus.

12.5.2 Spatial Linear Carrier Demodulation

These methods require the introduction of a large linear carrier in the x direction. The minimum magnitude of this carrier is such that the phase increases (or decreases) in a monotonic manner with x. This condition is necessary to avoid closed-loop fringes. This is possible if a tilt is introduced so that W' is always positive as in the plots in Fig. 12.12. Then the minimum slope is zero. So,

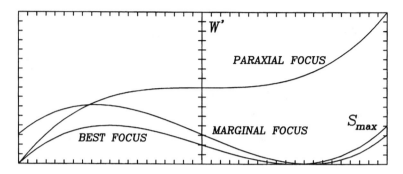

Figure 12.12 Effect on the radial wavefront slope of introducing tilt in a wavefront until the derivative of the wavefront is positive everywhere.

ideally, a tilt larger than this must be used, but this is the minimum value. Three focus possibilities exist as follows.

Paraxial focus. If a tilt is introduced at the paraxial focus in order to introduce the linear carrier, the maximum local wavefront slope is increased by a factor of 2, reducing the local minimum fringe η period to 0.5. A demodulation of these fringes with a spatial carrier may be performed, but this is not the ideal amount of defocusing to have the maximum possible local minimum fringe period and obtain the maximum testing asphericity capacity.

Best focus. If a tilt is introduced at the best focus, we obtain the maximum possible local minimum fringe period attainable with a linear carrier ($\eta = 2$), as shown in Fig. 12.13. This is the ideal configuration to analyze the fringe pattern with a modulated linear carrier.

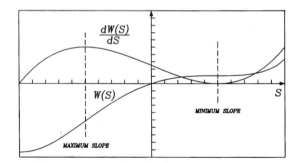

Figure 12.13 A wavefront and its radial slope at the best focus position, showing where the minimum slope occurs.

Table 12.1 Relative Minimum Fringe Periods for Several Wavefronts and Three Interferometric Analysis Methods

Interferometric analysis method	Wavefront focus	Wavefront tilt	Relative minimum fringe period
Phase shifting techniques	Paraxial	None	1.0
	Best	None	**4.0**
	Marginal	None	2.6
Spatial carrier demodulation	Paraxial	Yes	0.5
	Best	Yes	**2.0**
	Marginal	Yes	1.3
Circular carrier demodulation	Marginal	None	**2.6**

Marginal focus. If the proper tilt is introduced at the marginal focus, a linear carrier demodulation scheme may be used. However, this is not the ideal configuration for this method. The local minimum fringe period η is now equal to 1.2.

12.5.3 Circular Carrier Demodulation

With circular carrier demodulation, no tilt is introduced, because the circular symmetry must be preserved. A focus term must be selected such that the phase monotonically increases (or decreases) from the center toward the edge of the interferogram. From the three focus positions described (paraxial, best, and marginal), only the marginal focus position is acceptable as a minimum. Ideally, a defocusing larger than this amount should be used.

Marginal focus. At the marginal focus the wavefront radial slope does have any sign changes along an interferogram semidiameter. Thus this is the configuration to use with a radial carrier modulation. The local minimum fringe period η is equal to 2.6.

The results for the three interferogram analysis methods and different focus positions are summarized in Table 12.1.

12.6 MULTIPLE WAVELENGTH INTERFEROMETRY

In phase shifting interferometry, the phase is calculated modulo 2π, so there is a phase wrapping in the calculation of the phase. To be able to unwrap the phase, the phase between two adjacent measured points in the interferogram must be smaller than 2π. This is a limitation to the maximum wavefront slope and hence to the maximum asphericity being measured.

Wyant (1971), Polhemus (1973), Wyant *et al.* (1984), Cheng and Wyant (1984), Creath *et al.* (1985), Creath and Wyant (1986), and Gushov and Solodkin (1991) have studied the problem of phase determination when two or more different wavelengths are used.

If two different wavelengths λ_a and λ_b are simultaneously used, the wavetrain is modulated as in Fig. 12.14, with a group length λ_{eq} given by

$$\lambda_{eq} = \frac{\lambda_a \lambda_b}{|\lambda_b - \lambda_a|} \tag{12.26}$$

Wyant (1971) described two methods using two wavelengths. In the first method, a photographic recording of an interferogram is taken with one wavelength. Then another interferogram is formed with the second wavelength, placing the photograph of the first interferogram over the second. In this manner, a moiré image is formed from the photograph of one interferogram and the real-time image of the second. High frequencies of this moiré image are then filtered out with a pinhole.

In the second method the two interferograms are taken simultaneously, one on top of the other, by illuminating with the two wavelengths. The high spatial frequencies of the resulting moiré image are again filtered with a pinhole.

Polhemus (1973) described a real-time two-wavelength interferometer using a TV camera to detect the moiré pattern. Figure 12.15 shows the Moiré pattern obtained by the superposition of two different wavelengths, after low pass filtering. The resulting pattern is the image of an interferogram taken with the equivalent wavelength.

Cheng and Wyant (1984), Creath *et al.* (1985), and Creath and Wyant (1986) implemented phase shifting interferometers using two wavelengths. Two separate wrapped phase maps are obtained by taking two independent sets of measurements using each of the two wavelengths. We assume that the Nyquist limit has been exceeded because of the high wavefront asphericity. With one wavelength the phase unwrapping would be impossible, but it can be done with the two wavelengths. The two wavefront deformations are different if the scale is the phase, because the wavelengths are different. However, they must be equal if the OPD (optical path difference) is used instead of the phase. Thus, we have

$$OPD_a(x, y) = OPD_b(x, y) \tag{12.27}$$

We may also write

$$OPD_a(x, y) = \left(\frac{\phi_a(x, y)}{2\pi} + m_a \right) \lambda_a \tag{12.28}$$

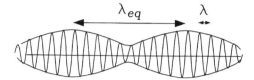

Figure 12.14 Wavetrain formed by two wavelengths.

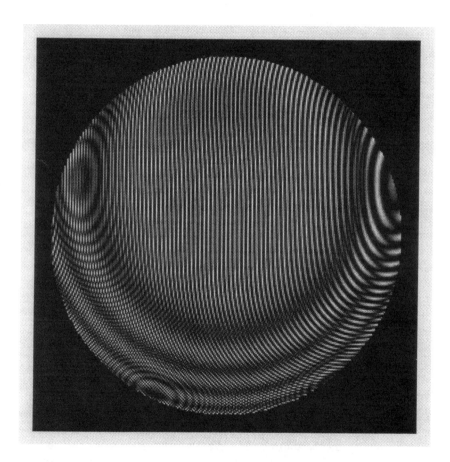

Figure 12.15 Moiré pattern formed by two interferograms taken with two different wavelengths. First wavelength $\lambda_a = 0.633$ and second wavelength $\lambda_b = 0.594$. Equivalent wavelength $\lambda_{eq} = 9.714$.

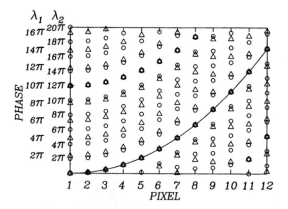

Figure 12.16 Graphical illustration of the phase unwrapping procedure, using two wavelengths, with $\lambda_1 = 1.2\lambda_2$ and $\lambda_{eq} = 6\lambda_1 = 5\lambda_2$.

and

$$\text{OPD}_b(x, y) = \left(\frac{\phi_b(x, y)}{2\pi} + m_b\right)\lambda_b \tag{12.29}$$

where m_a and m_b are integers. Thus, using Eq. 12.27 we have

$$\left(\frac{\phi_a(x, y)}{2\pi} + m_a\right)\lambda_a = \left(\frac{\phi_b(x, y)}{2\pi} + m_b\right)\lambda_b \tag{12.30}$$

We have one equation with two unknowns, m_a and m_b. The system may be solved if we assume that the difference of order numbers between two adjacent pixels is the same for both wavelengths. This hypothesis is valid if the asphericity is not extremely high. Thus, we may obtain

$$\Delta\text{OPD}_{n+1} = \frac{1}{2\pi}(\Delta\phi_{(n+1)a} - \Delta\phi_{(n+1)b})\lambda_{eq} \text{ if } \lambda_b > \lambda_a \tag{12.31}$$

The OPD values for all pixels in a row may be obtained if we take $\text{OPD}_1 = 0$. Figure 12.16 illustrates the phase unwrapping procedure using two different wavelengths with a ratio of 6 to 5. The only possible valid points when unwrapping the wavefront are the thick circles, where the two wavelengths coincide. The result is that even with subsampling, the unwrapping presents no ambiguities.

Cheng and Wyant (1985) enhanced the capability of two-wavelength interferometry by introducing a third wavelength. This allowed even steeper wavefront slopes to be measured.

REFERENCES

Cheng Y.-Y. and J. C. Wyant, "Two-Wavelength Phase Shifting Interferometer," *Appl. Opt.*, **23**, 4539–4543 (1984).

Creath K. and J. C. Wyant, "Direct Phase Measurement of Aspheric Surface Contours," *Proc SPIE.*, **645**, 101–106 (1986).

Creath K. and J. C. Wyant, "Aspheric Measurement Using Phase Shifting Interferometry," *Proc SPIE.*, **813**, 553–554 (1987).

Creath K., Y. -Y. Cheng, and J. C. Wyant, "Contouring Aspheric Surfaces Using Two-Wavelength Phase Shifting Interferometry," *Opt. Acta*, **32**, 1455–1464 (1985).

D?rband B. and H. J. Tiziani, "Testing Aspheric Surfaces with Computer Generated Holograms: Analysis of Adjustment and Shape Errors," *Appl. Opt.*, **24**, 2604–2611 (1985).

Dyson J., "Unit Magnification Optical System without Seidel Aberrations," *J. Opt. Soc. Am.*, **49**, 713–716 (1959).

Garc?a-M rquez J., D. Malacara, and M. Servín, "Limit to the Degree of Asphericity when Testing Wavefronts Using Digital Interferometry" *Proc. SPIE.*, **2263**, 274–281 (1995).

Gushov V. I. and Y. N. Solodkin, "Automatic Processing of Fringe Patterns in Integer Interferometers," *Opt. and Lasers in Eng.*, **14**, 311–324 (1991).

Jóźwicki R., "Telecentricity of the Interferometric Imaging System and its Importance in the Measuring Accuracy," *Optica Applicata*, **19**, 469–475 (1989).

Jóźwicki R., "Propagation of an Aberrated Wave with Nonuniform Amplitude Distribution and its Influence Upon the Interferometric Measurement Accuracy," *Optica Applicata*, **20**, 229–252 (1990).

Malacara D., Editor, *Optical Shop Testing*, John Wiley and Sons, New York, 1992.

Malacara D. and C. Menchaca, "Imaging of the Wavefront Under Test in Interferometry," *Proc. SPIE.*, **540**, 34–40 (1985).

Malacara, D., V. I. Vald, and M. Servín, "Spatial Carrier Analysis of Interferograms with Aspheric Wavefronts," *Proc. SPIE.*, **2340**, 190–201 *Interferometry 94*, Warsaw, Poland (1994).

Malacara-Hernández D., Z. Malacara-Hernández, and M. Servín, "Digitization of Interferograms of Aspheric Wavefronts," *Opt. Eng.*, **35**, 2102–2105 (1996).

Offner A. and D. Malacara, "Null Tests Using Compensators," in *Optical Shop Testing*, D. Malacara, Ed., John Wiley and Sons, New York, 1992.

Ono A. and J. C. Wyant, "Aspherical Mirror Testing Using a CGH with Small Errors," *Appl. Opt.*, **24**, 560–563 (1985).

Ono A., "Aspherical Mirror Testing with an Area Detector Array," *Appl. Opt.*, **26**, 1998–2004 (1987).

Polhemus C., "Two-Wavelength Interferometry," *Appl. Opt.*, **12**, 2071–2078 (1973).

Selberg L. A., "Interferometer Accuracy and Precision," *Proc. SPIE.*, **749**, 8–18 (1987).

Slomba A., F. and J. W. Figoski, "A Coaxial Interferometer with Low Mapping Distortion," *Proc. SPIE.*, **153**, 156–161 (1978).

Wyant J. C., "Testing Aspherics Using Two-Wavelength Holography," *Appl. Opt.*, **10**, 2113–2118 (1971).

Wyant J. C., B. F. Oreb, and P. Hariharan, "Testing Aspherics Using Two Wavelength Holography: Use of Digital Electronic Techniques," *Appl. Opt.*, **23**, 4020–4023 (1984).

Review Articles and Books

Briers J. D., "The Interpretation of Holographic Interferograms," *Opt. Quantum Electr.*, **8**, 469–501 (1976).

Creath K., "Phase-Measurement Interferometry Techniques," *Progress in Optics, Vol. 26* (1988).

Gåsvik K. J., *Optical Metrology*, 2nd Edition, John Wiley and Sons, New York, 1995.

Hariharan P., "Interferometric Metrology: Current Trends and Future Prospects," *Proc. SPIE*, **816**, 2–18 (1987).

Kreis T., "Quantitative Evaluation of Interference Patterns," *Proc. SPIE*, **863**, 68–76 (1987).

Kreis T. M., "Computer Aided Evaluation of Fringe Patterns," *Opt. and Lasers in Eng.*, **19**, 221–240 (1993).

Kujawinska M., "Expert System for Analysis of Complicated Fringe Pattern," *Proc. SPIE*, **1755**, 252–257 (1992).

Kujawinska M., "The Architecture of Multipurpose Fringe Pattern Analysis System," *Opt. and Lasers in Eng.*, **19**, 261–268 (1993).

Malacara D., *Optical Shop Testing*, 2nd Edition, John Wiley and Sons, New York, 1992.

Moore R. C., "Automatic Method of Real-Time Wavefront Analysis," *Opt. Eng.*, **18**, 461–463 (1979).

Osten W., R. Hofling, and J. Seadler, "Two Computer-Aided Methods for Data Reduction from Interferograms," *Proc. SPIE*, **863**, 105–113 (1987).

Reid G. T., "Image Processing Techniques for Fringe Pattern Analysis," *Proc. SPIE*, **954**, 468–477 (1988).

Reid G. T., "Automatic Fringe Pattern Analysis: A Review," *Opt. and Lasers in Eng.*, **7**, 37–68 (1986/87).

Robinson D. W. and D. C. Williams, ''Automatic Fringe Analysis in Double Exposure and Live Fringe Holographic Interferometry,'' *Proc. SPIE*, **599**, 134–140 (1985).

Robinson D. W. and G. T. Reid, Eds., *Interferogram Analysis: Digital Fringe Pattern Measurement Techniques*, Institute of Physics Publishing House, Briston, UK, 1993.

Shough D., ''Beyond Fringe Analysis,'' *Proc. SPIE*, **2003**, 208–223 (1993).

Schulz G. and J. Schwider, ''Interferometric Testing of Smooth Surfaces,'' *Progress in Optics, Vol. 13*, p. 431 (1976).

Schwider J., ''Advanced Evaluation Techniques in Interferometry,'' *Progress in Optics, Vol. 29*, p. 431 (1990).

Schmit J., K. Creath, and M. Kujawinska, ''Spatial and Temproal Phase-Measurement Techniques: A Comparison of Major Error Sources in One Dimension,'' *Proc. SPIE*, **1755**, 202–211 (1992).

Vlad V. I. and D. Malacara, ''Direct Spatial Reconstruction of Optical Phase from Phase-Modulated Images,'' *Prog. Opt.*, **23** (1994).

Wyant J. C. and K. Creath, ''Recent Advances in Interferometric Optical Testing,'' *Laser Focus/Electro-Optics*, November 1985, pp. 118–132.

Index